"The sea and land, and the peoples that span these lim[illegible] in isolation, both in academic scholarship and in prac[illegible] takes a widescreen, interdisciplinary approach to illum[illegible]ent socio-ecological processes, with keen attention to their equity implications."

Dr. Kenny Broad, *University of Miami*

"*Oceans and Society: An Introduction to Marine Studies* places the relationship between people and the ocean at its heart. The book echoes calls for interdisciplinary thinking, championing the development of innovative ways to better understand the relationships between people, ocean, and place, through useful case studies."

Dr. Emma McKinley, *Research Fellow, School of Earth and Environmental Sciences, Cardiff University & Chair of the Marine Social Sciences Network*

"As our climate changes and global populations reach eight Billion, we must improve ocean management. This book sets an important foundation for ocean management, as it introduces the breadth of marine studies, and connects oceans and society."

Dr. Quentin Hanich, *Ocean Nexus Chair in Fisheries Governance; Australian National Centre for Ocean Resources and Security (ANCORS), University of Wollongong*

"The challenges faced by our oceans are complex, multi-faceted and inter-related. Future ocean stewardship will rely on interdisciplinary approaches. This book is a timely compilation which will help guide students from diverse disciplinary backgrounds to an appreciation of the benefits and opportunities associated with engaging with human dimensions approaches."

Dr. Michelle Voyer, *Senior Research Fellow Australian National Centre for Ocean Resources and Security (ANCORS) at University of Wollongong*

"A must-read book for graduate students in the field of marine management. Understanding the 'ocean-society' intersection, from multiple perspectives and applied to multiple grand challenges, is essential to enable these next-generation marine managers to work collaboratively towards sustainable and just ocean futures."

Dr. Megan Bailey, *Associate Professor and Canada Research Chair, Marine Affairs Program, Dalhousie University*

"*Oceans and Society* is a valuable introduction to the human dimension of coastal and ocean issues, accessible to individuals with a variety of backgrounds united by their passion to sustain a healthy future for our ocean, coasts, and people."

Professor Jack Barth, *Executive Director, Marine Studies Initiative, Oregon State University*

Oceans and Society

This unique textbook presents an introduction to the interdisciplinary field of marine studies, exploring the dynamic relationship between people and the marine environment.

Emphasizing the human dimension of coastal and ocean issues, the book provides an innovative examination of the complex marine–human environment dynamics by drawing on social science and humanities approaches. Applying these interdisciplinary approaches, it addresses key challenges facing the marine environment, including changing climate, fisheries, aquaculture, marine pollution, energy production, and management of areas beyond national jurisdiction. While leading with a human dimension approach to these challenges, the chapters are all firmly grounded in foundational knowledge about coastal and ocean environments and processes. The textbook also includes examples of professional or academic areas of specialization within marine studies such as social and environmental justice, governance, global perspectives, traditional ecological knowledge and management, entrepreneurship, community development, conservation, and the blue economy. Ultimately, the book provides the first cohesive resource on marine studies to educate students, train interdisciplinary marine leaders, inspire new knowledge about people and the sea, generate innovative solutions for sustainable oceans, and build capacity for a new generation of marine-focused professionals.

Oceans and Society is essential reading for students on marine studies courses, as well as those studying marine governance, policy, conservation, and law more broadly. It will also be of great interest to students, researchers, and professionals interested in applying interdisciplinary approaches to environmental challenges.

Ana K. Spalding is an Associate Professor of Marine and Coastal Policy at Oregon State University, and Research Associate at the Smithsonian Tropical Research Institute and Coiba Research Station – AIP in Panama.

Daniel O. Suman is a Professor of Marine Policy and Coastal Management at the University of Miami's Rosenstiel School of Marine, Atmospheric, and Earth Science.

Earthscan Oceans

Coral Reefs
Tourism, Conservation and Management
Edited by Bruce Prideaux and Anja Pabel

Marine Extremes
Ocean Safety, Marine Health and the Blue Economy
Edited by Erika Techera and Gundula Winter

Marine Policy
An Introduction to Governance and International Law of the Oceans
2nd Edition
Mark Zacharias and Jeff Ardon

Conflicts over Marine and Coastal Common Resources
Causes, Governance and Prevention
Karen A. Alexander

Marine and Fisheries Policies in Latin America
A Comparison of Selected Countries
Edited by Manuel Ruiz Muller, Rodrigo Oyandel and Bruno Monteferri

Oceans and Society
An Introduction to Marine Studies
Edited by A. K. Spalding and Daniel O. Suman

Transdisciplinary Marine Research
Bridging Science and Society
Edited by Sílvia Gómez and Vera Köpsel

For further details please visit the series page on the Routledge website: http://www.routledge.com/books/series/ECOCE

Oceans and Society

An Introduction to Marine Studies

Edited by
Ana K. Spalding and
Daniel O. Suman

Designed cover image: © Getty images

First published 2023
by Routledge
4 Park Square, Milton Park, Abingdon, Oxon OX14 4RN

and by Routledge
605 Third Avenue, New York, NY 10158

Routledge is an imprint of the Taylor & Francis Group, an informa business

British Library Cataloguing-in-Publication Data
A catalogue record for this book is available from the British Library

Library of Congress Cataloging-in-Publication Data
Names: Spalding, Ana K., editor. | Suman, Daniel O. 1950– editor.
Title: Oceans and society : an introduction to marine studies / Edited by Ana K. Spalding and Daniel O. Suman.
Description: New York, NY : Routledge, 2023. | Includes bibliographical references and index.
Identifiers: LCCN 2022037797 (print) | LCCN 2022037798 (ebook) | ISBN 9780367524883 (hardback) | ISBN 9780367524869 (paperback) | ISBN 9781003058151 (ebook)
Subjects: LCSH: Marine ecology. | Marine resources. | Marine resources conservation.
Classification: LCC QH541.5.S3 O233 2023 (print) | LCC QH541.5.S3 (ebook) | DDC 577.7—dc23/eng/20220831
LC record available at https://lccn.loc.gov/2022037797
LC ebook record available at https://lccn.loc.gov/2022037798

ISBN: 978-0-367-52488-3 (hbk)
ISBN: 978-0-367-52486-9 (pbk)
ISBN: 978-1-003-05815-1 (ebk)

DOI: 10.4324/9781003058151

Typeset in Goudy
by codeMantra

To our colleagues in Marine Studies/Marine Affairs and Policy who pioneered the Human Dimensions field.

To our students who are the future ocean leaders.

To Tutty and Rosita who have provided some calm during our writing and preparation of the book.

Contents

Editors Bios

Ana K. Spalding

Dr. Spalding is an Associate Professor of Marine and Coastal Policy at Oregon State University, and Research Associate at the Smithsonian Tropical Research Institute and Coiba Research Station – AIP in Panama. She has a PhD in Environmental Studies from the University of California, Santa Cruz, an MA in Marine Affairs and Policy from the University of Miami, and a BA in International Economics from the University of Richmond. She has published widely on the socio-environmental outcomes of lifestyle migration to Panama; on the linkages between land use and policy, property rights, and development; and, more broadly, on the evolution of marine policy and conservation in Panama and the United States. She is also fascinated by interdisciplinarity and collaboration as an academic endeavor, where it no longer represents an abstract concept, but instead has become a critical framework for addressing global environmental threats. Her current research includes the study of adaptive capacity to changing ocean conditions in resource-dependent communities in California, science-policy engagement related to ocean acidification on the West Coast of the United States, and assessments of the social outcomes of marine protected areas.

Daniel O. Suman

Dr. Suman is a Professor of Marine Policy and Coastal Management at the University of Miami's Rosenstiel School of Marine, Atmospheric, and Earth Science. His research and project areas focus on coastal management, adaptation to climate change and sea level rise, governance of marine resources and space, management of mangroves and coastal wetlands, and marine protected areas – particularly in Latin America and the Caribbean, but also worldwide. Suman earned a PhD in Oceanography from the Scripps Institution of Oceanography (University of California, San Diego); a law degree (JD) from the University of California, Berkeley; an MA in International Education and Latin American Studies from Columbia University; and a BA from Middlebury College. At the University of Miami, he has taught courses for over 30 years in Environmental Law, Environmental Planning, Coastal Management, Coastal Law, and Water Resources Policy.

Contributors

Laura Anderson
Local Ocean Seafoods, Newport, OR, USA

Daniel D. Benetti
University of Miami, Miami, FL, USA

Peter Betjemann
Oregon State University, Corvallis, OR, USA

Kelly Biedenweg
Oregon State University, Corvallis, OR, USA

Bradley Boovy
Oregon State University, Corvallis, OR, USA

Hilary S. Boudet
Oregon State University, Corvallis, OR USA

Susanne M. Brander
Oregon State University, Corvallis, OR, USA

Diane Brandt
Renewable Northwest, Corvallis, OR, USA

Samantha Chisholm Hatfield
Oregon State University, Corvallis, OR, USA

Lorenzo Ciannelli
Oregon State University, Corvallis, OR, USA and Stazione Zoologica Anton Dohrn, Ischia, Italy

Andrés M. Cisneros-Montemayor
Simon Fraser University, Vancouver, BC, Canada

Angela Clark-Hughes
University of Miami, Miami, FL, USA

Flaxen Conway
Oregon State University and Oregon Sea Grant, Corvallis, OR, USA

A. N. Doerr
Oregon State University, Newport, OR, USA

Kirsten Grorud-Colvert
Oregon State University, Corvallis, OR, USA

Harriet Harden-Davies
Nippon Foundation – University of Edinburgh, Edinburgh, Scotland, UK

Brian K. Haus
University of Miami, Miami, FL, USA

Caroline B. LaPorte
Little River Band of Ottawa Indians, Seminole Tribe of Florida and University of Miami, Miami, FL, USA

Marta Maria Maldonado
Oregon State University, Corvallis, OR, USA

Carrie Pomeroy
University of California, Santa Cruz, Santa Cruz, CA, USA

Bryson Robertson
Oregon State University, Corvallis, OR, USA

Daniel Rothan
University of Miami, Miami, FL, USA

Jason Scorse
Center for the Blue Economy and Middlebury Institute of International Studies, Monterey, CA, USA

Inara Scott
Oregon State University, Corvallis, OR, USA

Manoj P. Shivlani
University of Miami, Miami, FL, USA

Ana K. Spalding
Oregon State University, Corvallis, OR, USA and Smithsonian Tropical Research Institute, Panama City, Panama

John D. Stieglitz
University of Miami, Miami, FL, USA

Daniel O. Suman
University of Miami, Miami, FL, USA

Michael Touchton
University of Miami, Miami, FL, USA

Melissa Ward
San Diego State University, San Diego, CA, USA and Oxford University, Oxford, UK

Benjamin J. Wickizer
Oregon State University, Corvallis, OR, USA

Ana Zangroniz
University of Florida/IFAS Extension, Miami, FL, USA

Foreword

Relationships among people and oceans are complicated. People rely on the oceans and are influenced by them. They derive well-being from coastal and ocean spaces. And they also affect coastal and ocean health. It's not possible to fully understand the myriad connections among people and oceans through one disciplinary lens. The field of marine studies draws on multiple disciplines to better understand complex linkages among people and the oceans. The comprehensive multi-disciplinary approach at the core of the field of marine studies is critical to addressing some of society's most challenging issues related to water quality, multi-use, energy development, coastal access, environmental justice, and many others.

As the field of marine studies has gained attention over the years, the need for a holistic, multi-disciplinary text has emerged. Ana K. Spalding and Daniel O. Suman's *Oceans and Society: An Introduction to Marine Studies* fills that need. By drawing on multiple academic disciplines from across the social sciences and humanities, as well as traditional ecological knowledge, *Oceans and Society* equips students, coastal and ocean practitioners, and anyone else interested in the human dimensions of the oceans with valuable knowledge and practical tools to make sense of the way people relate to the coasts and oceans. This work demonstrates the importance of the social sciences and the power of the humanities to lend insights into how people think about and interact with coastal and ocean spaces.

As a university professor who has been teaching introductory marine studies for two decades, I know how beneficial it is for students to be able to draw on different disciplines and areas of expertise when tackling important – but challenging – issues like marine pollution, habitat decline, inequitable access, and climate change. I also recognize how difficult it can be to provide just the right amount of disciplinary depth and breadth so that students are adequately equipped to engage with complex coastal and ocean issues. By providing in one place important theories and tools from a variety of disciplines and valuable cases, like the harvesting of Manoomin (wild rice) by Indigenous communities for subsistence and ceremony or the international governance of the Arctic Ocean, this book is a welcome addition to the field of marine studies.

It is essential that current and future generations of coastal practitioners, policy makers, researchers, and coastal and ocean users approach complex coastal and ocean issues through a lens of multidisciplinarity. The field of marine studies in general, and this book in particular, prepares them to do just that.

Dr. Tracey Dalton
Professor of Marine Affairs
University of Rhode Island
Kingston, RI, USA

Preface

Daniel and I, together, have over 60 years of experience conducting research, consulting, and teaching about the interactions between oceans and society. Notably, during this time our emphasis has been on "society" in all its dimensions (e.g., political, cultural, economic, legal, etc.) within the marine space. However, the path to becoming marine social scientists was far from clear. Daniel has a background in oceanography, law, Latin American studies, and education; I have degrees in environmental studies, marine affairs and policy, and economics; and we both share a deep connection to the ocean and curiosity about the people who rely on it (and we are both originally from Panama!). The latter, unfortunately, is not enough to secure a job nor does it provide clarity on the types of skills or knowledge needed to work with oceans and people. Indeed, before I met Daniel almost 20 years ago (Daniel was my MA and PhD supervisor), I had asked myself this question many times – *how can I work on social issues, in marine areas, without pursuing a degree in marine biology?* Since then, Daniel and I have actively collaborated on various projects related to coastal zone management, policy, tourism and sustainability, ocean conservation, and now on this book. Over the years we have often discussed challenges we faced while doing this work, such as funding (it tends to be less compared to funding available for our natural and physical science counterparts), expectations (often colleagues will expect us to translate or communicate the natural or physical science to the public, instead of recognizing our work as generating its own type of information), perceptions (inclusion of marine social scientists as research collaborators as a last minute add-on to a project to satisfy the funders), and marine-specific social science training (it is straightforward to pursue a degree in economics, anthropology, business administration, etc.; yet not as easy to find opportunities for training in marine economics, marine anthropology, marine business, or other marine-focused social sciences). In response, we decided to put together this volume on what we are calling *Marine Studies*, an interdisciplinary field that centers the human dimension (social sciences and humanities) and is grounded in a solid understanding of the natural and physical processes that define oceans and coasts. It is our hope that the volume will contribute to current efforts to build capacity and foster a community of interdisciplinary thinkers who can overcome these challenges and become the next generation of oceans and society leaders.

The volume is designed in three parts to provide a deep and practical understanding of the complex social, cultural, historical, economic, and environmental issues faced by global coasts and oceans. Part I sets the stage and defines the field of Marine Studies; describes the basic, applied, and interdisciplinary fields that enable the study of the human dimension; and outlines the natural and physical processes that make the ocean a unique place. Part II is a survey of contemporary and emerging grand challenges faced by oceans and society, such as fisheries, aquaculture, marine pollution, climate change, energy production, and the proposed framework for governance of the 60% of the ocean that is found in areas beyond national jurisdiction, aka "High Seas". Notably, most chapters in this Part of the book are co-authored by a natural or physical scientist in partnership with a social science or humanities expert. This was intentionally done to invite readers to think critically about how we portray problems in the ocean – *Is the problem that there are less fish in the sea? Or is the problem that human activity has led to fewer fish and is, in turn, negatively affecting those who rely on fishing for their livelihoods?* Finally, we recognize that readers will have different professional and academic interests and worldviews. Thus, Part III showcases a suite of applied approaches or perspectives to help address these grand challenges, such as ocean governance, conservation, social justice, traditional ecological knowledge, community development, entrepreneurship, and development for the blue economy.

If you are an **instructor** wanting to use some or all of this volume in your classes in marine or environmental studies, we encourage you to think about the goal of each part of the book. While you may certainly pick and choose chapters from Parts II and III, we encourage you to carefully present all the content from Part I as a foundation for subsequent discussions of the grand challenges and approaches. This can provide students with some clarity on their disciplinary or interdisciplinary identity, effectively building a community of Marine Studies scholars and future practitioners. To work through the marine socio-environmental problems of Part II, we have added questions for reflection to each chapter as a starting point for class discussion. You may facilitate activities, such as think-pair-share, where students can talk with each other about the points they found most interesting about the chapter. We also realize that due to space limitations we were only able to include a few specific examples in each chapter. To overcome this, you might ask students to each bring a current event related to the chapter and critically assess the social and environmental nature of the event. In our classes, for instance, we have engaged students by asking them to select one of the grand challenges and explore its many dimensions throughout the length of the term, with the expectation that they will produce a final project that reflects their interests and experiences. We have also developed case studies around some of these topics, inviting the whole class to represent a different approach or perspective, essentially replicating a real-world scenario where decision-making requires effective collaboration and communication across often conflicting interests.[1] Finally, to actively engage students to think about how their own personal vocation and interests might shape their future careers, in addition to questions for reflection, all the chapters in Part III include a short section on professional pathways that hints at the type of disciplinary specialization students might need to focus on to be more prepared for a given career direction. These professional pathways may appear generic. Certainly, for instance, to become a marine conservation practitioner a student may pursue a degree in marine biology, public policy, environmental education, etc. However, they would probably also want to focus on learning about the types of organizations that do conservation work and specifically tailor their coursework, assignments, and networking to building expertise on the topic.

If you are a **student** or **practitioner** reading this book, we invite you to read it critically and thoughtfully. You might ask yourself – *What is the change I want to see for oceans and society, and how can the content of each chapter help me think about creative and innovative solutions?* You may also notice that some of the topics are presented in more than one chapter. In fact, you may even notice that the same topic is addressed differently across chapters (e.g., the links between fisheries and aquaculture in providing food from the sea; or the implications of promoting marine renewable energy in ocean spaces that are already experiencing competing uses, such as fisheries, aquaculture, and conservation; or the role of international agencies in managing and regulating the ocean space). This, again, is intentional to reflect the interests, perspectives, and worldviews of a diverse society. Chapter authors have a range of disciplinary and professional experiences, expertise, and identities that are reflected in their writing. In this sense, we invite you to not just look at the content of the chapters but also read about the authors and think about their contributions to Marine Studies.

We have enjoyed the process of designing the structure of the book, inviting author contributions, and thinking deeply about how to best use this material to train the next generation of ocean leaders. Ultimately, we hope we have provided a tool that will inspire a new narrative for the ocean as a coupled natural–human system that calls for interdisciplinary thinking and holistic approaches to innovative solutions. Increasingly, policymakers, resource managers, conservation organizations, and natural and physical scientists are finding that solutions to pervasive environmental problems, such as climate change, require this interdisciplinary approach, including social science and humanities perspectives. Students of Marine Studies are ideally suited to fill this demand for future interdisciplinary marine-focused professionals.

Ana K. Spalding, *Oregon State University and Smithsonian Tropical Research Institute*
Daniel O. Suman, *University of Miami*

Acknowledgments

We thank the many people who have assisted us in the preparation of this book. First, we recognize the authors of the chapters who gave their time writing and revising the essays, as well as the anonymous external chapter reviewers. Taylor & Francis editorial assistants John Baddeley and Katie Stokes provided guidance regarding the many publishing details and were sensitive to delays caused by the pandemic. We thank Dr. Jack Barth and Dr. Kristen Milligan, Executive and Associate Director, respectively, of Oregon State University's Marine Studies Initiative who provided enthusiastic support for this project, as well as partial funding for co-editor Spalding. Oregon State students Trisha Patterson and Ravyn Cervantes provided excellent assistance with copyediting. Co-editor Suman's students from the University of Miami's Ocean Policy class and teaching assistant Johanna Mead provided constructive criticism and feedback for the chapters. We also recognize Mead's assistance with the indexing. Finally, we thank our family and friends who offered us support and encouragement during book preparation.

Note

1 You can find excellent case study materials in the UC Press journal titled *Case Studies in the Environment*, and in the National Socio-Environmental Synthesis Center's (SESYNC) Case Study Collection, available online at: https://www.sesync.org/resources/case-study-collection.

Part I

Setting the Stage

Part I sets the stage for the rest of the book by introducing the field of marine studies, describing various human dimension disciplines, and introducing basic natural and physical processes that define the marine environment.

DOI: 10.4324/9781003058151-1

1 An Introduction to Marine Studies

A. K. Spalding

Introduction

Marine studies is an emerging interdisciplinary[1] field of study that explores the dynamic relationship between people and the marine environment. As the reach of anthropogenic activities expands from heavily populated coastal areas into remote regions of the ocean, the marine environment has become a literal, as well as symbolic, last frontier for exploration and exploitation. At the same time, there is a growing awareness of the unequal outcomes and unjust practices associated with this expansion of ocean-based activities (e.g., Büscher et al., 2017; Campbell & Gray, 2019; Fusco et al., 2022). Scientists, practitioners, and civil society have made it clear that social equity and, more specifically, ocean justice must accompany all thinking about the future of coastal communities, ocean industries, conservation, and the governance of ocean spaces and resources in areas beyond national jurisdiction. More than ever, unraveling these dynamics to better understand how people affect the ocean and, in turn, how the ocean affects humans is essential. The field of marine studies is ideally suited to take on this challenge by engaging a suite of human dimensions approaches and tools (described in more detail in Chapter 2 of this volume) and applying these to real-world problems (marine-specific problems are presented in more detail in the chapters included in Part II of this book). This chapter specifically explores the evolution of marine studies as an interdisciplinary field. It then highlights the importance of adopting a marine studies approach by asking the question: Why should we care? And it ends with a discussion about ways in which interdisciplinary research and training can embrace the applied nature of this field, as well as support capacity building for the next generation of ocean and coastal professionals.

People and the Sea

The attraction to the ocean is deeply embedded into who we are as humans. From evolutionary science debates about Darwin's writings on the origins of life to faith-based representations in the Bible of the sea as simultaneously a limitless source of food and a dangerous life-threatening force, our connections to the ocean run deep and are inextricably linked to our existence. Depictions, interpretations,

DOI: 10.4324/9781003058151-2

and representations of the human relationship to the ocean abound in literature (e.g., Herman Melville's *Moby-Dick* or Samuel Taylor Coleridge's *The Rime of the Ancient Mariner*), art (e.g., Katsushika Hokusai's *The Great Wave off Kanagawa*), and history (e.g., the Age of Exploration, known for European colonization of the Americas). Similarly, naturalists such as Charles Darwin and Edward Forbes, intrigued by the ocean and its creatures, developed their own relationship with the ocean – where scientists are observers, and the ocean is the observed – by systematically documenting and studying the ocean in all its complexities. In the purest example of interdependence, humans have long relied on food from the sea for their survival. Academics often describe these relationships through distinct disciplinary perspectives. For instance, anthropologists tell us about seafaring cultures of the Pacific (Hau'ofa, 2008), political economists explain the role of neoliberal ideology on current uses of the ocean (e.g., Mansfield, 2004), fisheries biologists assess the health of fish stocks (e.g., Grorud-Colvert & Sponaugle, 2011), and oceanographers describe the relationship between ocean currents, fisheries resources, and climate change (e.g., Pinsky et al., 2018). But, as Singh et al. (2021) point out, will understanding the ocean in this fragmented way really lead to "the ocean we want"?

Importantly, it is us, as humans, who continue to socially construct[2] the ocean and have done so over time through our attitudes, beliefs, and actions. Humans have created what is known today as the Anthropocene ocean: a geographical space in which human activities are undeniably driving observed physical and environmental changes (Spalding & de Ycaza, 2020). Intricately linked as a tightly coupled natural–human system, these environmental changes are affecting human societies that are, in turn, also experiencing rapid change. It becomes ever clearer that understanding, and making decisions based on, this dynamic relationship between people and nature has become one of the most challenging, yet critical, tasks of our time. So ... what if, by recognizing the ocean as a socially constructed and dynamic space, scholars and practitioners were able to move beyond disciplinary limitations to informing decision-making? What if, instead of working narrowly within academic and practical silos, scholars and practitioners adopted a more holistic and integrated approach to understanding the material (practical) and symbolic (socially constructed) drivers of change in the ocean, and made decisions accordingly?

In practice, humans are actively creating the Anthropocene ocean through policy, management, and behaviors that are, in turn, shaped by needs, culture, values, and unequal power structures. These actions result in persistent plastic pollution in the ocean (Jambeck et al., 2015), overexploitation of living marine resources (Food and Agriculture Organization of the United Nations, 2022), and rising sea levels and temperatures as a result of anthropogenic climate change, among others. Symbolically, or through narratives about the ocean, humans are also affecting the ocean by, for instance, supporting the dominant perception of the ocean as limitless or "too big to fail" (Lubchenco & Gaines, 2019) or by conceptualizing the ocean and its resources as "open for business" (Virdin et al., 2020). Simply put, how we understand, perceive, and use the ocean matters. If we care about the future of the ocean and its resources, we must act and think in more holistic, integrated, and sustainable ways.

It follows, then, that a holistic approach to scholarship and practice – one that explicitly and intentionally integrates politics with ecological outcomes, or links literary and historical accounts of travel with state-of-the-art oceanographic observations – is a possible way forward. As Berkes points out in the preface to his 2015 book *Coasts for People: Interdisciplinary Approaches to Coastal and Marine Resource Management*: "Addressing the real problems of the world requires crossing disciplinary boundaries and, ultimately, eliminating the divides between science and management, resource user and decision-maker, and different kinds of knowledge" (p. 12). While admittedly a lofty goal, marine studies seeks to do just that! The next sections of this chapter outline how this interdisciplinary field can help address the very real challenges faced by the Anthropocene Ocean through focusing on the human dimension of coastal and marine issues.

Marine Studies as an Interdisciplinary Field of Study

You may be wondering whether this interdisciplinary thinking is really new. For instance, is it not obvious that the economy affects fisheries? Or that how we feel about the ocean affects whether or not we are willing to support or respect conservation regulations? The simple answer is that, in effect, it is not groundbreaking; and yes, some academic disciplines have embraced interdisciplinarity since at least the middle of the 20th century (Sauer, 1956). For instance, geographers, human and cultural ecologists, and environmental studies scholars have long engaged in explorations of the relationship between people and nature. However, it is the intentional application of this thinking to the ocean that is, arguably, novel. Certainly, the expectation that the information obtained from these interdisciplinary studies be used to solve real-world problems is new. Perhaps a key distinctive feature of marine studies as a field is that issues are framed foremost by the human dimension – and that the human dimension is inclusive of humanities, as well as disciplinary and interdisciplinary social sciences. By putting people first, in all their diversity and complexity, marine studies scholarship is ideally suited to disentangle the uneven, and often unjust, outcomes of resource use. Specifically, it may provide insights into the social and political conditions that are more conducive to successful conservation outcomes, or it may even help discern the human drivers of and proposed solutions to climate change. This does not mean natural and physical science contributions, alone, are not important. Instead, it suggests the need for a shift in thinking toward considering people as an inextricable part of the story of environmental change and associated solutions. While a traditional ecologist might focus on advancing knowledge about the life history of a given species based on ecological theories of evolution, an ecologist with training in marine studies might expand her focus on the life history of said species to explicitly consider the relationship between fish and associated human uses over time. A marine studies specialist might further look at the broader political economy, the culture, or even human migration patterns to understand why and how humans use certain fish species, or actions that might support the sustainable management of fisheries and the livelihoods of people dependent on living marine resources. Furthermore, marine studies often calls for collaborations across individuals who, together, form interdisciplinary research teams.

So, if not new, what are the intellectual origins of marine studies, and how does it differ from marine science? The "marine" aspect of the field is self-explanatory and encompasses all issues related to oceans (including the water column and seafloor) and coasts (spaces and uses along the broadly defined land–sea interface). To understand the meaning of "studies", consider environmental studies, a discipline that has existed since the 1970s. An inherently interdisciplinary field of study, environmental studies examines, understands, and addresses environmental challenges from a range of perspectives, including social, political, and economic concerns – often using more than one perspective at a time. The field is grounded in human ecology, a school of thought proposed by Barrows in 1923 that represented a critique of environmental determinism.[3] Subsequent scholarship, known as cultural ecology, moved away from the idea that the relationship between humans and the environment could be explained using ecological concepts and moved toward explanations of the evolution of culture and human civilizations, using culture and history as elements that also influence cultural change over time (Steward, 1955). For instance, Sauer's (1956) work on fire and agriculture reinforced Barrows' (1923) critique by recognizing the role of human activity in shaping environments into what he called "cultural landscapes". This new thinking about people and the environment had closer ties to the social sciences, applying concepts from economics, history, sociology, and political science to particular situations and environmental conditions. A more recent critique of cultural ecology, related to its failure to understand or fully account for the complexities of human interactions with their environment, is known as political ecology. First used by Eric Wolf in 1972, political ecology was born from the recognition that in post-WWII societies, people and communities no longer lived in isolation from larger political or economic forces. Political ecology scholars acknowledge the need to integrate broader social, political, and economic contexts into socio-environmental research.

Thus, drawing on this long tradition, environmental studies emerged from concepts that have evolved over time to include various human dimensions disciplines, and it applies that lens to relevant environmental problems. Importantly, the natural and social sciences work together to explore the causes of and identify solutions to complex environmental problems. In sum, environmental studies scholars acknowledge that environmental degradation is inextricably linked to the human condition and is often characterized by unequal or unjust outcomes for vulnerable communities. The inclusion of this human dimension is, indeed, what differentiates the environmental *studies* from environmental *science*. While environmental scientists acknowledge the interactions between people and the environment, they often focus primarily on the scientific drivers of and answers to environmental problems.

Currently, almost a century after the emergence of the field of human ecology, we are living in a time when converging climate, public health, economic, and racial justice crises, brought to a head in 2020 with the Covid-19 pandemic, have highlighted that failures of collective and public policies around health and the environment have perpetuated individual and collective suffering. In 2018, more than 1 billion people around the world were living in poverty (i.e., living on less than $3.20 per day), most of them in Sub-Saharan Africa (World Bank, n.d.). Furthermore, an estimated 25% of assessed animal and plant species are threatened (Intergovernmental Science-Policy Platform on Biodiversity and Ecosystem Services, 2019). Thus, the legacy of interdisciplinary thinking around people and the environment, as applied to the grand challenges facing society, is, arguably, more important and relevant than ever.

While the interdisciplinary nature of environmental or marine studies is not new, it is, unfortunately, far from widespread – likely because this type of research is harder to conduct, assess, and fund (Nature, 2015). Within the marine space, in particular, this integration of disciplines and application to real-world problems lags behind interdisciplinary efforts in terrestrial areas. An exploration of the reasons for this is beyond the purview of this chapter. But suffice it to say that the interdisciplinary perspectives on marine grand challenges presented in Part II of this book, and the various human dimensions approaches highlighted in Part III, are important reminders of a much-needed shift in thinking toward inclusive science that goes beyond the use of increasingly complex and sophisticated tools for describing and modeling ocean conditions. Indeed, such a shift could be responsive to calls for environmental justice and equity and the solutions-oriented science needed to achieve an ocean space that reflects the diversity of needs and interests of marine actors, knowledge-holders, industry, and government (Singh et al., 2021).

Finally, you may have heard the ideas described above as falling under the purview of disciplines other than marine studies. Other names typically used to describe interdisciplinary approaches to marine issues include maritime studies, marine social science, marine or blue humanities, marine affairs, among others. Indeed, these are all terms broadly used to refer to the study of people and the ocean. While the term marine studies is used here as an all-encompassing term, it is important to also acknowledge the many historical and current traditions that have influenced and inform the various chapters of this book.

Changing Oceans and Societies: Why Should We Care?

We have hinted at the challenges faced by the ocean and the people who depend on it. These include climate change, pervasive plastic pollution, questions around conventional exploitation of fossil fuels and increased knowledge about renewable sources of energy from the ocean, food from the sea (wild-caught and farmed), and global negotiations around resource management in ocean areas beyond national jurisdiction (further described in Part II of this book). The pervasiveness of these challenges suggests that the time to change our approach to finding solutions is now. Three points make this a unique moment for finding solutions for people and the sea.

First, physically and environmentally, the imprint of humans is ever-present. The Anthropocene ocean puts us, as humans, closer than ever to being affected by and having the opportunity to shape ocean futures (Biermann, 2021). For instance, technology has allowed us to explore further and deeper into the ocean than ever before, while also increasing our ability to predict future ocean conditions and, thus, reduce risk at sea. Furthermore, recent natural and associated social disasters (e.g., the 2004 Indian Ocean Tsunami, recurring heat waves across Europe, wildfires in Australia and most of the US West Coast in 2020 and 2022, and the global health crisis from the Covid-19 pandemic) have become regular reminders that we are highly dependent on a functioning natural, social, and political environment.

Second, related climate and global health crises have exposed significant flaws in how we understand and address society's problems, emphasizing the vulnerability of certain sectors of society and the need to simultaneously consider cross-cutting issues. You may consider, for example, the link between fisheries, nutrition, poverty, and gender. Global wild-capture fisheries plateaued in the 1980s (Pauly, 2008), and climate change is predicted to further affect their availability and distribution (Pinsky et al., 2018). Nutrition and food security continue to be a high priority on the sustainable development agenda; in 2017, some 815 million people around the world went to bed hungry, and about 2 billion people lacked key micronutrients (Development Initiatives, 2017). Closely related to nutrition, in 2015 about 10% of the world's population lived in extreme poverty, meaning that they lacked access to basic needs such as health, sanitation, and education, among others. For the first time since the 1990s, that number increased as a result of the Covid-19 crisis (Lakner et al., 2021). Studies show that these changes in fisheries, nutrition, and poverty disproportionately affect women around the world. If we consider post-harvest processing, women represent 50% of the global fisheries workforce (Food and Agriculture Organization of the United Nations, 2016). However, the nature of their involvement suggests they receive lower returns than men, in part due to engagement in less profitable segments of the value chain, gendered divisions of labor, and patterns of access to and ownership of assets (WorldFish, 2016). The value of securing gender equality has been extensively studied, effectively demonstrating that supporting women's engagement in productive activities, such as fishing, has the potential to improve nutrition and other development indicators through their contributions to the overall wellbeing of the household (WorldFish, 2016). As with poverty, the Covid-19 crisis has rolled back recent gains for women's rights and equality (UN Women, 2020), and negatively affected the fishing industry as a whole (Bennett et al., 2020). Applying lessons from recent efforts to support women and girls is more important than ever, and it offers an opportunity to engage more broadly across sectors to address these intersecting challenges.

Third, from a policy and governance perspective, the complex nature of the problems faced by people and the ocean means that solutions increasingly require both social and ecological information to make better resource management and development decisions. The ocean has only recently emerged as a critical element of the global development agenda, creating opportunities for action. In 2015, the ocean secured its own Sustainable Development Goal, SDG 14: Life Under Water, within the UN's *2030 Agenda for Sustainable Development* (United Nations, 2015); and scholars and practitioners are increasingly working on emphasizing the intersecting goals and benefits across all 17 SDGs (e.g., links between sustainable fisheries and poverty reduction [WorldFish, 2016]). Additionally, the decade that started in 2020 has, in a short time, seen the emergence of global efforts such as the High-Level Panel for a Sustainable Ocean Economy, as well as a variety of national and regional-level efforts. Indeed, the UN has declared this the UN Decade of Ocean Science for Sustainable Development. These actions allow for increased attention to the natural, social, and physical aspects of the ocean to be placed on the global agenda and have the potential to raise important questions and hold global and national institutions accountable for addressing the intersectional nature of our ocean's grand challenges.

How Does a Marine Studies Approach Help Address Ocean and Society Problems?

To provide a deep and practical understanding of the complex social, cultural, historical, political, and economic character of the marine environment, it is necessary to center human dimensions research, training, and capacity building. This training must be grounded in a solid understanding of marine natural and physical processes. The interdisciplinary perspective of a marine studies approach can be applied to real-world problems such as climate change, fisheries, aquaculture, pollution, energy production, governance, and biodiversity loss. Furthermore, this approach provides the skills and creates opportunities for future professionals to engage in a diversity of careers related to social and environmental justice, policy, entrepreneurship, community development, support for Indigenous rights, and conservation.

Marine Studies Research

Social science research and a humanities approach to the study of the marine environment has traditionally been organized around distinct disciplines, instead of by its shared focus on the ocean and associated activities. Analogous research on land use, food, and agriculture, in contrast, has enjoyed a community of scholarship built on shared study sites, themes, topics, theories, methods, and approaches. However, this is changing for the marine space. For instance, the Manifesto for the Marine Social Sciences (Bavinck & Verrips, 2020), generated at the Centre for Maritime Research's 2019 MARE conference, is the first attempt at identifying marine and coastal topics that are relevant to social scientists. In the Manifesto, the authors identified urgent marine social science topics, suggestions for further research, and thoughts on how to apply new methodologies and approaches to the study of marine issues. The resulting vision for marine social sciences includes an expansion of theory and applied research opportunities, as well as marine-based thematic focal areas such as ocean politics, regional perspectives, gender and fisheries, and sustainable blue growth. Importantly, there is a need to further connect the social sciences, the humanities, and the natural and physical sciences. This is often limited by challenges such as the inclusion of social science as an afterthought to a project, a relative lack of funding for environmental social science and humanities, and the differences in jargon between natural and social scientists (Spalding & Biedenweg, 2017). However, the interdisciplinary nature of marine studies research may help to overcome these challenges and facilitate integration by centering the importance of understanding the environmental and physical characteristics of the ocean (see Chapter 3 of this volume). In other words, marine studies research can help to understand the broad social and environmental context of marine issues, with a vision to identify holistic and viable solutions to the most pressing problems faced by people and the sea.

Interdisciplinary Training and Workforce Capacity Building

The vision for marine studies research outlined above requires appropriate training and capacity building. A key critique of interdisciplinarity is that it focuses on breadth (knowing a little about many things) versus depth (knowing a lot about a few things). However, in the context of marine studies, breadth of knowledge might be an asset – especially if accompanied by an understanding of the type of social science or humanities expertise that is needed to address a given problem (if it's not your expertise, you can always call on someone with knowledge in that field! Indeed, collaboration is a key element of marine studies), the type of natural or physical science that would be most useful for a given issue, and the social and governance context within which said issue is occurring (International Ocean Institute, 2018). Chapter 2 (*Human Dimension Approaches to Marine Studies*) and Chapter 3 (*The Ocean – An Introduction to the Marine Environment*) in this volume offer this foundational knowledge, while subsequent chapters in Part II illustrate how that knowledge is applied to key socio-environmental issues. Furthermore, research shows that while disciplinary knowledge can be valuable, other capacities such as the ability to work collaboratively, tolerance and reflexivity, trust, and the ability to balance power dynamics within groups are essential elements of interdisciplinary training and capacity building for the marine workforce of the future (Blythe & Cvitanovic, 2020).

Conclusion: A Field Evolved

This chapter shows how marine studies has evolved to integrate humanities, natural, and social science disciplinary approaches in the context of real-world problems. Through marine studies research, training, and capacity building we can prepare future interdisciplinary marine leaders, inspire new knowledge about people and the sea, generate innovative solutions for sustainable ocean futures, and build capacity for a new generation of marine-focused professionals. Environmental careers are diverse, wide ranging, and rapidly growing! Chapters in Part III of this volume provide examples of how marine studies training can be used to address social justice, understand governance and decision-making, incorporate and support Indigenous rights and knowledge, build marine-themed businesses, work with coastal communities, or focus on conservation. Importantly, marine studies lays the foundation for you to learn more about the links between people and the sea and invites students to embrace interdisciplinarity and align research and training with pressing environmental and societal needs.

Notes

1 Rosenfield (1992) defines multidisciplinary and interdisciplinarity as the spaces in which teams of people with different disciplinary backgrounds work in parallel with no or some integration across fields, respectively. Transdisciplinarity goes beyond integration and is characterized by a shared goal or approach to address the societally relevant question or issue.
2 Social construction is a sociological theory that suggests that knowledge of or about objects, events, landscapes, or even other living beings is shaped by meanings placed on said objects, landscapes, or living beings by society (Andrews, 2012).
3 Environmental determinism is a highly debated and critiqued geographical concept that suggests all human activities and characteristics are determined by environmental conditions. The concept has a complex history, starting in the early 20th century. Currently, geographers generally accept that the human condition is affected by politics, the economy, and other social factors, in addition to the environment (Livingstone, 2011).

References

Andrews, T. (2012). What is social constructionism? *Grounded Theory Review, 11*(1). http://groundedtheoryreview.com/2012/06/01/what-is-social-constructionism/

Barrows, H. (1923). Geography as human ecology. *Annals of the Association of American Geographers, 13*(1), 1–14. https://doi.org/10.1080/00045602309356882

Bavinck, M., & Verrips, J. (2020). Manifesto for the marine social sciences. *Maritime Studies, 19*(2), 121–123. https://doi.org/10.1007/s40152-020-00179-x

Bennett, N. J., Finkbeiner, E. M., Ban, N. C., Belhabib, D., Jupiter, S. D., Kittinger, J. N., Mangubhai, S., Scholtens, J., Gill, D., & Christie, P. (2020). The COVID-19 pandemic, small-scale fisheries and coastal fishing communities. *Coastal Management, 48*(4), 336–347. https://doi.org/10.1080/08920753.2020.1766937

Berkes, F. (2015). *Coasts for people: Interdisciplinary approaches to coastal and marine resource management*. Routledge. https://doi.org/10.4324/9781315771038

Biermann, F. (2021). The future of 'environmental' policy in the Anthropocene: Time for a paradigm shift. *Environmental Politics, 30*(1–2), 61–80. https://doi.org/10.1080/09644016.2020.1846958

Blythe, J., & Cvitanovic, C. (2020). Five organizational features that enable successful interdisciplinary marine research. *Frontiers in Marine Science, 7*. https://doi.org/10.3389/fmars.2020.539111

Büscher, B., Fletcher, R., Brockington, D., Sandbrook, C., Adams, W. M., Campbell, L., Corson, C., Dressler, W., Duffy, R., Gray, N., Holmes, G., Kelly, A., Lunstrum, E., Ramutsindela, M., & Shanker, K. (2017). Half-earth or whole earth? Radical ideas for conservation, and their implications. *Oryx, 51*(3), 407–410. https://doi.org/10.1017/S0030605316001228

Campbell, L. M., & Gray, N. J. (2019). Area expansion versus effective and equitable management in international marine protected areas goals and targets. *Marine Policy, 100*, 192–199. https://doi.org/10.1016/j.marpol.2018.11.030

Development Initiatives. (2017). *Global nutrition report 2017: Nourishing the SDGs*. https://globalnutritionreport.org/documents/822/Global_Nutrition_Report_2017.pdf

Food and Agriculture Organization of the United Nations. (2016). *The state of world fisheries and aquaculture 2016: Contributing to food security and nutrition for all*. Retrieved July 14, 2022 from https://www.fao.org/3/i5555e/i5555e.pdf

Food and Agriculture Organization of the United Nations. (2022). *The state of world fisheries and aquaculture 2022. Towards blue transformation*. Retrieved July 19, 2022 from https://www.fao.org/documents/card/en/c/cc0461en/

Fusco, L. M., Knott, C., Cisneros-Montemayor, A. M., Singh, G. G., & Spalding, A. K. (2022). Blueing business as usual in the ocean: Blue economies, oil, and climate justice. *Political Geography*, 98, 102670. https://doi.org/10.1016/j.polgeo.2022.102670

Grorud-Colvert, K., & Sponaugle, S. (2011). Variability in water temperature affects trait-mediated survival of a newly settled coral reef fish. *Oecologia, 165*(3), 675–686. https://doi.org/10.1007/s00442-010-1748-4

Hau'Ofa, E. (2008). We are the ocean. In *We Are the Ocean*. University of Hawaii Press.

Intergovernmental Science-Policy Platform on Biodiversity and Ecosystem Services. (2019). Summary for policymakers of the global assessment report on biodiversity and ecosystem services (summary for policy makers). IPBES Plenary at its seventh session (IPBES 7, Paris, 2019). Zenodo. https://doi.org/10.5281/zenodo.3553579

International Ocean Institute. (2018). *The future of ocean governance and capacity development*. Brill. https://doi.org/10.1163/9789004380271

Jambeck, J. R., Geyer, R., Wilcox, C., Siegler, T. R., Perryman, M., Andrady, A., Narayan, R., & Law, K. L. (2015). Plastic waste inputs from land into the ocean. *Science, 347*(6223), 768–771. https://doi.org/10.1126/science.1260352

Lakner, C., Yonzan, N., Mahler, D. G., Aguilar, R. A. C., & Wu, H. (2021). *Updated estimates of the impact of COVID-19 on global poverty: Looking back at 2020 and the outlook for 2021*. World Bank. Retrieved July 14, 2022 from https://blogs.worldbank.org/opendata/updated-estimates-impact-covid-19-global-poverty-looking-back-2020-and-outlook-2021

Livingstone, D. N. (2011). Environmental determinism. In J. A. Agnew & D. N. Livingstone (Eds.), *The Sage handbook of geographical knowledge* (pp. 368–380). Sage Publications.

Lubchenco, J., & Gaines, S. D. (2019). A new narrative for the ocean. *Science, 364*(6444), 911–911. https://doi.org/10.1126/science.aay2241.

Mansfield, B. (2004). Neoliberalism in the oceans: "Rationalization," property rights, and the commons question. *Geoforum, 35*(3), 313–326. https://doi.org/10.1016/j.geoforum.2003.05.002

Nature. (2015). Why interdisciplinary research matters. *Nature, 525*, 305. https://doi.org/10.1038/525305a

Pauly, D. (2008). Global fisheries: A brief review. *Journal of Biological Research- Thessaloniki, 9*, 3–9.

Pinsky, M. L., Reygondeau, G., Caddell, R., Palacios-Abrantes, J., Spijkers, J., & Cheung, W. W. (2018). Preparing ocean governance for species on the move. *Science, 360*(6394), 1189–1191. https://doi.org/10.1126/science.aat2360

Rosenfield, P. L. (1992). The potential of transdisciplinary research for sustaining and extending linkages between the health and social sciences. *Social Science & Medicine, 35*(11), 1343–1357. https://doi.org/10.1016/0277-9536(92)90038-R

Sauer, C. O. (1956). The education of a geographer. *Annals of the Association of American Geographers*, *46*(3), 287–299. https://doi.org/10.1111/j.1467-8306.1956.tb01510.x

Singh, G. G., Harden-Davies, H., Allison, E. H., Cisneros-Montemayor, A. M., Swartz, W., Crosman, K. M., & Ota, Y. (2021). Will understanding the ocean lead to "the ocean we want"? *Proceedings of the National Academy of Sciences, 118*(5). https://doi.org/10.1073/pnas.2100205118

Spalding, A. K., & Biedenweg, K. (2017). Socializing the coast: Engaging the social science of tropical coastal research. *Estuarine, Coastal and Shelf Science, 187*, 1–8. https://doi.org/10.1016/j.ecss.2017.01.002

Spalding, A. K., & de Ycaza, R. (2020). Navigating shifting regimes of ocean governance: From UNCLOS to sustainable development goal 14. *Environment and Society, 11*(1), 5–26. https://doi.org/10.3167/ares.2020.110102

Steward, J. H. (1955). *Theory of culture change: The methodology of multilinear evolution.* University of Illinois Press.

United Nations. (2015). *Transforming our world: The 2030 agenda for sustainable development* (A/Res/70/1). https://sdgs.un.org/sites/default/files/publications/21252030%20Agenda%20for%20Sustainable%20Development%20web.pdf

UN Women. (2020). COVID-19 and its economic toll on women: The story behind the numbers. Retrieved July 14, 2022 from https://www.unwomen.org/en/news/stories/2020/9/feature-covid-19-economic-impacts-on-women

Virdin, J., Vegh, T., Jouffray, J. -B., Blasiak, R., Mason, S., Österblom, H., Vermeer, D., Wachtmeister, H., & Werner, N. (2021). The Ocean 100: Transnational corporations in the ocean economy. *Science Advances*, *7*(3), eabc8041. https://doi.org/10.1126/sciadv.abc8041

Wolf, E. (1972). Ownership and political ecology. *Anthropological Quarterly*, *45*(3), 201–205. https://doi.org/10.2307/3316532

World Bank. (n.d.). *Poverty and inequality platform.* Retrieved July 21, 2022, from https://pip.worldbank.org/home

WorldFish. (2016). *Why gender equality matters in fisheries and aquaculture.* Retrieved July 14, 2022 from https://www.worldfishcenter.org/pages/why-gender-equality-matters-fisheries-aquaculture/

2 Human Dimension Approaches to Marine Studies

K. Biedenweg, S. Chisholm Hatfield, and A. K. Spalding

Introduction

People interact with the ocean in many ways. Some earn a living by spending weeks at sea to commercially fish for species like halibut or salmon; some like to walk along the shore to observe wildlife and contemplate the expansiveness of the earth; a few design complex machines to capture wave energy that will support human societies; and many live lives so entwined with the ebbs and flows of tidal and marine life cycles they may not distinguish between the marine and human system. Although our interactions with the ocean are similar to our interactions with dry land, they are also unique. To date, the majority of humans are unable to live in or on the open ocean itself; yet we still affect it and are highly intrigued by it. In fact, the grand challenges introduced in Part II (i.e., fishing, aquaculture, pollution, climate change, energy production, and resource use in Areas Beyond National Jurisdiction [ABNJ]) are inherently human challenges, created by and affecting humans. The unique character of interactions between people and the marine environment suggests a need to better understand the social processes that influence, and are in turn influenced by, the marine system. The approaches we might use to study these social processes are broadly known as human dimensions.

The human dimension of marine studies refers to all the ways that we understand the past, present, and future of human interactions with the marine environment. Social science disciplines, for example, contribute to this understanding using a scientific approach to identify patterns and trends in human–ocean interactions. Disciplinary fields include psychology (the study of the human mind and behavior), sociology (the study of human society), and political science (the study of politics and power at national and international scales). The humanities, in contrast, use interpretive or non-scientific approaches to study and find meaning in human society and culture. Such disciplines include history, literature, philosophy, languages, religion, and the arts. Some fields, such as anthropology, have traditionally drawn from the methods of both the social sciences and the humanities, hence why a broadly defined human dimension approach activates our knowledge of both. For example, we might combine our understanding of historical uses of fish with our understanding of current market values for

DOI: 10.4324/9781003058151-3

fish to establish appropriate fishery policies. In this way, we think of the human dimensions of marine studies as scientific, analytical, interpretive, applied, and interdisciplinary.

We divide this chapter into three sections: social sciences, interdisciplinary social sciences, and humanities. We also provide insights to a field that is a critical component of marine studies but does not always fit well within Western scientific categories, which is Traditional Ecological Knowledge (TEK). We want to emphasize, however, that these sections are not critical distinctions, nor entirely agreed upon by all human dimension specialists. Rather, the groupings are meant to help you wrap your head around the diversity of human dimension fields and how they inform our ways of knowing about oceans and society. Moreover, the disciplines we describe do not represent an exhaustive list of the fields that inform the human dimensions of marine studies. In fact, there are more than we can adequately explore here. Yet we hope this introduction showcases how these disciplines inform marine studies and sets you up to engage with the rest of this book more fully.

What Is Traditional Ecological Knowledge?

TEK is a sub-discipline of ecology and retains strong methodology components used in anthropology, such as ethnography. TEK refers to the rich and longstanding knowledge held by traditional Indigenous communities about the ecosystems with which they have interacted, in some cases, for over hundreds of thousands of years. TEK is a Western term created to describe Native Indigenous Science and practices relating to environmental behaviors, and thus each Tribe has their own definition based on cultural adherences and Knowledge. TEK differs from other knowledges, like Farmers or Fishermen's Ecological Knowledge, in that it is usually at least four generations' worth of information orally passed down and applied to natural resource issues; indeed, it can comprise up to 12 or more generations' worth of Indigenous Knowledge. It is often associated with anthropology, but the study itself achieved recognition during the 1970s when ecologists expanded on anthropological methods of examining the environmental systems and Indigenous Science(s) of Native American and First Nations peoples. TEK is typically documented through stories, imagery, songs, and beliefs that are tied to cultural tenets, which are linked to a Tribe's direct relationship with natural resources. It is unlike Western science, which generally refers only to knowledge that is written, vetted collectively, and is holistic rather than based in taxonomic categorization. That said, TEK is a critical and important human dimension of marine studies because of its expansive history, the ethnographic information it provides, and the detailed descriptions of the interactions between humans and the environment. For example, Eastern African spiritual and religious practices and social taboos inform how people manage natural resources such as fisheries (Mathooko, 2005). Native fishermen have used their empirical knowledge, developed over millennia (Erlandson et al., 2009; Nelson, 2008), of tidal cycles, microclimates, and biological phenomena to inform when they would remove their gear from spawning grounds to ensure reproduction and, thus, long-term subsistence opportunities (Hoagland, 2017; Reynolds et al., 2013).

The Basic Social Sciences

Here, we dive into some of the social sciences that you may already be familiar with and explain how they can directly help us understand the interactions between humans and the marine system (see Table 2.1). Specifically, we showcase how social science disciplines could help us better understand and potentially address three of the six grand challenges highlighted in Part II of this book: fisheries, renewable energy, and pollution.

For instance, *sociology*, the study of how human societies are structured, develop, and function, includes many relevant subfields covered in this book, including ethnic studies, environmental justice, and community development. The breadth of theory in these fields can help us understand why it is that primarily minority groups throughout United States work in fishery processing, how the placement of a tidal energy turbine may disproportionately impact households in one community over another, and how failing infrastructure resulting from systemic neglect of stormwater and sewage systems in poor communities leads to ocean pollution that is, in turn, blamed on those disadvantaged communities. While marine issues are historically understudied from a sociological perspective, Longo and Clark (2016) make a compelling argument about the important role of marine sociology as an emerging area of study. For example, sociologists developed the concept of social metabolism to explore how society's drive to commodify fisheries resources ultimately creates a divide (or metabolic rift) between people and the resources upon which they depend (Clausen and Clark, 2005; Longo et al., 2015).

Anthropology is the study of human societies and their cultures. Anthropology helps us understand, for instance, cultural beliefs that influence when and what species can be fished, historical and culturally relevant sources of energy that could spur innovation in the much-needed transition to renewable energy, and the cultural drivers or responses to plastic waste and other forms of pollution. Cultural and social anthropologists usually spend extensive time in one community to conduct their research through a process called ethnography, whereas archaeologists (a subfield of anthropology) often spend their time in the field looking for relics and artifacts of prior human societies. A 1981 review of the anthropology of fishing described how anthropologists can use their studies to explore fisheries, life on ships, and marine archaeology (Acheson, 1981). While many marine archaeologists focus on shipwrecks to understand humans' historical experiences with the oceans, others have found human-made monuments they believe date back thousands of years to a time when the oceans were lower than today.

Psychology, the study of the human mind and behavior, is the foundation for understanding how people learn and make decisions. In marine studies, psychology can help us understand key aspects of ocean literacy (what people know about the ocean and their influence on it) (National Oceanic and Atmospheric Administration, 2020) and why people engage in ocean stewardship (specific behaviors intended to care for the ocean) (United National Global Compact, 2020). Psychology is one of the fields that can help us understand reasons why artisanal fishers choose certain gear types for fishing (e.g., beach seines vs. hook and line), public attitudes toward renewable energy, and the reasons why people do or do not use plastic straws that are known to pollute waterways. A recent article summarizing psychological research on plastic pollution, for example, described that visual imagery triggers emotions that help people form memories and motivate behavioral change (Pahl et al., 2017). The authors of the study suggest that graphic images on plastic products, such as those found on cigarette packages, could substantially reduce the purchase and use of plastics that result in ocean pollution.

Political science is the theory and practice of government and politics, including the distribution of power and application of processes to make decisions. While human actions may negatively impact the marine environment, in the Anthropocene – a proposed geological epoch in which human activity is understood to be the dominant driver of environmental change (Lewis and Maslin, 2015) – we are also tasked with deciding how it will be managed. The field of political science can help us understand how international laws, such as the United Nations Convention on the Law of the Sea (UNCLOS), interact with national laws and local policies to determine who has rights to use, access, and distribute marine resources or spaces. It also helps clarify how cross-boundary issues, such as migratory living marine resources (e.g., tuna), are regulated and managed individually by nations or collaboratively across nations and regions; how conflicts over the use of and access to resources and spaces in the ocean are resolved; and how governance regimes are shaped over time. For instance, using a global environmental politics approach, Spalding and deYcaza (2020) suggest that ocean governance has evolved from a process that sought to establish jurisdictional zones on the water, to a critical element of the global sustainable development agenda. In other words, political science shows us where and how people, the ocean, and its resources are situated within global environmental politics.

Possibly the most commonly recognized social science is *economics*. This discipline focuses on studying the production and distribution of limited goods and services. In the marine space, it can help us understand the market forces that influence whether commercial fishing fleets target halibut versus tuna, whether marine energy is a financially viable alternative to energy derived from oil and gas, and the costs of pollution clean-up versus pollution prevention. For example, economists might study how a growing marine renewable energy sector will result in job creation, job displacement, and the creation of new supply chains that capture expenditures (Kerr et al., 2014).

Table 2.1 Example questions that some fields of human dimensions can address to contribute to our understanding of grand challenges in marine studies

	Fisheries	*Renewable Energy*	*Pollution*
Sociology	What Social Groups Engage in Fish Processing?	What Communities Will Be Most Impacted by the Placement of a Tidal Turbine?	How Do Failing Stormwater Systems Influence Communities Differently?
Anthropology	What Cultural Beliefs Influence When and What Is Fished?	What Were Historical, Culturally Relevant Sources of Energy?	How Does Ocean Pollution Impact Human Genetic and Epigenetic Evolution?
Psychology	Why Do Artisanal Fishers Choose to Use Beach Seines vs. Hook and Line?	What Are Public Attitudes About Renewable Energy?	What Outreach Strategies Will Decrease the Use of Plastic Straws?
Political Science	What Laws Establish the Rights to Fisheries?	What Actors Are Engaging in Development of Renewable Energy Technologies, and How Do They Interact with Each Other?	How Do Cross-Boundary Issues Influence the Prevention and Mitigation of Ocean Pollution?
Economics	What Market Forces Influence Tuna Catch?	How Does the Cost of Renewable Energy Compare to Non-Renewable Energy?	Is the Cost of Pollution Clean-Up Higher Than the Cost of Pollution Prevention?
Geography	Where Do Fishers Fish?	Where Would Be the Most Appropriate Site for a Renewable Energy Project, Considering Human and Ecological Factors?	Which Human Communities Are Impacted by Pollution, and Are There Spatial Patterns to These Human Impacts?
Political Ecology	To What Extent Do International Laws Consider Local Cultures and the Influence of Power Dynamics in Fishery Market Chains?	What Global and National Factors Influence the Development of Renewable Energy Projects, and Whose Voices Remain Unheard?	How Does Tourism from Predominantly Developed Nations Influence Marine Pollution in Small Island Developing States (SIDS)?
Public Health	Are Fish Polluted to the Extent That They Affect the Health of Those Who Are Dependent on Eating Them?	What Are the Potential Public Health Impacts of a New Wind Turbine?	How Can We Support Communities Who Live in Marine Areas with Higher Pollution?
Development Studies	How Can Fisheries Contribute to Local Community Wellbeing and Human Development?	How Might International Agencies Focus Their Investments on Marine Renewable Energy as a Way to Increase Education in Vulnerable Coastal Communities?	How Is Marine Pollution Distributed Across the Globe and Who Is Most Impacted by It?

Interdisciplinary Social Sciences

While most universities have departments that match the descriptions of the social science fields listed in the previous section, the reality is that many social scientists use theories and methods from several disciplines. This happens to such an extent that we can identify consistent interdisciplinary social sciences. For example, when *economic* theories are integrated with ideas from *business*, the knowledge can inform our understanding of marine entrepreneurship. This, in turn, is informed by other interdisciplinary fields like *business administration* which includes accounting, marketing, finance, human resource management, and many other concentrations that apply the knowledge of psychology (human behavior), sociology (institutions and human interactions), and economics (distribution and production of goods and services) to promote healthy marine-focused businesses associated with activities such as fisheries, renewable energy, or pollution control. Examples of marine entrepreneurship can include swim and clothing companies that use post-recycled materials and focus on ocean-friendly themes and designs, reef-safe sunscreen businesses, and local and sustainable seafood restaurants. While not all ocean-related businesses will include a sustainability component (indeed, many large marine-based enterprises are extractive and destructive in nature), several have emerged from an interest in and commitment to ocean conservation and often draw from interdisciplinary social science knowledge areas by supporting local employment opportunities and other benefits to the local environment and communities.

Another common interdisciplinary field is *geography*, the study of how people relate to place and space. Because people, place, and space are all affected by politics, the economy, culture, climate, and other environmental variables, it is inherently interdisciplinary. Indeed, different branches or subfields within geography include physical, environmental, and human foci. Importantly, the way geographers explore these multiple dimensions represents a unique contribution to the understanding of people and the sea. Geographers, for instance, explore fishing activity in the context of Marine Spatial Planning, a process that visualizes the diversity of ocean uses, often using maps developed with ocean users, within a policy and governance framework for a given geographic space (Karnad and St. Martin, 2020). Much as a city planner might refer to land-use maps to establish zoning laws, marine spatial planning maps and supporting documents are frequently used to inform the establishment of new marine protected areas or other ocean use classifications.

Another example is a field that emerged from elements of anthropology and geography, called *political ecology*, that uses critical interdisciplinary methods to explore the political, economic, and social factors associated with environmental issues. Political ecologists acknowledge that interactions between people and the marine environment are driven by power dynamics within human and political systems, which can be based on socio-demographic categories like race, ethnicity, and socio-economic status. They look at the interactions between international, national, communal, and family-level institutions to understand how human–nature relationships became what they are. In the context of marine studies, political ecologists can explore the extent to which local fisheries markets are impacted simultaneously by cultural beliefs and international trade regulations, how global and national factors influence the development of renewable energy projects, and how local and international tourism and migration influence localized marine pollution. Susan Stonich's research in the Honduran Bay Islands (1999), for example, explored how increasing tourism and immigration impacted the Bay Islands of Honduras. She found that international tourism not only was increasing median household income for Bay Islanders but also increased housing costs. Moreover, the housing development that was stimulated by foreigners retiring to the islands negatively impacted the health of barrier reefs surrounding the islands due to the lack of local regulations that would require reef-friendly construction practices.

Public health is both an interdisciplinary social science, from a research perspective, and an application of various social sciences to real-world human health scenarios, such as the global COVID-19 pandemic. Scientists who inform public health might study the extent to which fisheries contribute to the daily caloric intake of social groups, how adoption of marine renewable energy might improve air quality through a reduction of CO_2 emissions, and how specific types of pollution might impact our ability to eat healthy seafood. For instance, a consortium of global scientists found that fish contribute key micronutrients, such as calcium, iron, zinc, and vitamin A, in countries that are most at risk for human malnutrition due to lacking these nutrients (Hicks et al., 2019). And in Washington State in the USA, public health officials realized that the allowed level of industrial pollution to the Puget Sound was disproportionately exposing communities with diets high in fish (mostly Native Americans) to carcinogenic compounds (Polissar et al., 2012). The latter study resulted in a reconsideration for how much pollution should be allowed to enter public waterways and an adjustment of a public health measure called the fish consumption rate.

Finally, *development studies* is another interdisciplinary field that integrates political, sociological, economic, and geographic perspectives, among others, to better understand economic and social development in the global south. The Wellbeing in Developing Countries (WeD) research program based in the UK, for example, has identified that a better framework for evaluating the success of fisheries in developing countries would consider the fishers' *material* wellbeing (the gear, natural environment, and food that the person can count on), their *relational* wellbeing (how they interact with government institutions, cultural norms, and family), and their *subjective* wellbeing (how they think they are doing) (Coulthard, 2013). This three-dimensional measure of fishery success represents a more holistic understanding of the human dimensions of fisheries compared to singular measures often grounded in exclusively capitalist values, such as income from a day's labor or the number of fish caught. Similarly, development studies can help us understand the holistic impacts of new marine renewable energy infrastructure on different communities in the developing world and on the uneven distribution of marine pollution across the globe.

The Humanities

Most people know the sea as much through the humanities as through science. The humanities study narratives and representations of ideas and events, based on histories, philosophies, and use of language to understand the human experience. We like to think of the humanities as approaches to the creation of meaning of the human experience. *History*, for example, interprets events through time that have shaped our current context. Depending on where you live, history lessons may place more focus on events associated with land, but marine studies also have fascinating histories that give context to the world we find ourselves in today. We read histories of ocean exploration, science, and naval warfare. Indeed, historians explore not only the sea's influence on our culture development but also cultural influences on how we have come to understand the ocean itself.

Within the humanities, the *visual and performing arts* are important lenses for the study of marine issues. Imagine your favorite ocean-related painting, sculpture, play, or movie. What ideas about the marine environment did it instill in you? How did it make you feel? Visual and performing arts help us create meaning and contemplate the world around us, often in very new ways. *Finding Nemo*, *The Meg*, and *20,000 Leagues Under the Sea* are completely different genres of movies (look them up if you're not familiar!), but each leaves us with both emotional and cognitive reactions to the ocean. As Steinberg (2001) reminds us in his book, *The Social Construction of the Ocean*, our knowledge and understanding of the oceans is deeply tied to the human experience.

Lastly, *philosophy and religion* explore morality and norms in society, affecting how we may interact with or understand the marine environment. From Indigenous spirituality guided by reciprocity with the natural world, to the Quran and the Bible, most major world philosophies speak to human–ocean interactions and provide norms to guide our behaviors. Indeed, we see how documents like the Pope's 2015 Encyclical (Francis, 2015) have influenced Catholic ideals of human responsibilities toward the environment, particularly as they relate to climate change. Lynn White's renowned 1967 essay, "The Historical Roots of our Ecological Crisis", published in the journal *Science*, also appeals to the "moral imperative" as an alternative to science and technology in addressing pressing environmental issues.

Human Dimension Methods and Products

The differences within and between the social sciences and humanities are often based on the ways that information is collected, analyzed, and shared. For instance, some social scientists use interviews with key informants to collect data on the causes of pollution. These interviews will result in written notes or audio recordings that are then qualitatively analyzed for themes. These themes are usually shared in longer written documents. Other social scientists, however, might use quantitative assessments, such as economic analyses of market values, to calculate the costs of renewable energy. These quantitative analyses are more likely to be displayed in tables and graphs populated by numbers and dollar figures. Policy analyses may collect data from various texts and also analyze them for important themes and contributions that might be summarized in a written document meant for publication. Visual arts result in physical objects made from any number of materials. And the performing arts may not have any tangible product at all; instead, it might leave a memory you associate with watching a wonderful musical. These different products may sometimes make it difficult to integrate the human dimensions with the ecological dimensions of the ocean, but they are all equally important in understanding marine studies. You might say that the human dimensions are as complex as the humans we are trying to understand.

Putting Them All Together

Although we described several human dimension fields as though they were distinct, the reality is that they overlap substantially. As discussed in Chapter 1, the field of marine studies is inherently interdisciplinary and the human dimensions, as described in this chapter, represent the various social disciplines within it. We would not think about ocean literacy on its own, generally. We would think about how ocean literacy should inform policy development to improve renewable energy, and how we can use art to amplify existing knowledge from history and marine ecology to enhance ocean literacy. It then becomes obvious that there are many ways to contribute to the marine problems of our time through a degree related to the human dimensions of marine studies. Each social science or humanities field provides a specific contribution to our understanding of humans and oceans. And the combination of all these fields provides a broader, more holistic picture of marine studies. The rest of the book will dive deeper into these current issues and how the human dimension approach can be applied to build our knowledge and develop future marine-related careers.

References

Acheson, J. M. (1981). Anthropology of fishing. *Annual Review of Anthropology*, 275–316. https://doi.org/10.1146/annurev.an.10.100181.001423

Clausen, R., & Clark, B. (2005). The metabolic rift and marine ecology: An analysis of the ocean crisis within capitalist production. *Organization & Environment*, *18*(4), 422–444. https://doi.org/10.1177/1086026605281187

Coulthard, S. (2013). What does the debate around social wellbeing have to offer sustainable fisheries? *Current Opinion in Environmental Sustainability*, *4*(3), 358–363. https://doi.org/10.1016/j.cosust.2012.06.001

Erlandson, J. M., Rick, T. C., & Braje, T. J. (2009). Fishing up the food web? 12,000 years of maritime subsistence and adaptive adjustments on California's channel islands. *Pacific Science*, *63*(4), 711–724. https://doi.org/10.2984/049.063.0411

Francis, P. (2015). *Laudato Si': On Care for Our Common Home* [Encyclical].

Hicks, C. C., Cohen, P. J., Graham, N. A. J., Nash, K. L., Allison, E. H., D'Lima, C., Mills, D. J., Roscher, M., Thilsted, S. H., Thorne-Lyman, A. L., & MacNeil, M. A. (2019).

Harnessing global fisheries to tackle micronutrient deficiencies. *Nature*, *574*, 95–98. https://doi.org/10.1038/s41586-019-1592-6

Hoagland, S. J. (2017). Integrating traditional ecological knowledge with western science for optimal natural resource management. *IK: Other Ways of Knowing*, *3*(1), 1–15. https://doi.org/10.18113/P8ik359744

Karnad, D., & St. Martin, K. (2020). Assembling marine spatial planning in the global south: International agencies and the fate of fishing communities in India. *Maritime Studies*, *19*(3), 375–387. https://doi.org/10.1007/s40152-020-00164-4

Kerra, L. W., Colton, J., Conway, F., Hulle, A., Johnson, K., Jude, S., Kannen, A., MacDougall, S., McLachlan, C., Potts, T., & Vergunst, J. (2013). Establishing an agenda for social studies research in marine renewable energy. *Energy Policy*, *67*, 694–702. https://doi.org/10.1016/j.enpol.2013.11.063

Lewis, S. L., & Maslin, M. A. (2015). Defining the anthropocene. *Nature*, *519*(7542), 171–180. https://doi.org/10.1038/nature14258

Longo, S. B., & Clark, B. (2016). An ocean of troubles: Advancing marine sociology. *Social Problems*, *63*(4), 463–479. https://doi.org10.1093/SOCPRO/SPW023

Longo, S. B., Clausen, R., & Clark, B. (2015). *The tragedy of the commodity: Oceans, fisheries, and aquaculture*. Rutgers University Press.

National Oceanic and Atmospheric Administration. (2020). *Ocean literacy: The essential principles and fundamental concepts of ocean sciences for learners of all ages*. https://oceanliteracy.unesco.org/wp-content/uploads/2020/09/OceanLiteracyGuide_V3_2020-8x11-1.pdf

Mathooko, J. M. (2005). Application of traditional ecological knowledge in the management and sustainability of fisheries in East Africa: a long-neglected strategy? *Hydrobiologia*, *537*, 1–6. https://doi.org/10.1007/s10750-004-2788-8

Nelson, M. K. (Ed.). (2008). *Original instructions: Indigenous teachings for a sustainable future*. Simon and Schuster.

Pahl, S., Wyles, K. J., & Thompson, R. C. (2017). Channeling passion for the ocean towards plastic pollution. *Nature Human Behaviour*, *1*, 697–699. https://doi.org/10.1038/s41562-017-0204-4

Polissar, N. L., Neradilek, M., Aravkin, A. Y., Danaher, P., & Kalat, J. (2012). Statistical analysis of national and Washington State fish consumption data. *Prep. Wash. State. Dep. Ecol.*

Reynolds, N. D., & Romano, M. D. (2013). Traditional ecological knowledge: Reconstructing historical run timing and spawning distribution of Eulachon through tribal oral history. *Journal of Northwest Anthropology*, 47(1), 47–70.

Spalding, A. K., & de Ycaza, R. (2020). Navigating shifting regimes of ocean governance: From UNCLOS to sustainable development goal 14. *Environment and Society*, *11*(1), 5–26. https://doi.org/10.3167/ares.2020.110102

Steinberg, P. E. (2001). *The social construction of the ocean* (Vol. 78). Cambridge University Press.

Stonich, S. (1999). *The other side of paradise: Tourism, conservation, and development in the Bay Islands*. Cognizant LLC.

3 The Ocean

An Introduction to the Marine Environment

K. Grorud-Colvert and M. Ward

Introduction

Seen from space, the earth is a blue planet. The ocean covers 71% of the planet's surface (Reid, 2016), making the earth unique throughout our solar system. The ocean forms the foundation for life on our planet and sustains life in many ways. Traditionally, the earth was considered to have multiple "oceans". Here, we explore the fact that it can be considered a single global ocean, with effectively no lines or barriers subdividing it. The ocean's five basins – the Arctic, Pacific, Indian, Southern, and Atlantic Basins – are biologically, chemically, physically, and politically connected (Laffoley et al., 2020). This global ocean drives the earth's essential cycles, and its underlying chemistry, biology, and physics produce the patterns in marine resources upon which human communities have come to rely, linking together places, species, and peoples across the world.

The ocean sustains humanity by providing resources, supporting human health and livelihoods, and underpinning cultural identity through a unique set of environmental processes that we describe in this chapter. We also acknowledge the dual reality that certain groups and policies have had disproportionate impacts on the ocean through exploitation, overuse, and other unsustainable practices. We look toward policy, management, and local-to-global leadership to foster ocean-based solutions, with the goal of stopping, decreasing, and even reversing these impacts and inequities through creative approaches. Although humans serve as drivers of change, receivers of that change, and mediators of solutions, in this chapter we center the ocean itself and its processes. This ocean-based understanding, which you will continue to cover in other, more specialized courses, is the foundation for the particularities of the human dimensions of the marine environment. Inquiry into this holistic system is what we call marine studies.

This chapter focuses on the marine environment and is by no means comprehensive – the ocean is vast. Volumes have been written on each of the topics in this chapter. This summary gives a 3,000-foot overview of the marine environment, enabling readers to learn more about ocean systems and humanity's interactions with them.

DOI: 10.4324/9781003058151-4

The Ocean Drives Essential Earth System Cycles

The ocean works on a planetary level to make life on earth possible. For example, life requires water, and the oceans contain 97% of the water on earth (Baumgartner et al., 1975). Life also requires oxygen, and the microscopic plants ("phytoplankton") that teem through the upper layers of the ocean are photosynthetic powerhouses, producing over half of the earth's atmospheric oxygen (Sekerci & Petrovskii, 2015). Oxygen produced by these phytoplankton diffuses from the ocean water across the sea surface and into the atmosphere, where it is cycled into the total oxygen budget of the planet (Laffoley et al., 2020).

Temperature Regulation

Earth's proximity to the sun makes life as we know it possible. However, without the ocean, earth would be much hotter and much less hospitable. Solar radiation does not reach the planet evenly given its axial tilt and seasonal cycles, requiring heat transport in order to support life. Without this heat transport, the poles would be unbearably cold while the equator unbearably hot. Water has a higher capacity to hold heat than air does – the ocean can absorb as much as 97% of the solar radiation that hits the earth's surface (Bigg et al., 2003). As a result, the ocean provides the essential service of transporting and redistributing heat around the globe, driving and sustaining a myriad of earth system cycles including both weather (short-term changes in atmospheric conditions) and climate (long-term weather patterns in a region measured over years and decades).

Ocean Circulation

The ocean stores and moves heat via a conveyor-belt-like system called thermohaline circulation, which is driven by the fact that cold, salty water is denser than warm freshwater. The poles receive less direct sunlight due to the round shape of our planet, leading to colder temperatures, which in turn produce colder water. The cold, dense water then sinks, and can take upwards of 1,000 years to journey along the ocean basins to the tropics and back to the poles on this deep ocean conveyor belt. As the water moves toward the tropics, it slowly warms and becomes less dense. As a result, when paired with the right physical conditions, these deep waters rise to the surface, bringing with them nutrients that were previously inaccessible to organisms living in the shallow, light-penetrating "photic zone". These nutrients fuel rich communities of marine life across the globe.

Water Cycling

The ocean's exchange of heat, gasses, and water with the atmosphere plays a key role in the water or hydrologic cycle (Bigg et al., 2003). Heating of surface waters along the equator leads to greater evaporation, and subsequently greater humidity, precipitation, and formation of storms. These storms are then carried away from the equator and around the globe by an intricate system of winds. As water evaporates from the ocean, it is cycled such that 78% of all rain on the earth falls on the ocean (Baumgartner et al., 1975). The evaporated water also falls as rain on land and returns to the ocean via rivers, groundwater flow, and melting ice sheets and glaciers (Intergovernmental Panel on Climate Change [IPCC], 2019).

Nutrient Cycling

In addition to transporting heat, the ocean transports and cycles essential nutrients. Nutrients can enter the ocean from the land or through deposition from the atmosphere. The nutrients – such as nitrogen, phosphorus, silica, and iron – ultimately sink deep into the water, where eventually they may be brought to the surface by deep water upwelling. Coastal upwelling occurs when winds blow nearly parallel to a coastline, and the force of the wind, combined with the earth's rotational force (called the Coriolis effect), pushes surface waters offshore and deeper, colder, nutrient-rich waters are pulled up to replace them (see Figure 3.1). This regular pulse of nutrient-rich water into coastal zones creates highly productive ecosystems. It allows marine plants (including phytoplankton) to use these fresh nutrients for photosynthesis and forms the basis for marine food webs.

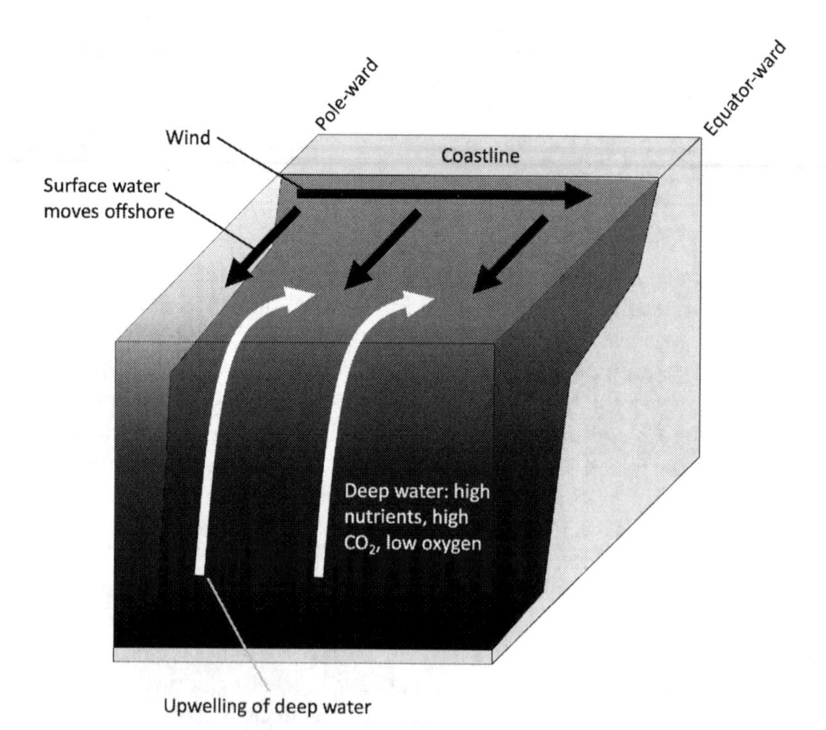

Figure 3.1 Coastal upwelling.

Carbon Cycling

The process of photosynthesis by marine plants (including phytoplankton) also absorbs CO_2. Over 50% of carbon fixation on our planet is accomplished by marine plants (Falkowski & Raven, 2013). This process sequesters carbon – another benefit the ocean provides as part of the global carbon cycle (see Figure 3.2). The ocean is a biological pump that removes CO_2 from the atmosphere and transfers and stores carbon during this process. When marine plants die, or are eaten by marine animals that die, they sink toward the ocean floor. This sinking brings organic carbon and nutrients from the surface ocean into deeper, aphotic zones, where bacteria, zooplankton, and other heterotrophic organisms can consume them. In addition to supporting life in the deep ocean, a fraction of this sinking, carbon-based material is also deposited in the sediments of coastal and deep seas. This carbon becomes buried in ocean sediments, removing carbon from the atmosphere on geological timescales, and providing an invaluable service to our climate. However, this process is being disrupted by excess levels of atmospheric carbon as the ocean removes at least 1 megaton of human-made CO_2 from the atmosphere every hour (Sabine et al., 2004) and has absorbed an estimated one-third of the CO_2 emitted since the Industrial Revolution (Gattuso et al., 2015; IPCC, 2019). It is the largest store of carbon on earth.

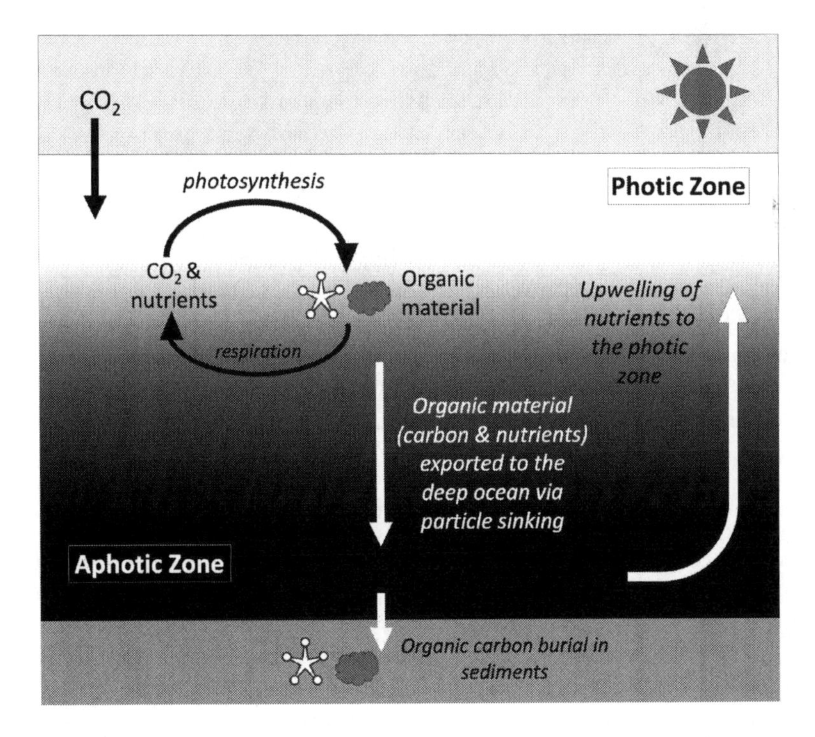

Figure 3.2 Ocean carbon cycle.

The Ocean Supports a Diversity of Marine Life

Life in the ocean is incredibly diverse. Scientists estimate there are about 2.2 million multi-celled species (eukaryotes) that live in the ocean, everything from snails to seagrass to squid to corals to whales. Yet about 91% of those species are still undiscovered by science (Mora et al., 2011). Nonetheless, the species we have discovered exhibit a range of adaptations that make them well-suited to life underwater in the ocean habitats they occupy. For example, mobile organisms, such as finfish and marine mammals, have evolved advanced body types and prey detection capabilities such as echolocation which allow them to thrive across wide expanses of ocean. Conversely, stationary (sessile) organisms have evolved to take advantage of food delivered by passing currents through filtration or tentacle-like appendages. A vast diversity of organisms and survival strategies exist in the ocean's fluid environment, and a growing number of scientific discoveries provide information about these species. Much less is known about the diversity of single-celled organisms (prokaryotes), such as bacteria and viruses, which make up the majority of life in the ocean by weight (Louca et al., 2019). Estimates of the number of microbial species in the ocean vary widely because it is mind-boggling to quantify; the number of species could range between 1×10^6 and 1×10^{27} (Louca et al., 2019; Mora et al., 2011).

Because the ocean is fluid and subsequently more open to movement by organisms, many populations of marine species have very large distributions and connect many different regions and ecosystems with their movements. Species contribute to ocean connectivity in many ways. Some may be widespread because they spend their entire life as plankton, carried by ocean currents. Others may spend their later life on the ocean floor but as plankton are transported away from their parents by ocean currents while they are very young. They may be widely distributed because they can swim great distances as adults. Regardless, populations of marine species are much more broadly distributed than terrestrial species as a whole (Gagné et al., 2020). This has many consequences, for example, it presents a challenge when managing fisheries that do not respect political boundaries and move in and out of country's waters, each with their own set of national rules and regulations for catch. One country's fishery may hinge on another country's adequate fishery management because the stock comes from upstream waters that are not their own.

The Ocean Bears the Brunt of Human Activities

For generations, humanity has relied on what seemed to be the ocean's unlimited bounty. The idea that the ocean is "too big to fail" – resources are unlimited and resilient – seemingly persists to this day as more intense and unsustainable uses of the ocean continue to increase (Lubchenco & Gaines, 2019). Accelerated climate change, caused by humans, is reworking our earth systems, including those supported by the ocean. This section summarizes four main earth system processes and how they are affected by growing environmental disruption.

The Climate Is Affected

There is no doubt that climate change is occurring, and we will continue to see its effects at an accelerating pace (Masson-Delmotte et al., 2019). Here is just part of the laundry list of anthropogenic impacts currently affecting the earth climate system:

- It is estimated that humans had caused warming of approximately 1.0°C by 2017 compared to levels before the Industrial Revolution, leading to more frequent and intense extreme weather events such as storms, fires, floods, and droughts over the past 50 years alone (Intergovernmental Science-Policy Platform on Biodiversity and Ecosystem Services [IPBES], 2019).
- We are seeing seasonal changes in sea ice that are likely unprecedented based on the last 1,000 years (IPCC, 2019). This is leading to thinning of Arctic Sea ice and amplified warming in the Arctic, where surface air temperature likely increased by more than double the global average over the past two decades (IPCC, 2019).
- The global average sea level has risen between 16 and 21 cm in the last 120 years (IPBES, 2019).
- Oxygen concentrations in the upper 1,000 m of the ocean have decreased between 0.5 and 3.3% (IPCC, 2019).
- Marine heat waves have doubled in frequency and have become longer-lasting, more intense, and more extensive (IPCC, 2019). This has led to impacts on single species – such as widespread bleaching of corals – and entire ecosystems, where shifts in food webs produce effects across species groups (Smale et al., 2019).
- The ocean has absorbed approximately 30% of anthropogenic CO_2 emissions. This absorption has led to reduced seawater pH over more than 95% of the ocean's surface area – a process termed "ocean acidification" (IPCC, 2019). These conditions decrease coral reef accretion rates; prevent larval molluscs, like oysters, from creating their shells; and also threaten organisms called pteropods or sea butterflies – tiny sea snails that play a major role as a food source for many ocean species (Kroeker et al., 2010).

Biodiversity Is Affected

All of these impacts from climate change also affect the biology of ocean organisms and the biodiversity that the ocean supports. For example, since the 1870s about half of the live coral cover on earth has been lost; this loss is continuing and accelerating in recent decades due to the effects of climate change (IPBES, 2019; IPCC, 2019). However, climate change is only the second largest impact on marine biodiversity; overexploitation of marine resources, primarily through fishing, is the single greatest contributor to the decline in biodiversity in the ocean (IPBES, 2019; Rogers et al., 2020). Adding to these impacts are those from land- and sea-based pollution and changes in land and sea use, including coastal development (IPBES, 2019). Moreover, these impacts do not typically occur in a vacuum but are cumulative; combined stress of myriad impacts is affecting at least 66% of the ocean's area (Halpern et al., 2015).

It is difficult to confirm how many marine species have become extinct over the past 500 years; at present 20 extinct marine species are known, but this is likely an underestimate because some marine organisms have not been observed for decades and may already be extinct (McCauley et al., 2015). Ocean species at risk of extinction or globally threatened include 25% of sharks, rays, and marine mammals (Dulvy et al., 2014; International Union for the Conservation of Nature, 2019). Populations of large wild species, such as tuna and other vertebrates, have declined over the last 50 years (IPBES, 2019), and since 1970 oceanic sharks and rays have declined globally by 71% (Pacoureau et al., 2021).

Even as species decline, stressors continue, including coastal development, use of biological resources, pollution, effects of transportation, invasive species, climate change, and other human disturbances (Rogers et al., 2020). More than 1 million square kilometers of the ocean floor are subject to bottom trawling each year (Amoroso et al., 2018). Deep sea mining is becoming more prevalent due to the increasing demand for raw materials and decreasing terrestrial resources (Miller et al., 2018) – the impacts are projected to be extremely severe (Niner et al., 2018).

Resources Are Affected

The loss in biodiversity also hints at the subsequent decrease in the productivity that it supports. Changes in the abundance and distribution of species that act as key resources to human populations also occur. The ranges of many marine species are projected to shift at the global scale by 2055, with an average increase of 30–70% of species moving to high latitudes and up to 40% of species moving out of tropical latitudes (Cheung et al., 2010). Predictions indicate that the largest decreases in animal biomass due to climate change will be at the middle latitudes, where many nations currently depend on seafood and fisheries (Lotze et al., 2019).

Overfishing has also compromised ocean resources. A massive increase in the size of the global fishing fleet from 1.7 to 3.7 million vessels occurred between 1950 and 2015 (Rousseau et al., 2019). While improved technology has led to exponential increases in fishing effort, catch per unit effort (CPUE) is declining due to depleted stocks of fish (Rogers et al., 2020). A conservative estimate suggests that unassessed fisheries produce 23% of the global fish catch (Costello et al., 2012), highlighting the issue of Illegal, Unreported, and Unregulated (IUU) fishing and the reality that many projections for fisheries may underestimate the true impact of global fishing.

People Are Affected

The impacts from climate change, overfishing, pollution, and other anthropogenic stressors accumulate over time, posing immense risks to human populations. For example, the low-lying coastal zones most in danger of sea level rise are currently home to 680 million people (nearly 10% of the global population in 2010), and of those, 65 million people live in small island developing states (SIDS) (IPCC, 2019).

A decrease in the availability of resources upon which humans rely also compromises our food security, health, and well-being (Österblom et al., 2020). Additionally, impacts are often distributed unequally – across place, time, and segments of society. Impacts based on the depletion and disruption to global climate, biodiversity, the natural functioning of ecosystems, and the contributions that nature makes to people are projected to be greatest in areas of the world with many of the most disadvantaged communities in the world (IPBES, 2019).

A Functioning Ocean Sustains Humanity

The total value of key assets provided by the ocean is estimated to be US $24 trillion annually, and the services that derive from these assets may add an additional $1.5 trillion to $6 trillion per year (Hoegh-Guldberg, 2015; Lillebø et al., 2017; Organization for Economic Cooperation and Development [OECD], 2016). However, some ocean services transcend economics, and monetary evaluations should not supersede the non-monetary and non-material benefits the ocean provides (Allison et al., 2020). These benefits, often referred to as nature's contributions to people or ecosystem services (Allison et al., 2020; Millennium Ecosystem Assessment, 2005) can be categorized in four types: Supporting, Regulating, Provisioning, and Cultural. Below we discuss what the ocean provides within each of these categories.

Supporting

The ocean's ecosystem services support life on the planet and all of the resulting benefits this biodiversity provides. All of the earth system processes discussed above fit the definition of supporting services. The ocean cycles carbon, nutrients, and water at the global level. These nutrients fuel primary production, which in turn fuels global food webs and transfers biological energy across the earth. Formation of foundational habitats, such as coral reefs, kelp forests, oyster reefs, and seagrass meadows, is another supporting service the ocean provides. For example, coral reef habitats support approximately 25% of the species in the ocean, with some 830,000 multi-cellular species estimated on coral reefs worldwide (Fisher et al., 2015). All of these foundational habitats support biodiversity by providing protection for the early life stages of marine species in nursery grounds, and also at other key life stages (Stuchtey et al., 2020).

Productive areas of high biodiversity can help to sustain surrounding marine environments, as well. For instance, they can export adult or larval animals to neighboring habitats ("spillover effect"), or act as storehouses of genetic diversity from the genetic library of these many species, contributing to improved resilience to changing ocean conditions (Blasiak et al., 2020). These areas are prime candidates for consideration as fully to highly protected MPAs when the other social and ecological enabling conditions are in place (Grorud-Colvert et al., 2021; Jones et al., 2018).

Regulating

Ocean services are vital to life on earth, via carbon sequestration and storage, temperature moderation, and balancing of extreme weather events. As water flows from land into the ocean, coastal habitats are an important source of water purification. For example, mangrove forests, wetlands, and oyster reefs can filter out pollutants and excess nutrients and trap sediments, leading to clearer and cleaner nearshore waters. These coastal habitats are also excellent examples of first-line defense against storm surges; they not only decrease the impacts of storm-surge waves on erosion and property damage but also prevent significant economic losses from large storms, such as hurricanes (Del Valle et al., 2020). Worldwide, coastal habitats provide protection for hundreds of millions of people (Stuchtey et al., 2020).

Provisioning

The ocean is a provider of food, medicines, biochemicals, and genetic resources, all deriving from its rich biodiversity. Food from the sea is an important source of protein and key nutrients – such as omega-3 fatty acids and iodine – for more than 3.3 billion people (Food and Agriculture Organization of the United Nations, 2020). This food provisioning is truly global; coral reefs alone benefit people from more than 100 countries by providing food and livelihoods (Masson-Delmotte et al., 2019). Some of these livelihoods are built on more than extraction of these marine species. Biodiverse places and habitats are significant attractions for tourism; another benefit from coral reefs – tourists visiting reefs globally contribute US $11.5 billion to the global economy (Masson-Delmotte et al., 2019). In fact, the global economy depends on the ocean through the livelihoods it supports. It provides jobs for hundreds of millions of people working in fishing, mariculture and aquaculture, shipping and ports, tourism, offshore energy, pharmaceuticals and cosmetics, and other industries (Stuchtey et al., 2020; Teh & Sumaila, 2013).

The ocean is also a source of many products that we use in our daily lives. Over 34,000 marine natural products have been discovered (MarinLit, 2020) – including marine drugs, nutraceuticals, cosmetics, aquaculture and new food products, and bulk chemicals (Blasiak et al., 2020). These products are made possible by the vast genetic and organismal diversity found in the ocean (think of those 2.2 million species of eukaryotes alone). Medicines may be found in unusual places. For example, Remdesivir, an antiviral that is derived from sea sponges, was approved by the U.S. Food and Drug Administration on May 1, 2020, as a treatment for the COVID-19 coronavirus (Taglialatela-Scafati, 2021). Not only does the ocean directly provide goods and services, but it also provides a means for their transportation across the globe. Around 90% of the world's goods are shipped on the ocean (Stuchtey et al., 2020).

Cultural

When thinking about ecosystem services, it can be tempting to try and ascribe monetary value to them. Indeed, we started off this section with an estimate of the trillions of dollars' worth of assets the ocean provides. However, there will always be key services from the ocean that transcend economics. Many of these are the cultural benefits, spanning spiritual, recreational, educational, aesthetic, and inspirational ways people interact with the ocean via their identities, activities, and ways of knowing.

Billions of people living along the coast ground their cultural and spiritual lives in the ocean (Stuchtey et al., 2020), not to mention all those who hold historical, aspirational, or artistic connections with the ocean without living near it. Many people relate to the sense of freedom, adventure, and awe that the ocean inspires (Allison et al., 2020). These and other cultural services are integrally connected with human values and behavior and influenced by social, economic, and political contexts (Millennium Ecosystem Assessment, 2005). Some marine species have been important to Indigenous peoples since time immemorial, and many species today bring personal enjoyment and subsistence to a diversity of peoples and communities. The ocean's habitats are important for both livelihoods and recreation. Cultural services represent many associated human value systems, underscoring the need to recognize local, Indigenous, and traditional knowledge, as well as academic knowledge of its functions and benefits (Allison et al., 2020).

Paths forward also require protecting these cultural services – peoples' ways of life and their health and well-being. Ecosystem services must be framed as such to underscore the interconnectedness of the ocean and people (Allison et al., 2020). Diverse stakeholders and rights-holders provide a full picture of needs and values and can support equitable distribution of the benefits the ocean provides (Österblom et al., 2020). The ocean of the future can be a place of prosperity for all, only if changes are undertaken equitably and inclusively. Material and monetary benefits from the ocean should not occur at the expense of the non-monetary and non-material benefits (Allison et al., 2020). Those who have benefited from the unsustainable use of the oceans since the start of industrialization bear differential responsibility to act; deliberate attention must be paid to benefits for women and others who are underserved and underrepresented (Österblom et al., 2020).

Conclusion

The ocean has always supported humanity and must continue to do so at even greater levels, and with more equitably distributed benefits. There is no single approach for sustaining a healthy, biodiverse, productive ocean. However, key to doing so is developing a solid understanding of the natural and physical processes of the ocean system, such as those described early in this chapter. While subsequent chapters in this volume focus on the human dimension, there is an opportunity to integrate the variety of people's perspectives, expertise, lived experiences, leverage points, and differential capacities to support change with this knowledge of the marine environment. These opportunities for integration and interdisciplinarity are a major focus of this book.

Questions for Reflection

1. How are humans affected by the following earth systems and how have humans affected these systems?
   a. Hydrologic cycle
   b. Nutrient cycling
   c. Temperature regulation
   d. Carbon cycle
2. Nature's contributions to people are extensive. What are some cultural services from the ocean that have influenced your life? Discuss the pros and cons of assigning economic values to the services that you identified.

References

Allison, E. H., Kurien, J., & Ota, Y. (2020). *The human relationship with our ocean planet*. World Resources Institute. https://www.oceanpanel.org/blue-papers/Human-RelationshipwithOurOceanPlanet

Amoroso, R. O., Pitcher, C. R., Rijnsdorp, A. D., McConnaughey, R. A., Parma, A. M., Suuronen, P., Eigaard, O. R., Bastardie, F., Hintzen, N. T., Althaus, F., Baird, S. J., Black, J., Buhl-Mortensen, L., Campbell, A. B., Catarino, R., Collie, J., Cowan, J. H., Durholtz, D., Engstrom, N., … Jennings, S. (2018). Bottom trawl fishing footprints on the world's continental shelves. *Proceedings of the National Academy of Sciences, 115*(43), E10275–E10282. https://doi.org/10.1073/pnas.1802379115

Baumgartner, A., & Reichel, E. (1975). *The world water balance: Mean annual global, continental and maritime precipitation, evaporation and run-off*. Elsevier Scientific Publishing Company.

Bigg, G. R., Jickells, T. D., Liss, P. S., & Osborn, T. J. (2003). The role of the oceans in climate. *International Journal of Climatology, 23*(10), 1127–1159. https://doi.org/10.1002/joc.926

Blasiak, R., Wynberg, R., Grorud-Colvert, K., Thambisetty, S., Bandarra, N. M., Canário, A. V. M., da Silva, J., Duarte, C. M., Jaspars, M., Rogers, A., Sink, K., & Wabnitz, C. C. C. (2020). The ocean genome and future prospects for conservation and equity. *Nature Sustainability, 3*(8), 588–596. https://doi.org/10.1038/s41893-020-0522-9

Cheung, W. W. L., Lam, V. W. Y., Sarmiento, J. L., Kearney, K., Watson, R., Zeller, D., & Pauly, D. (2010). Large-scale redistribution of maximum fisheries catch potential in the global ocean under climate change. *Global Change Biology, 16*(1), 24–35. https://doi.org/10.1111/j.1365-2486.2009.01995.x

Costello, C., Ovando, D., Hilborn, R., Gaines, S. D., Deschenes, O., & Lester, S. E. (2012). Status and solutions for the world's unassessed fisheries. *Science, 338*(6106), 517–520. https://www.science.org/doi/10.1126/science.1223389https://www.science.org/doi/10.1126/science.1223389

Del Valle, A., Eriksson, M., Ishizawa, O. A., & Miranda, J. J. (2020). Mangroves protect coastal economic activity from hurricanes. *Proceedings of the National Academy of Sciences, 117*(1), 265–270. https://doi.org/10.1073/pnas.1911617116

Dulvy, N. K., Fowler, S. L., Musick, J. A., Cavanagh, R. D., Kyne, P. M., Harrison, L. R., Carlson, J. K., Davidson, L. N., Fordham, S. V., Francis, M. P., Pollock, C. M., Simpfendorfer, C. A., Burgess, G. H., Carpenter, K. E., Compagno, L. J., Ebert, D. A., Gibson, C., Heupel, M. R., Livingstone, S. R., ... White, W. T. (2014). Extinction risk and conservation of the world's sharks and rays. *ELife, 3*, e00590. https://doi.org/10.7554/eLife.00590

Falkowski, P. G., & Raven, J. A. (2013). *Aquatic photosynthesis*. Princeton University Press. https://press.princeton.edu/books/paperback/9780691115511/aquatic-photosynthesis

Fisher, R., O'Leary, R. A., Low-Choy, S., Mengersen, K., Knowlton, N., Brainard, R. E., & Caley, M. J. (2015). Species richness on coral reefs and the pursuit of convergent global estimates. *Current Biology, 25*(4), 500–505. http://dx.doi.org/10.1016/j.cub.2014.12.022

Food and Agriculture Organization of the United Nations. (2020). *The state of world fisheries and aquaculture 2020: Sustainability in action*. FAO, Rome. https://doi.org/10.4060/ca9229en

Gagné, T. O., Reygondeau, G., Jenkins, C. N., Sexton, J. O., Bograd, S. J., Hazen, E. L., & Van Houtan, K. S. (2020). Towards a global understanding of the drivers of marine and terrestrial biodiversity. *PLOS ONE, 15*(2). https://doi.org/10.1371/journal.pone.0228065

Gattuso, J.-P., Magnan, A., Billé, R., Cheung, W. W. L., Howes, E. L., Joos, F., Allemand, D., Bopp, L., Cooley, S. R., Eakin, C. M., Hoegh-Guldberg, O., Kelly, R. P., Pörtner, H.-O., Rogers, A. D., Baxter, J. M., Laffoley, D., Osborn, D., Rankovic, A., Rochette, J., ... Turley, C. (2015). Contrasting futures for ocean and society from different anthropogenic CO_2 emissions scenarios. *Science, 349*(6243). https://www.science.org/doi/10.1126/science.aac4722

Grorud-Colvert, K., Sullivan-Stack, J., Roberts, C., Constant, V., Horta e Costa, B., Pike, E. P., Kingston, N., Laffoley, D., Sala, E., Claudet, J., Friedlander, A. M., Gill, D. A., Lester, S. E., Day, J. C., Gonçalves, E. J., Ahmadia, G. N., Rand, M., Villagomez, A., Ban, N. C., ... Lubchenco, J. (2021). The MPA guide: A framework to achieve global goals for the ocean. *Science, 373*(6560), eabf0861. https://www.science.org/doi/10.1126/science.abf0861

Halpern, B. S., Frazier, M., Potapenko, J., Casey, K. S., Koenig, K., Longo, C., Lowndes, J. S., Rockwood, R. C., Selig, E. R., Selkoe, K. A., & Walbridge, S. (2015). Spatial and temporal changes in cumulative human impacts on the world's ocean. *Nature Communications*, 6(1), 7615. https://www.nature.com/articles/ncomms8615

Hoegh-Guldberg, O. (2015). *Reviving the ocean economy: The case for action -2015*. World Wide Fund for Nature, Gland, Switzerland.

Intergovernmental Panel on Climate Change. (2019). *IPCC Special Report on the Ocean and Cryosphere in a Changing Climate*. Geneva, Switzerland: [H.-O. Pörtner, D. C. Roberts, V. Masson-Delmotte, P. Zhai, M. Tignor, E. Poloczanska, K. Mintenbeck, A. Alegría, M. Nicolai, A. Okem, J. Petzold, B. Rama, N.M. Weyer (Eds.)]. https://doi.org/10.1017/9781009157964

Intergovernmental Science-Policy Platform on Biodiversity and Ecosystem Services. (2019). *Global assessment report on biodiversity and ecosystem services of the Intergovernmental Science-Policy Platform on Biodiversity and Ecosystem Services*. IPBES secretariat, Bonn, Germany: E. S. Brondizio, J. Settele, S. Díaz, and H. T. Ngo (Eds.). https://doi.org/10.5281/zenodo.3831673

International Union for the Conservation of Nature. (2019). *The IUCN red list of threatened species*. Gland, Switzerland.

Jones, K. R., Klein, C. J., Halpern, B. S., Venter, O., Grantham, H., Kuempel, C. D., Shumway, N., Friedlander, A. M., Possingham, H. P., & Watson, J. E. M. (2018). The location and protection status of Earth's diminishing marine wilderness. *Current Biology, 28*(15), 2506–2512.e3. https://doi.org/10.1016/j.cub.2018.06.010

Kroeker, K. J., Kordas, R. L., Crim, R. N., & Singh, G. G. (2010). Meta-analysis reveals negative yet variable effects of ocean acidification on marine organisms. *Ecology Letters, 13*(11), 1419–1434. https://doi.org/10.1111/j.1461-0248.2010.01518.x

Laffoley, D., Baxter, J. M., Amon, D. J., Claudet, J., Hall-Spencer, J. M., Grorud-Colvert, K., Levin, L. A., Reid, P. C., Rogers, A. D., Taylor, M. L., Woodall, L. C., & Andersen, N. F. (2020). Evolving the narrative for protecting a rapidly changing ocean, post-COVID-19. *Aquatic Conservation: Marine and Freshwater Ecosystems, n/a*(n/a), 1–23. https://doi.org/10.1002/aqc.3512

Lillebø, A. I., Pita, C., Garcia Rodrigues, J., Ramos, S., & Villasante, S. (2017). How can marine ecosystem services support the Blue Growth agenda? *Marine Policy, 81*, 132–142. https://doi.org/10.1016/j.marpol.2017.03.008

Lotze, H. K., Tittensor, D. P., Bryndum-Buchholz, A., Eddy, T. D., Cheung, W. W. L., Galbraith, E. D., Barange, M., Barrier, N., Bianchi, D., Blanchard, J. L., Bopp, L., Büchner, M., Bulman, C. M., Carozza, D. A., Christensen, V., Coll, M., Dunne, J. P., Fulton, E. A., Jennings, S., … Worm, B. (2019). Global ensemble projections reveal trophic amplification of ocean biomass declines with climate change. *Proceedings of the National Academy of Sciences, 116*(26), 12907–12912. https://doi.org/10.1073/pnas.1900194116

Louca, S., Mazel, F., Doebeli, M., & Parfrey, L. W. (2019). A census-based estimate of Earth's bacterial and archaeal diversity. *PLOS Biology, 17*(2), e3000106. https://doi.org/10.1371/journal.pbio.3000106

Lubchenco, J., & Gaines, S. D. (2019). A new narrative for the ocean. *Science, 364*(6444), 911–911. https://www.science.org/doi/10.1126/science.aay2241

MarinLit. (2020). *A database of marine natural products literature*. http:// pubs.rsc.org/marinlit/

Masson-Delmotte, V., Zhai, P., Portner, H., Roberts, D., Skea, J., Shukla, P. R.,… & Waterfield, T. (2019). *Global warming at 1.5 degrees C: An IPCC Special Report*. Intergovernmental Panel on Climate Change. https://www.ipcc.ch/sr15/

McCauley, D. J., Pinsky, M. L., Palumbi, S. R., Estes, J. A., Joyce, F. H., & Warner, R. R. (2015). Marine defaunation: Animal loss in the global ocean. *Science, 347*(6219). https://www.science.org/doi/10.1126/science.1255641

Millennium Ecosystem Assessment. (2005). *Ecosystems and human well-being: Synthesis report*. Island Press.

Miller, K. A., Thompson, K. F., Johnston, P., & Santillo, D. (2018). An overview of seabed mining including the current state of development, environmental impacts, and knowledge gaps. *Frontiers in Marine Science, 4*. https://doi.org/10.3389/fmars.2017.00418

Mora, C., Tittensor, D. P., Adl, S., Simpson, A. G. B., & Worm, B. (2011). How many species are there on earth and in the ocean? *PLOS Biology*, 9(8), e1001127. https://doi.org/10.1371/journal.pbio.1001127

Niner, H. J., Ardron, J. A., Escobar, E. G., Gianni, M., Jaeckel, A., Jones, D. O. B., Levin, L. A., Smith, C. R., Thiele, T., Turner, P. J., Van Dover, C. L., Watling, L., & Gjerde, K. M. (2018). Deep-sea mining with no net loss of biodiversity - An impossible aim. *Frontiers in Marine Science, 5*. https://doi.org/10.3389/fmars.2018.00053

Organization for Economic Cooperation and Development. (2016). *The ocean economy in 2030*. OECD Publishing.

Österblom, H., Wabnitz, C., Tladi, D., Allison, E., Arnaud-Haond, S., Bebbington, J., Bennett, N., Blasiak, R., Boonstra, W., Choudhury, A., Cisneros-Montemayor, A., Daw, T., Fabinyi, M., Franz, N., Harden-Davies, H., Kleiber, D., Lopes, P., McDougall, C., Resosudarmo, B., & Selim, S. (2020). *Towards ocean equity*. High Level Panel for a Sustainable Ocean Economy.

Pacoureau, N., Rigby, C. L., Kyne, P. M., Sherley, R. B., Winker, H., Carlson, J. K., Fordham, S. V., Barreto, R., Fernando, D., Francis, M. P., Jabado, R. W., Herman, K. B., Liu, K.-M., Marshall, A. D., Pollom, R. A., Romanov, E. V., Simpfendorfer, C. A., Yin, J. S., Kindsvater, H. K., & Dulvy, N. K. (2021). Half a century of global decline in oceanic sharks and rays. *Nature, 589*(7843), 567–571. https://www.nature.com/articles/s41586-020-03173-9

Reid, P. C. (2016). Ocean warming: Setting the scene. In D. Laffoley & J. M. Baxter (Eds.), *Explaining ocean warming: Causes, scale, effects and consequences* (pp. 17–45). IUCN, Gland, Switzerland.

Rogers, A., Aburto-Oropeza, O., Appeltans, W., Assis, J., Ballance, L., Cury, P., Duarte, C., Favoretto, F., Kumagai, J., Lovelock, C., Miloslavich, P., Niamir, A., Obura, D., O'Leary, B., Reygondeau, G., Roberts, C., Sadovy, Y., Sutton, T., Tittensor, D., & Velarde, E. (2020). *Critical habitats and biodiversity: Inventory, thresholds and governance.* https://www.researchgate.net/publication/341219468_Critical_Habitats_and_Biodiversity_Inventory_Thresholds_and_Governance

Rousseau, Y., Watson, R. A., Blanchard, J. L., & Fulton, E. A. (2019). Evolution of global marine fishing fleets and the response of fished resources. *Proceedings of the National Academy of Sciences, 116*(25), 12238–12243. https://doi.org/10.1073/pnas.182034411

Sabine, C. L., Feely, R. A., Gruber, N., Key, R. M., Lee, K., Bullister, J. L., Wanninkhof, R., Wong, C. S., Wallace, D. W. R., Tilbrook, B., Millero, F. J., Peng, T.-H., Kozyr, A., Ono, T., & Rios, A. F. (2004). The oceanic sink for anthropogenic CO_2. *Science, 305*(5682), 367–371. https://www.science.org/doi/10.1126/science.1097403

Sekerci, Y., & Petrovskii, S. (2015). Mathematical modelling of plankton–oxygen dynamics under the climate change. *Bulletin of Mathematical Biology, 77*(12), 2325–2353. https://doi.org/10.1007/s11538-015-0126-0

Smale, D. A., Wernberg, T., Oliver, E. C. J., Thomsen, M., Harvey, B. P., Straub, S. C., Burrows, M. T., Alexander, L. V., Benthuysen, J. A., Donat, M. G., Feng, M., Hobday, A. J., Holbrook, N. J., Perkins-Kirkpatrick, S. E., Scannell, H. A., Sen Gupta, A., Payne, B. L., & Moore, P. J. (2019). Marine heatwaves threaten global biodiversity and the provision of ecosystem services. *Nature Climate Change,* 9(4), 306–312. https://doi.org/10.1038/s41558-019-0412-1

Stuchtey, M. R., Vincent, A., Merkl, A., Bucher, M., Haugan, P. M., Lubchenco, J., & Pangestu, M. E. (2020). *Ocean solutions that benefit people, nature, and the economy* (High Level Panel for a Sustainable Ocean Economy). World Resources Institute. www.oceanpanel.org/ocean-solutions

Taglialatela-Scafati, O. (2021). New hopes for drugs against COVID-19 come from the sea. *Marine Drugs, 19*(2), 104. https://doi.org/10.3390/md19020104

Teh, L. C. L., & Sumaila, U. R. (2013). Contribution of marine fisheries to worldwide employment. *Fish and Fisheries, 14*(1), 77–88. https://doi.org/10.1111/j.1467-2979.2011.00450.x

Part II
Grand Challenges

Part II introduces six contemporary topics, or "Grand Challenges," faced by the Ocean and the people who depend on it: fisheries, aquaculture, pollution, climate change, marine renewable energy, and governance in areas beyond national jurisdiction. Each chapter presents an integrated human, natural, and physical science perspective on the six topics.

DOI: 10.4324/9781003058151-5

4 Wild Capture Fisheries

Processes, Technology, and Solutions

F. Conway and L. Ciannelli

Wild capture fisheries may appear to be simple endeavors at first glance. Nevertheless, fishery systems across the globe are extremely complex and cover subsistence, recreational, and commercial fisheries. Often these fishery sectors interact or overlap. This chapter focuses largely on commercial fisheries.

Commercial fisheries in the Global North and South might share species or harvest methods, but they differ significantly. Artisanal or small-scale fisheries in Bangladesh are quite different from "developed-world" fisheries in Scotland or other countries with a long history of commercial harvest and access to scientific information and technological advances. This chapter will focus largely on the Global North.

Integrated and thoughtful approaches to fisheries can be truly complex. Management must consider the amounts of various living marine resources, where fish are caught and sold, and who the fishers are, as well as the ultimate consumers. Important questions center on the sustainability of the capture, as well as the equitable allocation and use of resources and generated revenue. Fisheries must be considered as a coupled natural–human (CNH) system. This chapter provides a quick glimpse into this topic.

We pose several questions. How might one really *know* about seafood and the marine environment?[1] How does what one *believes* they know influence one's perception of seafood-related stakeholders[2] and how they engage with and manage marine living resources? Does taking a coupled systems approach to the fisheries lead to creative and thoughtful ways to improve resource use and conservation? This chapter addresses these questions. We believe that knowledge about people cannot exist without knowledge of marine living resources and their habitats. Conversely, knowledge about the sea and its resources is incomplete without knowledge of the involved actors.

This chapter begins with a brief mention of the scope of fisheries, followed by an introduction of the CNH dimension of fisheries, and finally, presents an overview of some traditional and novel challenges that impact the resources and well-being of connected coastal communities. Three case studies illustrate processes, technologies, and solutions, and how fishing sectors (stakeholders and communities) cooperate with managers and scientists to solve problems and identify opportunities that address new challenges.

DOI: 10.4324/9781003058151-6

Fisheries Provide Meals, Jobs, and Much More

For many people, seafood – captured and cultivated – may mean little more than a meal. Seafood is indeed an important source of food. Global per-capita human consumption of wild-caught and farmed living marine resources has more than doubled since the 1960s (from 9.1 to 20.5 kg/yr), mostly from increased aquaculture production. About 87% of all fisheries production (wild-caught and farmed) is used for human consumption.

The economical and nutritional benefits of seafood capture are illustrated by recent US and global statistics. In the US, the total harvest of wild seafood was 4.5 million metric tons in 2017, valued at US$5.4 billion (National Marine Fisheries Service [NMFS], 2018). Table 4.1 provides a summary of the most harvested species by weight in the US.

Globally, about 85 million metric tons of wild seafood were landed in 2018 (Food and Agriculture Organization of the United Nations [FAO], 2020). Finfish constitute 85.5% of the global catch and are composed of small pelagic fish (anchovies and sardines), gadids (walleye pollock, cod, haddock), and scombrids (tuna, mackerel). Invertebrate species (squids, crabs, shrimps) include the remainder (FAO, 2020).

Fisheries also provide jobs. Worldwide, more than 40 million workers are employed in the commercial seafood sector. This does not account for secondary employment, such as seafood marketing. In the US, 168,000 people are directly employed by fisheries, and when considering seafood imports and the entire chain of seafood production from harvest to retail, the number increases to 1,200,000 people (NMFS, 2018).

Seafood global harvest has remained constant since the mid-1980s and may increase slightly during the next few decades as some stocks rebuild and previously underexploited stocks are targeted. In 2017, the percentage of fish and invertebrate stocks harvested at or below biologically sustainable levels was 65.8%. This has steadily decreased from 90% in the early and mid-1970s. Despite increasing fishing efforts, some bright spots exist. For example, fishing pressure on many

Table 4.1 The most harvested fish species in the US in 2017 (NMFS, 2018)

Species	*Where Harvested*	*Metric Tons and Percentage of Total Landings*
Walleye Pollock (*Gadus chalcogrammus*)	Bering Sea and Gulf of Alaska	1.5 million metric tons or 33% of total landings
Atlantic Menhaden (*Brevoortia tyrannus*)	Western Atlantic and Gulf of Mexico	641,000 metric tons or 15% of total landings
Pacific Hake (*Merluccius productus*)	US West Coast	351,000 metric tons or 8% of total landings
Pacific Cod (*Gadus macrocephalus*)	Gulf of Alaska and Bering Sea	300,000 metric tons or 6.7% of total landings
Pink Salmon (*Oncorhynchus gorbuscha*)	Alaska	225,000 metric tons or 5% of total landings

intensely managed fish stocks has decreased, and their biomass has increased, underscoring the importance of developing and implementing successful policies for the conservation and sustainable use of marine resources.

Other factors, such as technological advances in distribution, processing and waste reduction, expansion of markets, and awareness of health benefits from seafood consumption, also play important roles. Nevertheless, great variations exist throughout the world. African countries have the lowest per-capita consumption at about 10 kg/yr, despite more than 15% of global landings originating from that continent's Exclusive Economic Zones (EEZs). Regions such as Russia, Japan, Western Europe, and North America have the highest per-capita seafood consumption. These disparities in capture and consumption of seafood raise concerns for global food security and underscore the need for policies aimed at equitable distribution of economic and health benefits of marine resources.

Fisheries Are a Coupled Natural–Human System

Seafood capture is tightly linked with the human systems that depend on it for nutrition, jobs, recreation, and cultural activities. Just like captured seafood influences human economies and cultural activities, humans impact populations of living marine resources and their habitats, primarily through commercial and recreational harvests, as well as through coastal and ocean development. These two-way interactions between seafood and humans constitute the building blocks of a CNH system.

Several natural processes affect the production of wild capture fisheries in the ocean. First, fish production is linked to primary production of the ocean. The most productive areas of the ocean are the coastal waters over the continental shelves and the eastern boundaries coastal upwelling systems of the ocean (Chavez & Messié, 2009). About 20 million metric tons of wild seafood, or 20% of the global catch, are annually harvested from coastal upwelling areas, despite occupying <1% of the global ocean surface (Pauly & Christensen, 1995). Apex and mid-food web predators can also affect the available biomass of harvested fish via predator–prey interactions. Lastly, processes that depend on the size of the targeted populations (e.g., disease transmission, food limitation) can also affect fish abundance. This density-dependent control (DingsØr et al., 2007) is an important mechanism that many models utilize for managing fish populations. From an ecological perspective, mortality is affected by predation and starvation, and is highest during early life stages when individuals are vulnerable to starvation, physiological stress, and a wide range of predators (Bailey & Houde, 1989).

Climate also impacts fisheries. One major climate driver of fisheries production is the El Niño–Southern Oscillation (ENSO), defined as the difference in sea level pressure between the Western and Eastern Pacific regions (Timmermann et al., 2018). During positive phases of ENSO, the strong equatorial easterly trade winds decrease in intensity or reverse in direction, causing a reduction of the tropical upwelling, and consequently strong reductions of primary productivity throughout the Eastern Tropical Pacific. Low primary production impacts the availability of Peruvian anchoveta – the most harvested marine fish species on earth. Other commercially important species, such as sardines and jack mackerel, of the Eastern Tropical Pacific are also negatively impacted during ENSO events. Strong El Niño phases (1992, 1998, 2015) have caused noticeable reductions of global harvest.

A general principle of fisheries production is linked to the life history of the species; short-lived species can quickly rebound from heavy declines, while long-lived species will take longer to rebound from low stock size. However, long-lived species reproduce multiple times during their life cycle and thus can endure short periods of unfavorable environmental variability better than short-lived species, which only reproduce once or few times during their life cycle and are therefore more susceptible to environmental variations. These basic principles are used to manage fish populations. However, uncertainties related to assessment of stock status, the effect of density-dependent dynamics, trophic interactions, and climate and environmental variability present challenges for fisheries management. Additional difficulties for management are synergistic effects between natural variability (i.e., linked with climate or local environmental variability) and human harvest (Brander, 2010).

The history of fishery science and the development of new fisheries are closely related to technological developments. From the fisheries perspective, technological advances are exemplified by larger and better-equipped vessels and fishing gear that increase catch efficiency, as well as those that reduce bycatch and increase maritime safety. From the fishery science perspective, new technologies provide new and more data for better assessment of the resources, habitats, and oceanographic conditions (e.g., acoustic surveys, autonomous vehicles, new environmental sensors, GPS). Development of faster computing power allows for more sophisticated tactical and strategic models for fisheries management. The history of many fisheries worldwide reflects this duality of technology, as a driver for depletion and a catalyst for management and conservation. The latter half of the 20th century was characterized by a rapid development of new vessels and gear, which allowed fleets to fish more intensely, further from shore, and in deeper waters (Frank et al., 2018). An exponential decline of large predators worldwide occurred as a result of fisheries harvest from the 1950s to the 1980s (Myers & Worm, 2003). From that point on, the global harvest has leveled off. Slight increases of global catch in recent years may be a promising indication that new technologies, coupled with practical management and policies, are yielding the desired effects (FAO, 2020).

The development of commercial fisheries and fishery science and technology is guided by governance (policy and management). Policies created to balance the conservation and use of fishery resources have evolved during the past decades, and throughout much of the world sustainability has become the goal (FAO, 1995; 2015). Box 4.1 describes fishery legislation in the US that has evolved to become ever more sustainable.

BOX 4.1 US Fisheries Policy and Management

The Magnuson–Stevens Fisheries Conservation and Management Act (MSFCMA), approved in 1976 and significantly modified in 1996, is the main tool for managing fisheries within the US. Prior to the MSFCMA, unregulated fishing 12 miles offshore by foreign fleets raised concerns (Weber, 2002). Additionally, the 1970s began to experience a paradigm shift, which initiated a move away from a belief in the ocean's limitless bounty (Weber, 2002). US jurisdiction was extended to 200 nautical miles, and eight regional fishery management councils were created.

Three major revisions have occurred since the initial implementation of the MSFCMA: The Sustainable Fisheries Act of 1996 and the Reauthorization Acts of 2006 and 2007 (NMFS, 2019). Fundamental goals of the MSFCMA and of subsequent management strategies are to maintain sustainable and healthy fisheries, prevent fishery collapses, reduce bycatch, and support the rebuilding of depleted stocks – without causing undue harm to fishing communities. Thus, the Act attempts to balance conservation and sustainable use of fishery resources.

Humans do not manage the fish. Rather, they manage the intersection of fish and people. It is helpful to start with some clarification of terms. Although recreational fishing is an important economic activity in much of the world, we use the terms "fishing" or "fishing industry" to refer to the commercial harvest of wild capture fisheries.

The commercial fishing industry looks quite different around the world, even in the Global North. We draw on several examples for comparative purposes: Artisanal fishing in Italy (Box 4.2), Greenland fisheries (Box 4.5), and groundfish fisheries on the US West Coast (discussed in depth later in the chapter). Nevertheless, some similarities exist. Most fishing occurs near coastal communities (geographic communities). Fishing businesses, often called "the fishing communities" (communities of interest or occupation), are comprised of individuals who operate at a business level (often based on family ties). These fishing communities share their own values and perceptions while being part of the larger fishing industry and community of place (Conway et al., 2002). Local, state, and national institutions influence fishing communities, profoundly impacting the ecological, social, cultural, and political dynamics, resilience, and community wellbeing.

Known and Novel Challenges

Throughout the world, fisheries face both known and novel challenges. Some challenges result from environmental changes and uncertainties, while others stem from the increasing demand for seafood, advances in technologies, and the difficulties of enforcing distance-water fisheries on the high seas. Addressing these challenges with adaptive and sustainable approaches must guide the fishery sector.

Known Challenges

Challenges in the natural dimension affecting wild capture fisheries' CNH systems include declining fish stocks due to overexploitation and changing environmental conditions (Frank et al., 2016), bycatch of depleted species, and habitat degradation through coastal developments and bottom trawling. Some of these impacts can be remediated through implementation of policies allowing for the rebuilding of depleted stocks and restoration of degraded habitats. For example, area and seasonal closures, limits on effort and catch, and marine protected areas are successful strategies for rebuilding many stocks and preserving essential fish habitats. Bycatch reduction gears have drastically limited the harvest of seabirds, sea turtles, marine mammals, and other depleted stocks. Restoration of mangrove forests, seagrass beds, and coral reefs helps to improve nursery and foraging grounds for many marine resources.

The human dimension presents known challenges, as well. Global seafood markets are a known human dimension challenge. Over 90% of seafood consumed in the US is imported (Greenberg, 2014). At the same time, much of the seafood harvested in the US is marketed globally. Thus, trade issues or the recent global pandemic can greatly impact seafood harvesters and processors (White et al., 2020). The US has attempted to use trade embargos on the import of seafood products (e.g., on Mexican tuna imports) to encourage nations to adopt less harmful fishing practices that have resulted in legal claims and heightened international tensions.

Conflicts commonly arise between different fishery sectors regarding allocation of the catch and coastal dockage space. Classic conflicts often occur between industrial fishing fleets and small-scale (artisanal) fisheries (e.g., in Peru/Chile – Arellano & Swartzman, 2010), recreational fishers and small-scale fishers (e.g., on Ischia Island, Italy – Box 4.2; Figure 4.1), as well as between small-scale fishers and coastal tourism developments (e.g., in the Florida Keys – Schittone, 2001).

BOX 4.2 Artisanal Fishery of Ischia, Italy – Conflicts between Artisanal Fisheries and Tourism

Artisanal fisheries dominate the entire Italian fishing fleet; as of 2019 there were 11,984 licensed fishing vessels, 67.9% of which were composed of small vessels (<12 m in length), operating within a few miles from shore. While the number of larger vessels has declined by nearly 20% during the last 20 years, small fishing boats have remained nearly constant. About 175,000 metric tons of fish were landed in 2019, producing an income of about 900 million euros. Small fishing vessels produced 12% of the landings and 17% of the revenue (Ministry of Forestry and Agriculture [Italy], 2020).

Italian artisanal fisheries are threatened by several factors: overfishing, illegal fishing, pollution, coastal development, climate change, and competing interests and uses of coastal habitats. The artisanal fishery in Ischia Island, seaward of Naples, exemplifies these issues. An author of this chapter (Lorenzo Ciannelli) was born and raised on Ischia and grew up experiencing first-hand the relationships between a flourishing tourism industry and a dwindling artisanal fishery. The island's artisanal industry is very diversified and utilizes fixed bottom trammel nets for the capture of Mediterranean hake, shelf flatfish, Norway lobster, and red prawns, and pelagic nets for the capture of medium and large pelagic, such as tunas, bonito, dolphinfish, amberjack, and swordfish (Silvestri, 2003). While technology has certainly contributed to the development of faster, motorized, well-equipped, and safer vessels, day-to-day operations linked to the capture (when, where, and how-to) and marketing of the fish have not changed, as they are rooted in traditions that are centuries old and orally transmitted between generations. For example, visual navigation via alignment of landmarks (churches, trees, promontories) is the primary way by which fishermen locate productive habitats and avoid obstacles for their nets. Trammel nets are hand-dropped daily in the afternoon and retrieved the following morning. Fishermen return to the docks by late morning to sell their catch directly to local consumers.

Ischia's beautiful shorelines, green mountains, delicious food, hot springs, and ancient buildings and streets make the island a popular tourist destination. The tourism industry began to develop in Ischia in the 1950s, and the livelihoods of increasing numbers of local people have become dependent on tourism. Tourism in Ischia generates an estimated 250 million euros annually and employs nearly 7,000 people, many of whom are young. These numbers vastly exceed those involved in local artisanal fishing. Tourism is vital to the local communities. However, it demands space and uses of coastal habitats – demands that may threaten the very livelihoods of the local fishing industry (Figure 4.1). This situation raises questions about fair valuation and equitable allocation of ecosystem services. These themes are not unique to Ischia and exemplify the complex, multifaceted nature of artisanal fisheries and conflicts with tourism's increasing demands on coastal resources.

Figure 4.1 Image of the bay in the Lacco Ameno municipality on Ischia Island, Italy. The image shows many anchored recreational vessels. Excessive mooring and anchoring in these coastal habitats have negative impacts on the ecosystems, by dragging anchors and chains over bottom habitats. Small fishing vessels compete for space with larger numbers of recreational vessels. Photo credit: L. Ciannelli.

The US has had limited access to commercial harvest of fish for several decades. A range of access-related management rules for commercial fisheries extends from some simple restricted permit programs and limited-entry systems to more detailed transferable-quota programs, commonly called "catch shares" (Pooley, 1998) (Box 4.3). Limited entry refers to regulations that cap the total number of fishers, fishing vessels, or equipment in a fishery (Rettig & Ginter, 1978). Catch share management strategies allocate a specific portion of the total allowable catch to individuals, cooperatives, or communities.

BOX 4.3 Catch Shares

Catch shares in the US are a form of rights-based management – not a full property right, but rather a revocable privilege to land an assigned percentage of the total quota. Each catch share recipient must cease fishing upon reaching the allocation (Morrison, 2019). Catch share programs use individual allocations, as well as allocations to groups and entities. These include Individual Transferable Quotas (ITQs), granted to both individuals and entities, such as processors (Fina, 2011; Matulich, 2009), fishers organized as corporations and limited liability corporations, fishing cooperatives, and fishing communities.

Broadly speaking, catch shares have resulted in consolidation, changed the dynamics between fishers and supply chains, altered the employment structure of crews, increased financial barriers, and, therefore, reduced entry opportunities and resulted in disproportionate losses of fishing rights for small vessels and communities (Carothers, 2010). These fishery management tools may improve safety (Pfeiffer & Gratz, 2016) but also result in a loss of access, exposure, and opportunity for younger generations (Cramer et al., 2018; McCay et al., 1995). This may be one factor causing "the graying of the fleet" (Box 4.4).

BOX 4.4 Graying of Fisheries

In many countries the average age of fishers is increasing. Iceland has seen increases in the average age of fishermen to 60 years old (Nielsen et al., 2017). East Coast, West Coast, and Alaska fishing communities in the US are also experiencing this reality. The need to better understand the concept of graying has been increasingly recognized in recent years (Cramer et al., 2018; Donkersloot & Carothers, 2016; Ringer et al., 2018; White, 2015). A 2014 study by Russell et al. interviewed participants in the US fishing industry on the West Coast from Seattle, Washington, to Monterey, California, concluding the existence of graying of the fleet. One concern is that as fishermen retire and leave the industry, the majority of catch rights once held by the community will evaporate, causing a loss of individual and community income, identity, and sustainability. Commercial fishing is central to promotion of social relationships and cultural values (Cramer et al., 2018; Donkersloot & Carothers, 2016). A decrease in the entry of young people into the fishing sector would alter many social and cultural aspects that are the heart and soul of coastal communities.

Novel Challenges

New challenges are appearing that we have yet to fully understand, and therefore, have implemented limited management actions to address. For example, global warming and a changing climate alter species distribution (Pinsky et al., 2013). Range shifts of marine fish are now impacting coastal fishing communities (Rogers et al., 2019). Fish species can adapt to global warming by shifting their spatial distribution, leading to a redistribution of global catches, with loss at the tropics and gains at higher latitudes (Cheung et al., 2010). However, as illustrated in the case of the Greenland fisheries (Box 4.5), uncertainty surrounds these processes, in part due to questions of how productivity of fish – and their predators and food sources – will change with global warming. Fish and marine invertebrates have complex life cycles and multiple life stages, and the fitness of the entire population depends on the most sensitive life stages when faced with climate variability (Ciannelli et al., 2014a). Typically, early life stages (eggs, larvae, and juvenile stages) are most sensitive to changing environmental conditions. Of all stages in a species life cycle (Dahlke et al., 2018), these are the least sampled and understood. Additional challenges include the impact of ocean acidification and deoxygenation on fish and invertebrate populations (Keller et al., 2017). These factors will produce winners and losers, possibly depending on sensitive life history stages.

BOX 4.5 Greenland Fisheries – Managing Fisheries during Climate Change

Fisheries are an important economic and cultural resource for Greenland, accounting for more than 90% of its exports (FAO, 2004). Greenland's waters extend 200 nautical miles from shore and are divided into east, west, and south regions. In each region, commercial fisheries are managed as offshore and coastal fisheries. The coastal fleet includes vessels less than 75 gross tonnage (GT) operating within three nautical miles from the coast. The offshore fleet of larger boats (>75 GT) operates beyond three miles. Over 1,700 small boats (of less than 6 m) fish in coastal areas. In recent years, the coastal fleet has landed more than half of all cod caught in Greenland, as well as doubled the halibut caught by the offshore fleet (FAO, 2004). Economically and culturally, coastal fleets are the drivers of Greenland fisheries.

The most important coastal fisheries include halibut, shrimp, cod, capelin, and lumpfish. Offshore fisheries land shrimp, halibut, cod, redfish, and pelagic species, such as herring, mackerel, blue whiting, and capelin. Cod fisheries exemplify the common development cycle. Developed in the mid-1950s, they peaked a decade later. However, in the early 1990s the west and south Greenland fisheries collapsed due to a combination of overfishing and adverse climatic conditions (Bonanomi et al., 2015). Since 2000, the cod landings have increased, mostly in the east, although not to 1960 levels. Currently, halibut also show signs of overfishing, as evidenced by decreased size at age class, as well as age at maturity. Concerns also surround the inshore Greenland halibut fishery exceeding the total allowed catch (TAC) and the impact of trawling on benthic habitats (Long & Jones, 2021).

Greenland fisheries are projected to grow in the coming years as some commercially harvested species shift their ranges northward, and fish populations located at the northern edge of their distribution range increase productivity. The Greenland government has expanded the ability to harvest, assess, and manage its fisheries in the changing Arctic (Ministry for Foreign Affairs, 2020). However, while climate change may improve the fisheries of the most mobile species (e.g., mackerel and capelin) due to range shifts, it also presents many uncertainties. How global warming and ice melting will affect local ocean productivity is a significant unknown (Meire et al., 2017). Even for pelagic species, great uncertainty surrounds the multitude of processes that affect their distribution. Will fishery managers be able to effectively regulate the increasing fishing effort as fish biomass changes with the warming climate? Can Greenland successfully control illegal fishing?

New challenges also affect local economies (Magel et al., 2020). The global pandemic has certainly impacted fishing communities and the fishery sector. Examples include outbreaks at seafood processing plants (Global Seafood Alliance, 2020) and limits to global markets (particularly China) during the peak season for the crab fishery. It is unclear how fishing fleets will adapt to these kinds of novel changes (Selden et al., 2020), but many scenarios are possible. More extreme weather events brought by climate change can create hardships for small-vessel fleets, while larger vessels are able to adapt by staying out longer, fishing in more distant regions, or coping with the delays in other ways (Kuonen et al., 2019).

Scientists still do not know the full extent to which species can adapt to changes in water temperature and other oceanographic conditions. In the face of these uncertainties, a wise course of action is to provide opportunities for adaptation of wild populations, and therefore, conserve the diversity of strategies that species exhibit (Schindler et al., 2010). Uncertainty and unknowns lead to challenges for science, practice, and governance. Creativity and flexibility in ocean governance will be increasingly important because of the complexity, the cumulative impacts of challenges like global climate change (e.g., shifting species distribution, ocean acidification, and hypoxia), emerging market incentives (e.g., third-party certification and traceability), pollution (e.g., microplastic impacts in wild and cultivated seafood), and the multiple interests (marine renewable energy, fiber optic cables, shipping, and tourism) in marine space. The coupled system (CNH) approach that considers the vulnerability, resilience, and adaptive capacity *of the entire marine system* can lead to creative and thoughtful ways to respond (Cinner & Barnes, 2019).

Novel Approaches

The study of the CNH system of wild capture fisheries requires interdisciplinary and transdisciplinary investigative approaches, given the multifaceted nature of the system and the strong connection between the resource and its users (Ciannelli et al., 2014b). An interdisciplinary approach to research includes multiple disciplines and strives to develop new investigative tools to address the problem (Paterson et al., 2010). This is necessary, but not sufficient, because we also must consider different types of knowledge and uses that relate to this system (Beaudreau & Levin, 2014). Strong engagement of all resource users is also necessary to ensure that the investigative approach includes different types of knowledge (e.g., local ecological and experiential knowledge), and that research questions and possible management solutions are relevant to a variety of stakeholders.

US West Coast Groundfish

We close this chapter with a discussion of the US West Coast Groundfish fishery that provides a multi-decadal example of novel (current and future) challenges and approaches to a complex fisheries system.

About 500,000 metric tons of wild fish and invertebrates are caught and sold annually on the US Pacific Coast from California to Washington. This catch accounts for about $1 billion in sales and creates more than 230,000 jobs related to seafood capture, processing, and consumption (NMFS, 2018). A large fraction by weight (38%) of the wild catch is groundfish, a complex of over 90 species of fish that spend most of their life cycles near or on the seabed. For management purposes, groundfish are grouped into four main categories: rockfish (mostly of the genus *Sebastes*, nearly 60 species), flatfish (20 species), cartilaginous fish (skate and sharks), and groundfish (such as sablefish, Pacific hake or whiting, cabezon, and lingcod).

Commercial fishing vessels utilizing bottom or midwater trawl gear catch most of the groundfish. Other vessels operate using fixed gear (longline, pots) or hook and line (particularly close to shore). In the Pacific Coast region, about 30 vessels operate midwater trawls targeting Pacific hake, and another 70 vessels capture groundfish using bottom trawls (Figure 4.2). The number of bottom trawlers has decreased substantially over time. By 2019 only half the number of vessels were operating, as compared to two decades earlier, and only one-quarter the number that were operating in the late 1980s (NMFS, n.d.). Major policy changes discussed in Box 4.1 have led to fleet consolidation.

The history of the US West Coast non-whiting groundfish fishery resembles that of other commercial fisheries in the US and abroad (see Greenland, Box 4.5). Fisheries start with an initial development and expansion phase, undergo a period of sustained high catch followed by a steady decline, and then move into a period of reconstruction. Environmental and climate variability adds uncertainty (Figure 4.3).

After the adoption of the 200-mile US Fishery Conservation Zone, the participation of fishing nations (Russia, China, Japan) continued to occur through joint venture systems. However, by the early 1980s, foreign fisheries were no longer able to participate due to the overcapitalization and increasing capacity of the US domestic fleet. The NMFS determined that US fishing vessels possessed such great

Figure 4.2 US fishing vessels in Newport, Oregon, that operate bottom trawl gear. Photo credit: L. Ciannelli.

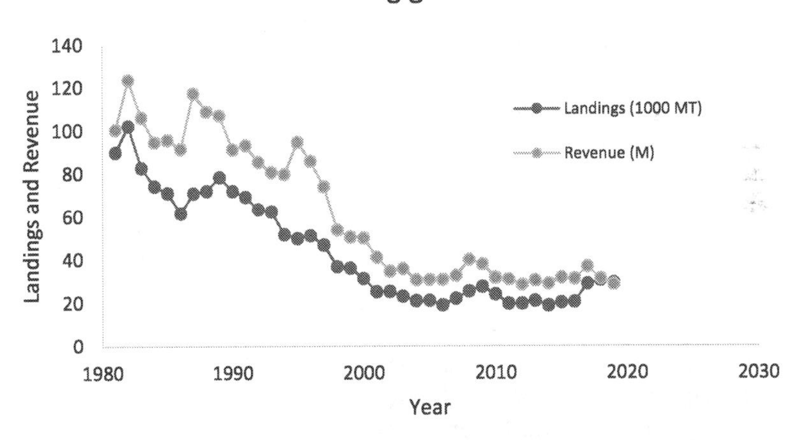

Figure 4.3 Landings (thousands of metric tons) and revenue (millions of dollars) of non-whiting groundfish in the US West Coast bottom and midwater trawl fishery. Data on landings and revenue are from the Pacific Fishery Information Network (PacFIN, https://reports.psmfc.org/pacfin/f?p=501:1000:16204540881702::NO:::). Revenue data are expressed in December 2019 US dollars adjusted for inflation using the seasonally adjusted Consumer Price Index for All Urban Consumers from the US Bureau of Labor Statistics (https://beta.bls.gov/dataViewer/view/timeseries/CUSR0000SA0).

fishing potential that there was no room for foreign vessels in this fishery. The groundfish harvest continued to rise and peaked in the early 1980s (Figure 4.3). By the early 2000s, nine species of groundfish were declared overfished (seven types of rockfish, lingcod, Pacific hake). The US Secretary of Commerce issued an official "disaster declaration" for the West Coast groundfish fishery in 2000 (Conway & Shaw, 2008).

During this difficult period, fishery scientists and managers created plans to rebuild eight different stocks of important groundfish species. Quotas were drastically reduced, and nets were modified to minimize contact with benthic habitat. Major spatial restrictions (area closures called Rockfish Conservation Areas or RCAs) were designed to prohibit trawlers from the core depth zone with highest density of adults of the overfished groundfish stocks. The goal of these measures was to relieve fishing pressure on those stocks and commence rebuilding. A 2006 amendment to the Pacific Coast Groundfish Fishery Management Plan designated areas as Essential Fish Habitat (EFH), strengthening the RCAs.

Since the 2000 disaster declaration, improved fisheries science, industry compliance, innovative technological and regulatory advances, and improved recruitment resulted in the recovery of most of the formerly depleted stocks. Rebuilding of this groundfish fishery appears to have been successful. With the lifting of some regulatory restrictions, new opportunities may exist for groundfish harvest – although shifting market demands may present new challenges.

Conclusion

Natural and social scientists, managers, and industry practitioners together must address biological, environmental, economic, and social challenges in collaborative teams that not only include different disciplines but also recognize different types of knowledge. Scientists must be comfortable working with managers and stakeholders to collectively develop research priorities and incorporate local ecological knowledge into scientific understanding. NOAA has recently developed new priorities for addressing the impacts of climate change on marine fisheries. Agency scientists cannot do this alone, however, due to time limitations and their immediate tactical needs (e.g., completing surveys, conducting stock assessments) that continuously obviate deeper evaluation of the data and climate models/scenarios. Survey data tend to be underutilized, and many fishery management agencies must coordinate with academic teams to enhance the utility of existing information and promote innovation through targeted research. Research partnerships with agencies are mutually beneficial because they expose students and early career scientists from academia to unique professional and research opportunities (e.g., internships at agency labs, seagoing opportunities on research vessels). New areas of investment in research and education promote convergence training: graduate education that merges across disciplines and provides experiential learning opportunities for students to prepare them for diverse career pathways (Ciannelli et al., 2014b). Teams of students can work collaboratively to address applied problems developed by stakeholders and academics, such as bioeconomic impacts of species range shifts (Magel et al., 2020), evaluation of socio-ecological success of marine reserves (Wilson et al., 2021), and communication of uncertainty in marine weather forecasts (Kuonen et al., 2019).

Questions for Reflection

1. Are wild capture fisheries more than just food?
2. What are some of the natural, human, and technological challenges facing wild capture fisheries?
3. What are some of the similarities and differences of wild capture fisheries across the Global North and between fisheries in the Global North and the Global South?
4. What are the educational needs to sustainably manage fisheries?

Notes

1 Ocean currents, temperatures, habitats, and their relationships.
2 Stakeholders include seafood harvesters or growers, seafood businesses, and the coastal communities where people involved in fisheries live and work, as well as policymakers and fishery managers who implement the policies.

References

Arellano, C. E., & Swartzman, G. (2010). The Peruvian fishery: Changes in patterns and distribution over time. *Fisheries Research*, *101*(3), 133–145. https://doi.org/10.1016/j.fishres.2009.08.007

Bailey, K. M., & Houde, E. D. (1989). Predation on eggs and larvae of marine fishes and the recruitment problem. In J. H. S. Blaxter & A. J. Southward (Eds.), *Advances in marine biology* (Vol. 25, pp. 1–83). Academic Press. https://doi.org/10.1016/S0065–2881(08)60187-X

Beaudreau, A. H., & Levin, P. S. (2014). Advancing the use of local ecological knowledge for assessing data-poor species in coastal ecosystems. *Ecological Applications*, *24*(2), 244–256. https://doi.org/10.1890/13–0817.1

Bonanomi, S., Pellissier, L., Therkildsen, N. O., Hedeholm, R. B., Retzel, A., Meldrup, D., Olsen, S. M., Nielsen, A., Pampoulie, C., Hemmer-Hansen, J., Wisz, M. S., Grønkjær, P., & Nielsen, E. E. (2015). Archived DNA reveals fisheries and climate induced collapse of a major fishery. *Scientific Reports*, *5*(1), 1–8. https://doi.org/10.1038/srep15395

Brander, K. (2010). Impacts of climate change on fisheries. *Journal of Marine Systems*, *79*(3–4), 389–402. https://doi.org/10.1016/j.jmarsys.2008.12.015

Carothers, C., Lew, D. K., & Sepez, J. (2010). Fishing rights and small communities: Alaska halibut IFQ transfer patterns. *Ocean Coastal Management*, *53*(9), 518–523. https://doi.org/10.1016/j.ocecoaman.2010.04.014

Chavez, F. P., & Messié, M. (2009). A comparison of eastern boundary upwelling ecosystems. *Progress in oceanography*, 83(1–4), 80–96. https://doi.org/10.1016/j.pocean.2009.07.032

Cheung, W. W. L., Lam, V. W. Y., Sarmiento, J. L., Kearney, K., Watson, R. E. G., Zeller, D., & Pauly, D. (2010). Large-scale redistribution of maximum fisheries catch potential in the global ocean under climate change. *Global Change Biology*, *16*(1), 24–35. https://doi.org/10.1111/j.1365–2486.2009.01995.x

Ciannelli, L., Bailey, K., & Olsen, E. M. (2014a). Evolutionary and ecological constraints of fish spawning habitats. *ICES Journal of Marine Science, 72*(2), 285–296. https://doi.org/10.1093/icesjms/fsu145

Ciannelli, L., Hunsicker, M., Beaudreau, A., Bailey, K., Crowder, L. B., Finley, C., Webb, C., Reynolds, J., Sagmiller, K., Anderies, J. M., Hawthorne, D., Parrish, J., Heppell, S., Conway, F., & Chigbu, P. (2014b). Transdisciplinary graduate education in marine resource science and management. *ICES Journal of Marine Science, 71*(5), 1047–1051. https://doi.org/10.1093/icesjms/fsu067

Cinner, J. E., & Barnes, M. L. (2019). Social dimensions of resilience in social-ecological systems. *One Earth, 1*(1), 51–56. https://doi.org/10.1016/j.oneear.2019.08.003

Conway, F. D. L., Gilden, J., & Zvonkovic, A. (2002). Communication, power and innovation in fishing families and communities. *Fisheries, 27*(10), 20–29.

Conway, F., & Shaw, W. (2008). Socioeconomic lessons learned from the response to the federally declared West Coast groundfish disaster. *Fisheries, 33*(6), 269–277. https://doi.org/10.1577/1548–8446-33.6.269

Cramer, L. A., Flathers, C., Caracciolo, D., Russell, S. M., & Conway, F. (2018). Graying of the fleet: Perceived impacts on coastal resilience and local policy. *Marine Policy, 96*, 27–35. https://doi.org/10.1016/j.marpol.2018.07.012

Dahlke, F. T., Butzin, M., Nahrgang, J., Puvanendran, V., Mortensen, A., Pörtner, H. -O., & Storch, D. (2018). Northern cod species face spawning habitat losses if global warming exceeds 1.5°C. *Science Advances, 4*(11), eaas8821. https://doi.org/10.1126/sciadv.aas8821

DingsØr, G. E., Ciannelli, L., Chan, K. -S., Ottersen, G., & Stenseth, C. N. (2007). Density dependence and density independence during the early life stages of four marine fish stocks. *Ecology*, 88(3), 625–634. https://doi.org/10.1890/05–1782

Donkersloot, R., & Carothers, C. (2016). The graying of the Alaskan fishing fleet. *Environment: Science and Policy for Sustainable Development*, 58(3), 30–42. https://doi.org/10.1080/00139157.2016.1162011

Fina, M. (2011). Evolution of catch share management: Lessons from catch share management in the North Pacific. *Fisheries*, 36(4), 164–177. https://doi.org/10.1080/03632415.2011.564509

Food and Agriculture Organization of the United Nations. (1995). *Code of conduct for responsible fisheries*. www.fao.org/3/a-v9878e.pdf

Food and Agriculture Organization of the United Nations. (2004). *Fishery country profile - Greenland*. https://www.fao.org/fishery/docs/DOCUMENT/fcp/en/FI_CP_GL.pdf

Food and Agriculture Organization of the United Nations. (2015). *Voluntary guidelines for securing sustainable small-scale fisheries in the context of food security and poverty eradication*. www.fao.org/3/a-i4356en.pdf).

Food and Agriculture Organization of the United Nations. (2020). *The state of world fisheries and aquaculture 2020. Sustainability in action*. https://doi.org/10.4060/ca9229en

Frank, K. T., Petrie, B., Leggett, W. C., & Boyce, D. G. (2016). Large scale, synchronous variability of marine fish populations driven by commercial exploitation. *Proceedings of the National Academy of Sciences, 113*(29), 8248–8253. https://doi.org/10.1073/pnas.1602325113

Frank, K. T., Petrie, B., Leggett, W. C., & Boyce, D. G. (2018). Exploitation drives an ontogenetic-like deepening in marine fish. *Proceedings of the National Academy of Sciences, 115*(25), 6422–6427. https://doi.org/10.1073/pnas.1802096115

Global Seafood Alliance. (2020, June 3). *Crisis response: How is the seafood supply chain responding to the COVID-19 pandemic?* [Video]. YouTube. https://www.youtube.com/watch?v=c52DAIy9Phw

Greenberg, P. (2014). *American catch: The fight for our local seafood*. Penguin Press.

Keller, A. A., Ciannelli, L., Wakefield, W. W., Simon, V., Barth, J. A., & Pierce, S. D. (2017). Species-specific responses of demersal fishes to near-bottom oxygen levels within the California Current large marine ecosystem. *Marine Ecology Progress Series, 568*, 151–173. https://doi.org/10.3354/meps12066

Kuonen, J., Conway, F., & Strub, T. (2019). Relating ocean condition forecasts to the process of end-user decision making: A case study of the Oregon commercial fishing community. *Marine Technology Society Journal, 53*(1), 53–66. https://doi.org/10.4031/MTSJ.53.1.1

Long, S., & Jones P. J. (2021). Greenland's offshore Greenland halibut fishery and role of the Marine Stewardship Council certification: A governance case study. *Marine Policy, 127*, 104095. https://doi.org/10.1016/j.marpol.2020.104095

Magel, C. L., Lee, E. M. J., Strawn, A. M., Swieca, K., & Jensen, A. D. (2020). Connecting crabs, currents, and coastal communities: Examining the impacts of changing ocean conditions on the distribution of U.S. West Coast Dungeness crab commercial catch. *Frontiers in Marine Science, 7*, 401. https://doi.org/10.3389/fmars.2020.00401

Matulich, S. C. (2009). The value of individual processing quota in the Alaska red king crab fishery: A preliminary analysis. *Marine Resource Economics, 24*(2), 187–193. https://doi.org/10.1086/mre.24.2.42731379

McCay, B. J., Creed, C. F., Finlayson, A. C., Apostle, R., & Mikalsen, K. (1995). Individual transferable quotas (ITQs) in Canadian and US fisheries. *Ocean & Coastal Management, 28*(1–3), 85–115. https://doi.org/10.1016/0964–5691(95)00068-2

Meire, L., Mortensen, J., Meire, P., Juul-Pedersen, T., Sejr, M. K., Rysgaard, S., Nygaard, R., Huybrechts, P., & Meysman, F. J. R. (2017). Marine-terminating glaciers sustain high productivity in Greenland fjords. *Global Change Biology, 23*(12), 5344–5357. https://doi.org/10.1111/gcb.13801

Ministry of Forestry and Agriculture (Italy). (2020). *Annual report on the efforts made by Italy in 2019 to achieve a sustainable balance between fishing capacity and fishing opportunities in compliance with art 22 Reg.* (CE) n.1380/2013.

Ministry for Foreign Affairs (Government of Iceland). (2020, December 20). *Greenland and Iceland in the New Arctic: Recommendations of the Greenland Committee appointed by the Minister for Foreign Affairs and International Development Co-operation*. https://www.government.is/library/01-Ministries/Ministry-for-Foreign-Affairs/PDF-skjol/Greenland-Iceland-rafraen20–01-21.pdf

Morrison, W. (2019, April 8). *National Marine Fishery Service's catch share policy*. NMFS. https://media.fisheries.noaa.gov/dam-migration/01–121.pdf

Myers, R. A., & Worm, B. (2003). Rapid worldwide depletion of predatory fish communities. *Nature, 423*(6937), 280–283. https://doi.org/10.1038/nature01610

National Marine Fisheries Service. (n.d.). *Fisheye: Fisheries economic explorer*. https://data-explorer.northwestscience.fisheries.noaa.gov/fisheye/

National Marine Fisheries Service. (2018). *Fisheries economics of the United States, 2016* (NMFS-F/SPO-18). U.S. Dept. of Commerce & NOAA. https://www.fisheries.noaa.gov/resource/document/fisheries-economics-united-states-report-2016

National Marine Fisheries Service. (2019). *Laws & policies: Magnuson-Stevens Act*. https://www.fisheries.noaa.gov/topic/laws-policies#magnuson-stevens-act

Nielsen, M., Nielsen, A. H. R., Waldo, S., Blomquist, J., Asche, F., Bergesen, O., Viðarsson, J. R., Sigurðardóttir, S., & Sveinþórsdóttir, R. (2017). *Employment and salary of Nordic coastal fishermen*. Nordisk Ministerråd. https://doi.org/10.6027/TN2017–558

Paterson, B., Isaacs, M., Hara, M., Jarre, A., & Moloney, C. L. (2010). Transdisciplinary co-operation for an ecosystem approach to fisheries: A case study from the South African sardine fishery. *Marine Policy, 34*(4), 782–794. https://doi.org/10.1016/j.marpol.2010.01.019

Pauly, D., & Christensen, V. (1995). Primary production required to sustain global fisheries. *Nature, 374*(6519), 255–257. https://doi.org/10.1038/374255a0

Pfeiffer, L., & Gratz, T. (2016). The effect of rights-based fisheries management on risk taking and fishing safety. *Proceedings of the National Academy of Sciences, 113*(10), 2615–2620. https://doi.org/10.1073/pnas.1509456113

Pinsky, M. L., Worm, B., Fogarty, M. J., Sarmiento, J. L., & Levin, S. A. (2013). Marine taxa track local climate velocities. *Science, 341*(6151), 1239–1242. https://doi.org/10.1126/science.1239352

Pooley, S. G. (1998). *Issues and options in designing and implementing limited access programs in marine fisheries* (NOAA-TM-NMFS-SWFSC-252). U.S. Dept. of Commerce & NOAA. https://repository.library.noaa.gov/view/noaa/3047/noaa_3047_DS1.pdf

Rettig, R. B., & Ginter, J. J. C. (Eds.). (1978). *Limited entry as a fishery management tool.* University of Washington Press.

Ringer, D., Carothers, C., Donkersloot, R., Coleman, J., & Cullenberg, P. (2018). For generations to come? The privatization paradigm and shifting social baselines in Kodiak, Alaska's commercial fisheries. *Marine Policy*, 98, 97–103. https://doi.org/10.1016/j.marpol.2018.09.009

Rogers, L. A., Griffin, R., Young, T., Fuller, E., St. Martin, K., & Pinsky, M. L. (2019). Shifting habitats expose fishing communities to risk under climate change. *Nature Climate Change*, 9(7), 512–516. https://doi.org/10.1038/s41558-019-0503-z

Russell, S., Sparks, K., Arias-Arthur, A., & Varney, A. (2014). *Pacific Coast groundfish trawl fishery social study*. Northwest Fisheries Science Center. https://www.fisheries.noaa.gov/west-coast/socioeconomics/west-coast-groundfish-trawl-fishery-social-study

Schindler, D. E., Hilborn, R., Chasco, B., Boatright, C. P., Quinn, T. P., Rogers, L. A., & Webster, M. S. (2010). Population diversity and the portfolio effect in an exploited species. *Nature*, *465*(7298), 609–612. https://doi.org/10.1038/nature09060

Schittone, J. (2001). Tourism vs. commercial fishers: Development and changing use of Key West and Stock Island, Florida. *Ocean & Coastal Management*, *44*(1–2), 15–37. https://doi.org/10.1016/S0964–5691(00)00078-8

Selden, R. L., Thorson, J. T., Samhouri, J. F., Bograd, S. J., Brodie, S., Carroll, G., Haltuch, M. A., Hazen, E. L., Holsman, K. K., Pinsky, M. L., Tolimieri, N., & Willis-Norton, E. (2020). Coupled changes in biomass and distribution drive trends in availability of fish stocks to US West Coast ports. *ICES Journal of Marine Science*, *77*(1), 188–199. https://doi.org/10.1093/icesjms/fsz211

Silvestri, G. (2003). *Lacco Ameno tuna and other fishing on the Island of Ischia.* Imagaenaria.

Timmermann, A., An, S.-I., Kug, J.-S., Jin, F. -F., Cai, W., Capotondi, A., Cobb, K. M., Lengaigne, M., McPhaden, M. J., Stuecker, M. F., Stein, K., Wittenberg, A. T., Yun, K.-S., Bayr, T., Chen, H.-C., Chikamoto, Y., Dewitte, B., Dommenget, D., Grothe, P. … Zhang, X. (2018). El Niño–Southern Oscillation complexity. *Nature*, 559, 535–545. https://doi.org/10.1038/s41586-018-0252–6.

US Bureau of Labor Statistics. (2022, April 18). *Occupational outlook handbook*. https://www.bls.gov/ooh/math/mathematicians-and-statisticians.htm

Weber, M. L. (2002). *From abundance to scarcity: A history of U.S. marine fisheries policy.* Island Press.

Wilson, M. N., Laufer, A. E., Howard, E. M., & Wong-Ala, J. A. (2021). Lessons from the trenches: Students' perspectives of their own marine transdisciplinary education. *Frontiers in Marine Science*, *7*, 592368. https://doi.org/10.3389/fmars.2020.592368

White, C. S. (2015). Getting into fishing: Recruitment and social resilience in North Norfolk's 'cromer crab' fishery, UK. *Sociologia Ruralis*, *55*(3), 291–308. https://doi.org/10.1111/soru.12101

White, E. R., Froehlich, H. E., Gephart, J. A., Cottrell, R. S., Branch, T. A., Bejarano, R. A., & Baum, J. K. (2020). Early effects of COVID-19 on US fisheries and seafood consumption. *Fish and Fisheries*, *22*(1), 232–239. https://doi.org/10.1111/faf.12525

5 Global Marine Aquaculture Development

J. D. Stieglitz, M. Touchton, D. D. Benetti, D. Rothen, A. Clark-Hughes, B. K. Haus, A. Zangroniz, and D. O. Suman

Continued global population growth has focused attention on improving the production, efficiency, and sustainability of food production systems to provide for food security. Capture fisheries are the last remaining industrialized animal protein production system still operating in a hunter-gatherer procurement strategy, while aquaculture represents a farming alternative whose growth can be attributed to several inherent advantages compared to capture fisheries. Aquaculture, or fish farming, has emerged as one of the most promising forms of food production for both animal and plant products and is located in both marine and coastal environments and inland waters.

Aquaculture currently provides about half the global aquatic animal production and produces more fish for human consumption than capture fisheries. The development of this rapidly growing activity faces several challenges, as well as many opportunities. While aquatic animals are a significant food source for billions of people, wild-caught (capture) production has plateaued and may decline in the coming decades. The Food and Agriculture Organization of the United Nations (FAO, 2022) estimates that in 2019 about 35.4% of global fish stocks were overfished (i.e., fished at biologically unsustainable levels), while 57.3% were fished sustainably. Capture fisheries offer little room for growth; only 7.2% of stocks are underfished (FAO, 2022). Climate change will exacerbate this trend, alter ranges of many stocks, and lead to uncertain ecosystem effects. Coastal wetland losses, as well as degradation of tropical coral reefs and coastal water quality, all adversely impact the quality and quantity of wild-caught species.

Residents of coastal communities may be particularly vulnerable to the decreasing availability of captured aquatic resources. For generations, communities have depended upon healthy and easily accessible local seafood. As these resources diminish in supply and safety, food security for residents of coastal communities may become more tenuous. Poverty levels in small coastal communities are high, particularly in developing countries, and any threat to food supply could have grave impacts on community livelihoods and well-being.

FAO (2022) reported that the 2020 global production of fisheries was 177.8 million tons. Of this total, capture fisheries contributed about 51% while aquaculture production accounted for 49%. However, since a portion (~20.4 million tons) of capture fisheries is utilized for non-food products, such as fishmeal and fish

DOI: 10.4324/9781003058151-7

oil for animal feeds, aquaculture provided 56% of seafood consumed by humans in 2020, compared to 44% from capture fisheries. The contribution from aquaculture production has steadily increased over the past four decades, while capture fisheries (marine and inland) have remained relatively constant (Table 5.1). Over 20 million people are directly engaged in aquaculture operations, compared with almost 60 million in capture fisheries (FAO, 2022). Asian countries dominate farming of aquatic animals and are responsible for about 89% of total production; China's share alone is almost 60%, and India, Bangladesh, Indonesia, Thailand, and Vietnam are significant producers.

We must distinguish between marine/coastal and inland farming activities, which generally represent marine species vs. freshwater species. The marine/coastal production of aquatic animals in 2020 was about 33.1 million tons, or only about 37.8%. Both aquaculture sectors (marine/coastal and inland) have experienced high growth in past decades (Table 5.2). Inland aquaculture utilizes a variety of techniques from tanks, earthen ponds, in-pond raceways, and net pens in inland water bodies, as well as integrated paddy rice-fish culture. Finfish dominate inland aquaculture production, particularly carp, tilapia, and catfish. Cultivation of marine species occurs in coastal ponds, lagoons, and tanks or in the open ocean (Figure 5.1). Mariculture operations may use floating or submersible net pens, lines, or rafts and often depend on seedstock that is produced in nurseries on land. Asian countries together with Latin American nations dominate coastal aquaculture activities.

Table 5.2 indicates the categories of aquatic animals (finfish, crustaceans, mollusks) grown in marine/coastal operations. Mollusks (oysters, scallops, mussels, clams) dominate in live weight with similar contributions of finfish (Atlantic salmon and others) and crustaceans (shrimp, prawns, crabs). Asian countries contribute 87% of finfish production, 89% of crustacean production, and 92% of mollusk production (FAO, 2022). We address the growth of Atlantic salmon and shrimp cultivation later in the chapter.

Table 5.1 World fisheries and aquaculture production in millions of tons live weight (FAO, 2022)

	1990s	*2000s*	*2010s*	*2020*
CAPTURE				
Inland	7.1	9.3	11.3	11.5
Marine/Coastal	81.9	81.6	79.8	78.8
Total Capture	88.9	90.9	91.0	90.3
AQUACULTURE				
Inland	12.6	25.6	44.7	54.4
Marine/Coastal	9.2	17.9	26.8	33.1
Total Farmed	21.8	43.4	71.5	87.5
TOTAL GLOBAL FISHERIES AND AQUACULTURE	110.7	134.3	162.6	177.8

Figure 5.1 Net pen used for offshore aquaculture production of cobia (*Rachycentron canadum*) at an offshore aquaculture farm off the coast of Panama. Photo credit: J. Stieglitz.

Table 5.2 Global inland and marine/coastal aquaculture production in thousands of tons live weight in 2020 (FAO, 2022)

Category	*Inland*	*Marine/Coastal*	*Total*
Finfish	49,120	8,341	57,461
Crustaceans	4,477	6,760	11,237
Mollusks	193	17,548	17,741
Other Aquatic Animals	594	469	1,062
Total	54,384	33,117	87,501

The rapid growth of global aquaculture in past decades is estimated to be 6.7% from 1990 to 2020 (FAO, 2022). Yet, concerns remain over balancing this growth with associated socio-environmental impacts. The question is no longer whether aquaculture is necessary but, rather, how to ensure sustainability.

Aquaculture has a bright future, and if growth trends continue, it will surpass the sum of capture fisheries production worldwide. By 2030, aquaculture should produce 106 million tons live weight, an increase of 21% over 2020 values (FAO, 2022). However, FAO predicts that growth rates will decelerate in the coming decade as countries increase their environmental oversight of farming activities.

Challenges and Advances

The growth in marine aquaculture development has resulted in many benefits – notably, increased consumer access to seafood – but growth has sometimes come at a cost. Balancing economic viability and environmental sustainability remains challenging. Utilization of technical advances, improved understanding of production hurdles, and implementation of sustainable measures to overcome challenges all facilitate marine aquaculture's role as leading producer of the world's seafood supply. The following sections discuss some of the primary challenges and advances.

Technical

Technical challenges are either biologically or otherwise associated with the engineering of growing and harvesting systems. From a biological standpoint, a principal challenge is identification of reliable sources of healthy, high-quality seedstock used for stocking grow-out systems. Major marine species groups, such as bivalves, crustaceans, fish, and macroalgae, all face species-specific technical challenges. Overcoming these requires research and development to resolve the biological and physiological demands of the species' life stages. Other challenges are associated with engineering aspects of grow-out systems for large-scale production. Technical developments are often difficult to implement because aquaculture enterprises range from large corporations with global operations to countless small-scale local farms. Such diversity presents social, economic, regulatory, and political complexities that can make large-scale adoption of technical advances difficult (Kumar et al., 2018).

Technical advances include increased efficiency and reduction of total environmental impact (Boyd et al., 2020; Lovatelli et al., 2013; Price et al., 2015). Innovations also occur on both the supply and demand sides (Kumar & Engle, 2016). Recent developments in genetics; animal health and welfare and disease control; life-cycle closure of many high-value species; improvements in recirculating aquaculture systems (RAS); improved feed efficiencies and new feed ingredients; and engineering technologies for open-ocean and land-based systems are all significant. These innovations offer the potential to improve the overall production efficiency of marine aquaculture.

Great progress has been made in domestication of major species, such as shrimp, tilapia, and salmon (Teletchea, 2015), leading to the availability of high-quality, genetically selected, improved seedstock. However, despite advances, the seedstock for many marine species is still sourced from the wild broodstock, leading to high variability in quality and overall aquaculture performance, reduced production efficiencies, and environmental impacts.

Land-based marine systems, such as RAS, coastal ponds, "flow-through"-type systems, as well as near-shore coastal net pens, are viable options for many commercially produced marine fish and crustaceans. However, in their current form these production systems cannot meet the expected demand of 40–50 million metric tons of additional seafood that aquaculture must supply by 2030 to meet the world's projected demand (World Bank, 2013). Effectively satisfying the projected seafood demand requires improvements in production efficiencies across multiple criteria (e.g., biological, economic, environmental, and operational). The open ocean environment, characterized by strong currents and deep water, permits greater assimilative waste capacity, representing one of the most promising frontiers for technical advancement in marine aquaculture (Benetti et al., 2010; Buck & Langan, 2017). Improved economic viability and environmental sustainability are inextricably tied to technical advances in the field.

Economic

Global marine aquaculture also faces numerous economic challenges because it is essentially a practice of refining relatively low-cost raw materials into higher value products. This refining process exists for both fed (fish, shrimp) and unfed (macroalgae, mollusks) species, with the primary difference that the latter typically utilize freely accessible inputs for growth. Profitability is achieved when the margins between production costs and revenue from outputs reach a point of overall economic viability. Numerous complexities associated with these basic principles can be tied to the fact that marine aquaculture and the seafood industry are global enterprises. Raw ingredients, equipment, seedstock, and other critical elements necessary to support individual farming operations are sourced globally. This complex supply chain leads to economic hurdles outside the control of any single operation. The global economy's tumultuous nature can lead to a farm being profitable one day, then losing money the next, simply because of currency exchange rates on the supply or demand side. Understanding the nuances of the value-chain is critical to distinguishing successful aquaculture ventures from failures. Well-informed producers attempt to minimize the risk of these economic uncertainties, often leading to eventual vertical integration[1] of aquaculture businesses (Olson & Criddle, 2008). By controlling many of the variables in the production process, operators attempt to minimize risk as much as possible. Such practices are not unique to marine aquaculture, yet the inherent factors associated with growing a live product for market adds a significant layer of complexity. Small-scale producers can be particularly vulnerable to financial risks given their narrow profit margins. Advances, including use of disease-resistant and specific pathogen-free seedstock, improved understanding of animal welfare and life support requirements, progress in animal health, and implementation of best management practices have all led to more sustainable marine aquaculture operations.

Marine aquaculture possesses many inherent economic advantages over wild-harvest seafood operations. Benefits stem from the availability of fresh aquaculture products on regular schedules, while wild harvest depends on the success or failure of a particular fishing season. Additionally, farm-raised seafood has a known history that can be easily traced, certified, and delivered to the consumer. Increasing awareness of human health and nutrition has led many consumers to seek seafood options that can be certified as free from certain contaminants.

Aquaculture can deliver these products through careful management of the entire production process and selection of feed inputs. Additionally, aquaculture permits high-demand seafood to be produced closer to markets, thereby potentially reducing the overall carbon footprint associated with global seafood production. As consumers increasingly understand aquaculture's benefits, the sector's growth should not directly compete with commercial wild-catch fisheries.

Environmental

Ocean aquaculture illustrates a classic case of "tragedy of the commons", whereby ocean space – a common-pool resource – is developed in what is deemed to be the most profitable manner regardless of the environmental impacts (e.g., water pollution, invasive species introductions, diseases, habitat degradation). Examples repeat across the major species sectors of marine aquaculture, including salmon and shrimp aquaculture. The key to achieving sustainability is identification of methods, species, systems, and management practices that allow for an appropriate balance of environmental protection and economic viability. In recent decades as the industry has matured, operations that have achieved this balance are those that typically attract the greatest investment. Even operations once deemed "traditional" and unable to improve sustainability indicators, such as pond-based marine shrimp and fish production, have seen significant improvements in overall operating efficiencies, profitability, social acceptance, and environmental sustainability (Boyd et al., 2020).

Legal/Regulatory

The ocean presents attractive investment options for aquaculture, often to the exclusion of equity and sustainability (Jouffray et al., 2020). Regulatory structures that grant licenses to aquaculture activities are viewed by some as privatization of a common-pool resource. Thus, perceptions of traditional users, particularly fishers, toward ocean aquaculture are often negative. Transparent and participatory governance from global to local levels is critical for addressing concerns about sustainability and equity. Marine spatial planning (ocean zoning) is one anticipatory planning technique that could assist the development of rational regulation of the suite of anthropogenic ocean activities, including aquaculture (Lester et al., 2018).

In many countries ocean aquaculture operations face numerous legal and regulatory obstacles. Framework legislation is essential for delegating institutional authority and responsibilities and clearly enunciating principles and objectives for this sector. Legislation should mandate the promulgation of regulations and development of a national aquaculture plan that establishes operational guidelines, permitting structures, and parameters for site selection; sets requirements for monitoring for disease, water quality, and escapees; and develops attractive parameters for leases. A country cannot expect to create a welcoming environment for investors without clear, well-structured legislation and implementing regulations. The permitting process itself may present significant disincentives for ocean aquaculture activities. A developer may have to obtain numerous permits from various agencies, confront high transactional costs in terms of time and money, and face an extremely high burden of proof that the activity will not cause significant environmental impacts. The result could be an environment creating uncertainty and disincentives for potential investors. Ocean aquaculture in the US is plagued by regulatory difficulties (Box 5.1).

BOX 5.1 Regulatory Challenges for US Ocean Aquaculture

Open ocean aquaculture in the US is illustrative of some of the complexities and challenges facing this sector (Lester et al., 2018). The Gulf of Mexico exemplifies the conflict among branches of the federal government over authority for ocean aquaculture. At least five federal agencies have permitting authority over different aspects of ocean aquaculture, and no streamlined coordinated permitting system currently exists. The current situation (July 2022) is one of great uncertainty for aquaculture investors. The Executive Branch has taken steps to facilitate ocean aquaculture. In 2016 the Gulf of Mexico Regional Fisheries Management Council adopted an Open Ocean Aquaculture Fisheries Management Plan (NMFS) that was subsequently approved by the NOAA Fisheries. Executive Order 13921 (May 7, 2020) attempted to promote US ocean aquaculture by mandating a streamlined permitting process that NMFS would coordinate, the creation of Aquaculture Opportunity Areas where farms would be located, and rapid environmental review. Commercial fishing and environmental groups challenged the authority of the NMFS under the Magnuson–Stevens Fishery Conservation and Management Act to regulate aquaculture activities and were ultimately successful in the US Circuit Court of Appeals (*Gulf Fishermens Association v. NMFS*) in August 2020. Legislation (Advancing the Quality and Understanding of American Aquaculture Act/AQUAA Act – S.3100) that would establish the regulatory framework for open ocean aquaculture has been introduced in Congress with bipartisan support, but the proposal has failed to advance in numerous sessions. By 2022, the absence of a consistent message from the three governmental branches left US open ocean aquaculture development in limbo. Nevertheless, despite the absence of a predictable regulatory framework, one net pen farm in the eastern Gulf of Mexico (Ocean Era, LLC) obtained all required permits and should begin operations to grow Almaco jack in late 2022.

Case Studies of Two Marine Species: Shrimp and Atlantic Salmon

Shrimp and Atlantic salmon are two stories of seafood farming that have entered Western markets and are among the largest farmed marine species. Both have experienced rapid growth and exhibit increasingly more sustainable social and environmental practices; however, there are still severe social and environmental impacts.

Shrimp Aquaculture

Two decades after World War II, Japanese fishery scientists succeeded in raising commercial quantities of kuruma shrimp, *Penaeus japonicus*, in captivity. Several years later, Filipino scientists introduced aquaculture techniques for giant black tiger shrimp, *Penaeus monodon*, and shrimp aquaculture began its rapid spread through Southeast Asia. By the 1970s, the boom in shrimp farming was the proof that the "Blue Revolution"[2] could boost production of seafood. Shrimp aquaculture operations expanded rapidly throughout many tropical developing countries that viewed high investment returns and expanding export markets for the product as stimuli for national development. By 2018, production surpassed 6.5 million tons of farmed shrimp (*P. vannamei* and *P. monodon*) with about 75% of production from Asia and 25% from Latin America (FAO, 2022). Post-larvae introduced to grow-out ponds are obtained from the wild (extensive, artisanal operations) or produced in hatcheries (semi-intensive or intensive operations). Shrimp is a gourmet export product providing significant returns for domestic and foreign investors, while typically offering little benefit for food security in producing countries.

Today, 12 of the top shrimp farming countries (Vietnam, China, Bangladesh, Ecuador, Indonesia, Myanmar, India, Mexico, Philippines, Thailand, Brazil, and Honduras) house almost 97% of the pond area (Boyd & McNevin, 2018), and all have experienced associated high rates of mangrove loss. Many of the extensive shrimp farms were built in mangrove ecosystems during the boom period, causing significant mangrove destruction. By 2001, aquaculture was responsible for 52% of global mangrove loss, while shrimp farming itself accounted for 38% of losses (Warne, 2011). Numerous incentives explain the historical siting of shrimp ponds in mangroves. Wild larvae ("seed") for stocking ponds were collected in mangrove channels and estuaries. Brackish water could be easily pumped from tidal channels; alternatively, water simply filled the ponds during high tides. Wastewater from ponds could be discharged into mangrove tidal channels. Transport of feed and shrimp harvest by boat was a feasible option. Most importantly, land was cheap, available, and usually publicly owned. Governments often undervalued mangrove ecosystem services and granted inexpensive concessions for use of the land and water (or failed to enforce prohibitions on removing mangroves).

Many governments and international financial institutions (e.g., World Bank, Asian Development Bank, Inter-American Development Bank) lacked appreciation of the ecosystem benefits of mangrove ecosystems and prioritized short-term economic gain, thus encouraging conversion of coastal wetlands to shrimp aquaculture ponds. A third of international fisheries assistance to the Philippines (approximately $1 billion) supported aquaculture projects. Primavera (2000) suggested that aquaculture was responsible for perhaps half of that country's mangrove losses. Governments granted concessions to develop mangrove forests through conversion to shrimp ponds for extended periods of time at low rents (Primavera, 2005). Armitage (2002) noted that governments and multilateral financial institutions, allied with well-connected entrepreneurs, facilitated construction of shrimp aquaculture ponds at the expense of traditional users of common-pool resources. The privatization or extended leasing of common-pool resources undermined community-based property rights and resource access and led to marginalization of numerous traditional mangrove users. Essentially, a multi-use public resource (mangrove forests) had become a single-use private asset (shrimp ponds). Shrimp ponds located in mangrove areas have also been responsible for serious social conflicts between pond operators and traditional users of mangrove ecosystems, including small-scale fishers, crab and clam collectors, and small-scale wood harvesters, many of whom live in extreme poverty. Shrimp pond operators have often blocked access routes on which traditional users depend to access public resources, sometimes leading to violent confrontations between local community members and shrimp pond guards.

The troubling situation of shrimp aquaculture has improved in recent decades due to several factors (Ashton, 2008). A growing number of operations depend on farm-raised broodstock and hatchery post-larvae – decreasing the dependency on wild caught larvae, as well as improving efficiency. There is an increased recognition that mangrove areas are suboptimal for shrimp pond location, and ponds are increasingly being built further inland. Other technological advances, such as the installation of recycled/closed water circulation systems, continuous aeration, improved artificial feeds with reduced percentages of fishmeal components, and the decreased use of antibiotics, have all lightened the ecological footprint of shrimp farms.

Vietnam and Indonesia have pioneered silvofishery models that integrate shrimp aquaculture into mangrove areas and embrace multiple uses of mangroves along with shrimp culture (Ahmed et al., 2018). The Mangroves and Markets (MAM) Project in Vietnam, supported by SNV Netherlands Development Organization, encourages a traditional type of shrimp pond that requires a minimum of 50% of mangrove cover. Shrimp produced from this program also receive certification that recognizes less harmful shrimp aquaculture techniques (SNV Netherlands Development Organization, n.d.).

Many countries have passed new legislation and adopted regulations that prohibit new concessions of shrimp ponds in mangroves. Implementation of these prohibitions can often be a challenge, however. Campaigns of international and domestic NGOs, as well as organized opposition from mangrove user groups, have also exerted pressure on governments and consumers (Suman, 2019). While mangrove and coastal wetlands protection continue to face serious challenges worldwide, recognition of the important role of mangrove biomass and sediments as sinks for CO_2 (blue carbon) is increasingly appreciated by the international community (Ahmed et al., 2018; Nellemann et al., 2009).

Certification programs involving collaboration of the industry and NGOs, described further below, also reduce the adverse environmental and social impacts of shrimp farming through consumer choice. In 1999 the Network of Aquaculture Centres in Asia-Pacific partnered with the World Bank, WWF, and later UNEP to form the Consortium on Shrimp Farming and the Environment – subsequently developing the International Principles for Responsible Shrimp Farming (FAO/NACA/UNEP/WB/WWF, 2006). The development of codes of best practice (i.e.,

Code of Conduct for Responsible Fisheries (FAO, 2011a) and *Technical Guidelines on Aquaculture Certification* (FAO, 2011b)), through collaborations of international organizations, academics, NGOs, and the industry itself, also evidences the positive evolution of this industry, success in the increasing siting of ponds away from mangroves, and the termination of support from international lending institutions (Ashton, 2008).

Atlantic Salmon Farming

Salmon farming began in the 1960s with the raising of juvenile Atlantic salmon (*Salmo salar*) in net pens in Norwegian fjords. Atlantic salmon was an ideal species for cultivation because of its large, nutrient-rich eggs and the ease with which it spawned and grew in captivity. In the early 1970s researchers began to conduct genetic-based selection of salmon from Norwegian rivers. This produced salmon varieties with high growth rates and distinct biologies, unofficially called "*Salmo domesticus*" (Greenberg, 2010; Gross, 1998). Norway maintains dominance in this sector today. Norwegian aquaculture industries exported their net pen farming methods and genetic stock to other cold-water regions with deep-water estuaries and fjords (southern Chile; the Maritime Provinces and British Columbia in Canada; Scotland), and production of farmed salmon exploded. Salmon farming typically involves two stages: (1) hatching from eggs and raising in freshwater tanks, and (2) transfer of juvenile salmon to net pens in the sea where they grow on pelletized feed and are harvested after several years.

Global production of farmed Atlantic salmon has increased ten times over since 1990, and by 2020, the global production was 2.72 million tons (FAO, 2022; Iversen et al., 2020). Five major producing countries accounted for over 95% of the farmed salmon production in 2015 (Norway, 55.3%; Chile, 25.4%; Scotland, 7.6%; Canada, 6%; Faroe Islands, 3.3%) (Iversen et al., 2020).

This fast-growing industry has experienced several environmental and social challenges. Atlantic salmon (*S. salar*) is native to the North Atlantic. Nevertheless, it has been introduced to Pacific areas (British Columbia, Canada, as well as southern Chile). Concern surrounds salmon escapees from net pens and their potential to compete with native species of fish, as well as the impact of antibiotics and other aquaculture therapeutants.

The high densities of farmed salmon have degraded water quality in numerous areas because of fish feces and unconsumed feed. Fjords or nearshore channels where cages are located have experienced reduced dissolved oxygen and algal blooms (eutrophication) (Salgado et al., 2015). Complaints by environmental groups and fishers often mention the inability of government authorities to monitor farm operations and determine carrying capacities (Salgado et al., 2015). Concerns also center on the use of lower trophic-level pelagic species (e.g., anchovies, herring, and sardines) that are reduced to fish meal and fish oil – both important components of feed for terrestrial and aquatic animal feeds, including salmon.

With increasing densities of farmed salmon, parasitic sea lice have become a rising problem in areas where salmon are farmed. Infestations of sea lice can reduce fish growth, increase fish mortality, and cause substantial financial losses for farms. Abolofia et al. (2018) estimate that the potential biomass growth lost per production cycle can range from 3.62 to 16.55% of farm revenue. In addition, the sea lice infestations may spill over to wild fish populations. Due to a staggering outbreak of infectious salmon anemia (ISA) viral disease in Chile beginning in 2007, salmon aquaculture failed to show growth from 2009 to 2012 (Abolofia et al., 2018; Asche et al., 2009). The Chilean salmon industry has recovered since the ISA disease outbreak due to important new knowledge and adaptive management (Alvial et al., 2012).

Marine concessions to salmon farming companies have caused spatial conflicts, particularly due to the displacement of fishers and what is often viewed as the privatization of a public resource. In southern Chile, artisanal fishers and Indigenous peoples consider that salmon farms have blocked their access to their traditional fishing areas (Salgado et al., 2015).

Much progress has been made in the past decade to reduce ecosystem impacts and increase efficiency. NGO campaigns and increased public awareness of problems emanating from salmon farms have led to the development of certification programs for "sustainable farms" and best management practices. The Aquaculture Stewardship Council developed an initial standard for salmon aquaculture in 2012, subsequently revised in 2017 (Aquaculture Stewardship Council [ASC], 2017), focusing on habitat protection, disease management, fish feed, introduction and management of non-native and genetically modified species, and social responsibility.

Technological advances have made salmon farming more economically efficient and more environmentally friendly (Kumar et al., 2018). Improved feed formulations produce more efficient feed conversion rates, and innovative feeding systems reduce feed waste. Advances in aquafeed nutrition have significantly reduced the amount of fish meal and oil in feeds for salmon and other carnivorous marine fish aquaculture species (Bendiksen et al., 2011; Jackson & Shepherd, 2012; Tacon & Metian, 2008) to the point where salmon aquaculture is a net producer of fish oil and protein (Aas et al., 2019; Crampton et al., 2010; Ytrestøyl et al., 2015).

Sea lice removal has improved with utilization of cleaner fish that eat sea lice, as well as warm water and freshwater treatments. Antibiotic usage has declined dramatically in recent years with improved biosecurity measures (Morrison & Saksida, 2013); this trend should continue with technical advances and improved farming practices (Bravo, 2012). Genetic selection is increasing yields and improving overall production efficiencies. Improved management requiring better spacing and implementation of carrying capacity standards attempt to minimize water pollution.

Closed-system salmon farming operations are increasing around the world. Some of these technologies may be placed in the water, while others are land-based, consisting of large RAS operations that have begun to raise salmon from egg to market size entirely in land-based tank systems (Kramer, 2015). While land-based salmon farming faces capital cost challenges, salmon RAS technologies may reduce water pollution, permit faster growth, eliminate conflicts over marine space, and produce salmon that earn the highest marks for sustainability (Kramer, 2015; The Explorer, n.d.).

Opportunities for the Future

The future is bright for the global marine aquaculture industry, with continuous progress being made in many areas. Opportunities exist for improving overall operating efficiencies, sustainability, and economic viability of new and existing operations. Some of the most promising opportunities follow.

Integrated Multi-Trophic Aquaculture (IMTA)

Most marine aquaculture systems utilize a single trophic level approach, whereby a single species of one trophic level is grown in isolation with little attention paid to upstream or downstream utilization of nutrient inputs used. Alternatively, IMTA combines culture of species in different trophic levels with complementary ecosystem functions, such as top trophic level salmon that require feed, lower trophic level seaweed that extract nutrients from seawater, and filter feeders (mussels, oysters, sea urchins, sea cucumbers) that consume particulate matter from unconsumed fed and fish feces – illustrated by Cooke Aquaculture IMTA in the Bay of Fundy, New Brunswick, Canada (Chopin, 2013; Granada et al., 2016; Greenberg, 2010). Research has shown that when conducted properly, IMTA techniques can recycle 35–100% of nutrients from fed aquaculture operations (Troell et al., 2009). IMTA represents a promising opportunity for improving the sustainability and economic viability of marine aquaculture and may even play a role in climate change mitigation (Duarte et al., 2017).

Restorative Aquaculture

Coastal aquaculture faces significant political and social acceptance challenges in many areas based on concerns over potential nutrient enrichment ("eutrophication") of coastal waters (Primavera, 2006; Ridler et al., 2007). In the US alone, 65% of the estuaries and coastal waters are severely degraded from excessive nutrient inputs. The eutrophication that plagues these waters leads to other negative conditions, such as harmful algal blooms (HABs) and low oxygen ("hypoxic") conditions that can kill seagrasses and fish. Use of sustainable culture of seaweeds and bivalves can extract nutrients from waters and restore the environment to a state in which wild species can thrive. This type of farming is known as "restorative aquaculture" (Alleway et al., 2019; Gentry et al., 2020; Theuerkauf et al., 2019). Additionally, by relieving pressure on capture fisheries, it allows coastal communities to remain economically productive, thereby increasing resiliency during periods of economic downturn and climate change.

Open Ocean Aquaculture

The open ocean or "offshore" environment represents one of the most promising areas for marine aquaculture development. Recent assessments indicate that sustainable marine aquaculture in only a fraction of the world's oceans could supply

Figure 5.2 Almaco jack (*Seriola rivoliana*) in an ocean net pen at a farming operation off the coast of Kona, Hawaii. Photo credit: J. Stieglitz.

the world's seafood demand, compared to the vast areas of the sea utilized to support wild-catch fisheries (Costello et al., 2020; Gentry et al., 2017). Offshore aquaculture efforts have focused on marine finfish and to a smaller extent on some macroalgae species. Technological advances in engineering of net pens and mooring systems to withstand the difficulties associated with location far offshore, as well as automation and artificial intelligence, combine to improve the operating efficiencies, environmental sustainability, and overall profitability (Figure 5.2).

Certification Programs

Certification of seafood products attempts to promote sustainable production and responsible farming by minimizing the adverse environmental and social impacts from aquaculture operations (Bush et al., 2013; FAO, 2011b). As a market-based tool, certification sets environmental, social, and safety standards for the industry, audits compliance, and labels products so that consumers can make value choices. Certification schemes compete for seafood suppliers that comply with standards via third-party certification, buyers that distribute to consumers, and consortia of environmental groups, academics, and international organizations that approve science-based standards. Certification has gained widest acceptance in Europe and North America, and although the percentage of world aquaculture production that is certified is small (4.6%), its coverage is growing.

The FAO approved its *Technical Guidelines on Aquaculture Certification* in 2011, establishing the framework for numerous certification programs. The FAO established an institutional procedure for certification programs: setting the standards, developing an accreditation system to assess a specific aquaculture process, and third-party certification of an aquaculture activity. FAO guidelines focus on four substantive criteria: animal health and welfare, food safety, environmental integrity, and socio-economic aspects.

Aquaculture certification is currently led by the Aquaculture Stewardship Council (ASC), the Global Aquaculture Alliance, the Aquaculture Certification Council (ACC), and the GLOBAL G.A.P. (Good Agricultural Practices) organization. The ASC standards are consistent with the FAO aquaculture guidelines. ASC issues certifications (assessed via third parties) for farms, suppliers, and products (ASC, n.d.). The Global Aquaculture Alliance (GAA, 2020) also coordinates a comprehensive third-party aquaculture certification program for the four stages of production (hatcheries, farms, feed, and processing plants) (GAA, 2020). GAA's Best Aquaculture Practices (BAPs) are seafood-specific certification programs that address FAO's four substantive criteria.

Certification programs are definite steps toward sustainable and responsible aquaculture operations, but they do have limitations (Parkes et al., 2010). Questions always remain about the accuracy of information. Consumers and distributors may be confused by contradictory information from different certification schemes. Sustainability is often narrowly defined and based on site-specific operations, ignoring cumulative and regional impacts from numerous fish farms on water quality and coastal habitats. Second, it is difficult for stakeholders in the Global South to participate (funds, language, communications, managerial capacity), and suppliers must bear many costs of certification programs (Mutersbaugh et al., 2005). Third, social issues (working conditions, worker rights, salaries) are difficult to certify, and reliance on national labor laws is not effective. Finally, small-scale producers may also experience great challenges to participate in certification programs.

Conclusions – Moving Forward

Coastal communities depend upon healthy, easily accessible, local seafood. Food security is threatened as these resources diminish in supply and safety due to overfishing, climate change, and environmental degradation. Human health concerns related to consumption of seafood from polluted coastal waters have led to increasing seafood imports to offset the lack of locally available products. As a result, many coastal communities are vulnerable to increases of non-communicable diseases as their diets shift from locally available seafood to imported, less nutritious, more heavily processed food sources (Hicks et al., 2019; Savage et al., 2019). Sustainable marine aquaculture will play a critical role in addressing these issues.

Aquaculture has emerged as the key to meeting the world's increasing demand for seafood. Even with improved fishery management, capture fisheries are incapable of satisfying seafood demand. In many cases, farm-raised seafood is of higher quality than wild-caught seafood both in terms of nutritional composition as well as freshness. Combined with advances in seafood traceability and consumer education, these factors have led many to seek sustainable marine aquaculture products to satisfy their seafood intake. These trends are expected to increase in the coming decades, providing abundant growth opportunities. When conducted in a responsible and sustainable manner, aquaculture offers great potential to enhance food security, improve coastal water quality, and build resilience in coastal communities. Through the advances detailed in this chapter, as well as resolution of public policy, economic, and social equity issues, marine aquaculture will continue its sustainable expansion to meet future seafood demand.

Questions for Reflection

1. What are the principal technical, economic, and regulatory challenges associated with the development of marine aquaculture?
2. In what ways can marine aquaculture help fill the gap between global supply and demand of seafood in environmentally and economically sustainable ways?
3. What have been some of the socio-environmental impacts of shrimp and Atlantic salmon aquaculture, and how are these impacts being addressed?
4. What are some of the limitations of aquaculture certification programs?

Notes

1 Vertical integration is the combination, in one business enterprise, of several stages of production.
2 Blue Revolution refers to the significant growth in global aquaculture production.

References

Aas, T. S., Ytrestøyl, T., & Åsgård, T. (2019). Utilization of feed resources in the production of Atlantic salmon (Salmo salar) in Norway: An update for 2016. *Aquaculture Reports, 15*, 100216. https://doi.org/10.1016/j.aqrep.2019.100216

Abolofia, J., Asche, F., & Wilen, J. E. (2017). The cost of lice: Quantifying impacts of parasitic sea lice on farmed salmon. *Marine Resource Economics, 32*(3). http://dx.doi.org/10.1086/691981

Ahmed, N., Thompson, S., & Glaser, M. (2018). Integrated mangrove-shrimp cultivation: Potential for blue carbon sequestration. *Ambio, 47*(4), 441–452. https://doi.org/10.1007/s13280-017-0946-2

Alleway, H. K., Gillies, C. L., Bishop, M. J., Gentry, R. R., Theuerkauf, S. J., & Jones, R. (2019). The ecosystem services of marine aquaculture: Valuing benefits to people and nature. *Bioscience, 69*(1), 59–68. https://doi.org/10.1093/biosci/biy137

Alvial, A., Kibenge, F., Forster, J., Burgos, J. M., Ibarra, R., & St-Hilaire, S. (2012). *The recovery of the Chilean salmon industry: The ISA crisis and its consequences and lessons.* Global Aquaculture Alliance, 85.

Armitage, D. L. (2002). Socio-institutional dynamics and the political ecology of mangrove forest conservation in Central Sulawesi, Indonesia. *Global Environmental Change, 12*(3), 203–217. https://doi.org/10.1016/S0959-3780(02)00023-7

Aquaculture Stewardship Council. (n.d.). *About the ASC.* https://www.asc-aqua.org/what-we-do/about-us/about-the-asc/

Aquaculture Stewardship Council. (2017, April). *ASC salmon standard v1.1.* https://www.asc-aqua.org/wp-content/uploads/2017/07/ASC-Salmon-Standard_v1.1.pdf

Asche, F., Hansen, H., Tveteras, R., & Tveteras, S. (2009). The salmon disease crisis in Chile. *Marine Resource Economics, 24*(4), 405–411. https://doi.org/10.1086/mre.24.4.42629664

Ashton, E. C. (2008). The impact of shrimp farming on mangrove ecosystems. *CAB Reviews: Perspectives in Agriculture, Veterinary Science, Nutrition and Natural Resources,* 3(003). https://doi.org/10.1079/PAVSNNR20083003

Bendiksen, E. Å., Johnsen, C. A., Olsen, H. J., & Jobling, M. (2011). Sustainable aquafeeds: Progress towards reduced reliance upon marine ingredients in diets for farmed Atlantic salmon (Salmo salar L.). *Aquaculture, 314*(1–4), 132–139. https://doi.org/10.1016/j.aquaculture.2011.01.040

Benetti, D. D., Benetti, G. I., Rivera, J. A., Sardenberg, B., & O'Hanlon, B. (2010). Site selection criteria for open ocean aquaculture. *Marine Technology Society Journal, 44*(3), 22–35. https://doi.org/10.4031/MTSJ.44.3.11

Boyd, C. E., D'Abramo, L. R., Glencross, B. D., Huyben, D. C., Juarez, L. M., Lockwood, G. S., McNevin, A. A., Tacon, A. G. J., Teletchea, F., Tomasso Jr., J. R., Tucker, C. S., & Valenti, W. C. (2020). Achieving sustainable aquaculture: Historical and current perspectives and future needs and challenges. *Journal of the World Aquaculture Society, 51*(3), 578–633. https://doi.org/10.1111/jwas.12714

Boyd, C. E., & McNevin, A. A. (2018). Land use in shrimp aquaculture. *World Aquaculture,* 49(1), 28–34.

Bravo, S. (2012). Environmental impacts and management of veterinary medicines in aquaculture: The case of salmon aquaculture in Chile. In M. G. Bondad-Reantaso, J. R. Arthur, & R. P. Subasinghe (Eds.), *Improving biosecurity through prudent and responsible use of veterinary medicines in aquatic food production* (FAO Fisheries and Aquaculture Technical Paper No. 547, pp. 11–24). FAO.

Buck, B. H., & Langan, R. (2017). *Aquaculture perspective of multi-use sites in the open ocean: The untapped potential for marine resources in the anthropocene.* Springer Nature. https://doi.org/10.1007/978-3-319-51159-7

Bush, S. R., Belton, B., Hall, D., Vandergeest, P., Murray, F. J., Ponte, S., Oosterveer, P., Islam, M. S., Mol, A. P. J., Hatanaka, M., Kruijssen, F., Ha, T. T. T., Little, D. C., & Kusumawati, R. (2013). Certify sustainable aquaculture? *Science, 341*(6150), 1067–1068. https://doi.org/10.1126/science.1237314

Chopin, T., MacDonald, B., Robinson, S., Cross, S., Pearce, C., Knowler, D., Noce, A., Reid, G., Cooper, A., Speare, D., Burridge, L., Crawford, C., Sawhney, M., Ang, K. P., Backman, C., & Hutchinson, M. (2013). The Canadian integrated multi-trophic aquaculture network (CIMTAN) - A network for a new era of ecosystem responsible aquaculture. *Fisheries,* 38(7), 297–308. https://doi.org/10.1080/03632415.2013.791285

Costello, C., Cao, L., Gelcich, S., Cisneros-Mata, M.Á., Free, C. M., Froehlich, H. E., Golden, C. D., Ishimura, G., Maier, J., Macadam-Somer, I., Mangin, T., Melnychuk, M. C., Miyahara, M., de Moor, C. L., Naylor, R., Nøstbakken, L., Ojea, E., O'Reilly, E., Parma, A. M. … Lubchenco, J. (2020). The future of food from the sea. *Nature,* 588(7836), 95–100. https://doi.org/10.1038/s41586-020-2616-y

Crampton, V. O., Nanton, D. A., Ruohonen, K., Skjervold, P. O., & El-Mowafi, A. (2010). Demonstration of salmon farming as a net producer of fish protein and oil. *Aquaculture Nutrition, 16*(4), 437–446. https://doi.org/10.1111/j.1365-2095.2010.00780.x

Duarte, C. M., Wu, J., Xiao, X., Bruhn, A., & Krause-Jensen, D. (2017). Can seaweed farming play a role in climate change mitigation and adaptation? *Frontiers in Marine Science, 4*, 100. https://doi.org/10.3389/fmars.2017.00100

FAO/NACA/UNEP/WB/WWF. (2006). *International principles for responsible shrimp farming.* Network of Aquaculture Centres in Asia-Pacific (NACA). https://enaca.org/?id=542#:~:text=The%20International%20Principles%20for%20Responsible%20Shrimp%20Farming%20provide, for%20a%20more%20sustainable%20development%20of%20shrimp%20farming.

Food and Agriculture Organization of the United Nations. (2011a). *Code of conduct for responsible fisheries.* https://www.fao.org/3/i1900e/i1900e00.htm

Food and Agriculture Organization of the United Nations. (2011b). *Technical guidelines on aquaculture certification.* https://www.fao.org/publications/card/en/c/0ec15d4a-6295-51a5-af16-5c724c62de25/

Food and Agriculture Organization of the United Nations. (2022). *The state of world fisheries and aquaculture 2022.* https://doi.org/10.4060/cc0461en

Gentry, R. R., Alleway, H. K., Bishop, M. J., Gillies, C. L., Waters, T., & Jones R. (2020). Exploring the potential for marine aquaculture to contribute to ecosystem services. *Reviews in Aquaculture, 12*(2), 499–512. https://doi.org/10.1111/raq.12328

Gentry, R. R., Froehlich, H. E., Grimm, D., Kareiva, P., Parke, M., Rust, M., Gaines, S. D., & Halpern, B. S. (2017). Mapping the global potential for marine aquaculture. *Nature Ecology & Evolution, 1*(9), 1317–1324. https://doi.org/10.1038/s41559-017-0257-9

Global Aquaculture Alliance. (2020). *Best aquaculture practices.* https://www.bapcertification.org

Granada, L., Sousa, N., Lopes, S., & Lemos, M. F. L. (2016). Is integrated multitrophic aquaculture the solution to the sectors' major challenges? – A review. *Reviews in Aquaculture*, 8(3), 283–300. https://doi.org/10.1111/raq.12093

Greenberg, P. (2010). *Four fish: The future of the last wild food.* Penguin Books.

Gross, M. R. (1998). One species with two biologies: Atlantic salmon (Salmo salar) in the wild and in aquaculture. *Canadian Journal of Fisheries and Aquatic Sciences, 55*(S1), 131–144. https://doi.org/10.1139/d98-024

Hicks, C. C., Cohen, P. J., Graham, N. A. J., Nash, K. L., Allison, E. H., D'Lima, C., Mills, D. J., Roscher, M., Thilsted, S. H., Thorne-Lyman, A. L., & MacNeil, M. A. (2019). Harnessing global fisheries to tackle micronutrient deficiencies. *Nature, 574*, 95–98. https://doi.org/10.1038/s41586-019-1592-6

Iversen A., Asche, F., Hermansen, Ø., & Nystøyl, R. (2020). Production cost and competitiveness in major salmon farming countries 2003–2018. *Aquaculture, 522*, 735089. https://doi.org/10.1016/j.aquaculture.2020.735089

Jackson, A., & Shepherd, J. (2012). The future of fishmeal and fish oil. In *Second international congress on seafood technology on sustainable, innovative, and healthy Seafood* (Vol. 189) (pp. 189–221). https://www.fao.org/3/i2534e/i2534e.pdf#page=199

Jouffray, J. -B, Blasiak, R., Norström, A. V., Österblom, H., & Nyström, M. (2020). The blue acceleration: The trajectory of human expansion into the ocean. *One Earth, 2*(1), 43–54. https://doi.org/10.1016/j.oneear.2019.12.016

Kramer, L. (2015, October 14). *Land-based salmon aquaculture: A future with potential.* SeaFood Source. Retrieved July 18, 2022 from https://www.seafoodsource.com/news/aquaculture/land-based-salmon-aquaculture-a-future-with-potential

Kumar, G. & Engle, C. R. (2016). Technological advances that led to growth of shrimp, salmon, and tilapia farming. *Reviews in Fisheries Science & Aquaculture, 24*(2), 136–152. https://doi.org/10.1080/23308249.2015.1112357

Kumar, G., Engle, C., & Tucker, C. (2018). Factors driving aquaculture technology adoption. *Journal of the World Aquaculture Society, 49*(3), 447–476. https://doi.org/10.1111/jwas.12514

Lester, S. E., Gentry, R. R., Kappel, C. V., White, C., & Gaines, S. D. (2018). Offshore aquaculture in the United States: Untapped potential in need of smart policy. *Proceedings of the National Academy of Sciences, 115*(28), 7162–7165. https://doi.org/10.1073/pnas.1808737115

Lovatelli, A., Aguilar-Manjarrez, J., & Soto, D. (2013). *Expanding mariculture farther offshore. Technical, environmental, spatial and governance challenges. FAO Technical Workshop, Orbetello, Italy, March 22–25, 2010* (No. 24). FAO Library.

Morrison, D. B., & Saksida, S. (2013). Trends in antimicrobial use in Marine Harvest Canada farmed salmon production in British Columbia (2003–2011). *The Canadian Veterinary Journal, 54*(12), 1160–1163. https://pubmed.ncbi.nlm.nih.gov/24293677

Mutersbaugh, T., Klooster, D., Renard, M. -C., & Taylor, P. (2005). Certifying rural spaces: Quality-certified products and rural governance. *Journal of Rural Studies, 21*, 381–388. https://doi.org/10.1016/j.jrurstud.2005.10.003

Nellemann, C., Corcoran, E., Duarte, C. M., Valdés, L., De Young, C., Fonseca, L., & Grimsditch, G. (Eds.) (2009). *Blue carbon: The role of healthy oceans in binding carbon: A rapid response assessment.* UNEP/Earthprint.

Olson, T. K., & Criddle, K. R. (2008). Industrial evolution: A case study of Chilean salmon aquaculture. *Aquaculture Economics & Management, 12*(2), 89–106. https://doi.org/10.1080/13657300802110687

Parkes, G., Young, J. A., Walmsley, S. F., Abel, R., Harman, J., Horvat, P., Lem, A., MacFarlane, A., Mens, M., & Nolan, C. (2010). Behind the signs: A global review of fish sustainability information schemes. *Reviews in Fisheries Science, 18*(4), 344–356. https://doi.org/10.1080/10641262.2010.516374

Price, C., Black, K. D., Hargrave, B. T., & Morris Jr., J. A. (2015). Marine cage culture and the environment: Effects on water quality and primary production. *Aquaculture Environment Interactions*, 6(2), 151–174. https://doi.org/10.3354/aei00122

Primavera, J. H. (2000). Development and conservation of Philippine mangroves: Institutional issues. *Ecological Economics, 35*(1), 91–106. https://doi.org/10.1016/S0921-8009(00)00170-1

Primavera, J. H. (2005). Mangroves, fishponds, and the quest for sustainability. *Science, 310*(5745), 57–59. https://doi.org/10.1126/science.1115179

Primavera, J. H. (2006). Overcoming the impacts of aquaculture on the coastal zone. *Ocean & Coastal Management, 49*(9–10), 531–545. https://doi.org/10.1016/j.ocecoaman.2006.06.018

Ridler, N., Wowchuk, M., Robinson, B., Barrington, K., Chopin, T., Robinson, S., Page, F., Reid, G., Szemerda, M., Sewuster, J., & Boyne-Travis, S. (2007). Integrated multi-trophic aquaculture (IMTA): A potential strategic choice for farmers. *Aquaculture Economics & Management, 11*(1), 99–110. https://doi.org/10.1080/13657300701202767

Salgado, H., Bailey, J., Tiller, R., & Ellis, J. (2015). Stakeholder perceptions of the impacts from salmon aquaculture in the Chilean Patagonia. *Ocean and Coastal Management, 118*(Part B), 189–204. https://doi.org/10.1016/j.ocecoaman.2015.07.016

Savage, A., McIver, L., & Schubert, L. (2020). Review: The nexus of climate change, food and nutrition security and diet-related non-communicable diseases in Pacific Island countries and territories. *Climate and Development, 12*(2), 120–133. https://doi.org/10.1080/17565529.2019.1605284

SNV Netherlands Development Organization. (n.d.). *MAM-II: Scaling up ecosystem-based adaptation in the Mekong Delta.* https://snv.org/project/mam-ii-scaling-ecosystem-based-adaptation-mekong-delta

Suman, D. (2019). Mangrove management: Challenges and guidelines. In G. M. E. Perillo, E. Wolanski, D. R. Cahoon, & C. S. Hopkinson (Eds.), *Coastal wetlands: An integrated ecosystem approach* (pp. 1055–1079). Elsevier.

Tacon, A. G., & Metian, M. (2008). Global overview on the use of fish meal and fish oil in industrially compounded aquafeeds: Trends and future prospects. *Aquaculture, 285*(1–4), 146–158. https://doi.org/10.1016/j.aquaculture.2008.08.015

Teletchea, F. (2015). Domestication of marine fish species: Update and perspectives. *Journal of Marine Science and Engineering, 3*(4), 1227–1243. https://doi.org/10.3390/jmse3041227

The Explorer. (n.d.). *Closed-system salmon farming protects the environment.* https://www.theexplorer.no/solutions/marine-donut---paving-the-way-for-sustainable-fishfarming/

Theuerkauf, S. J., Morris Jr., J. A., Waters, T. J., Wickliffe, L. C., Alleway, H. K., & Jones, R. C. (2019). A global spatial analysis reveals where marine aquaculture can benefit nature and people. *PloS One, 14*(10), e0222282. https://doi.org/10.1371/journal.pone.0222282

Troell, M., Joyce, A., Chopin, T., Neori, A., Buschmann, A. H., & Fang, J. -G. (2009). Ecological engineering in aquaculture — Potential for integrated multi-trophic aquaculture (IMTA) in marine offshore systems. *Aquaculture, 297*(1–4), 1–9. https://doi.org/10.1016/j.aquaculture.2009.09.010

Warne, K. (2011). *Let them eat shrimp: The tragic disappearance of the rainforests of the sea.* Island Press.

World Bank. (2013). *Fish to 2030: Prospects for fisheries and aquaculture* (Agriculture and Environmental Services Discussion Paper, No. 3). https://openknowledge.worldbank.org/handle/10986/17579

Ytrestøyl, T., Aas, T. S., & Åsgård, T. (2015). Utilisation of feed resources in production of Atlantic salmon (Salmo salar) in Norway. *Aquaculture, 448*, 365–374. https://doi.org/10.1016/j.aquaculture.2015.06.023

6 Marine Pollution

Science, Art, and the Accumulations of Time

S. Brander and P. Betjemann

The history of art and literature reveals that humans have long represented the oceans as seemingly boundless, large enough and deep enough to withstand any disturbance. As ocean temperatures rise and waters acidify – the effects of climate change – we now know that the ocean is not "boundless, endless, and sublime – the image of Eternity", as one 19th-century poet wrote. But the belief in the ocean's unaffectable immensity has had a very long shadow. As recently as 1974, *Scientific American* published an article by the deep-sea explorer Willard N. Bascom, who argued that waste disposal in the ocean, if "thoughtfully controlled", was a plausible way for humans to get rid of pollutants (Bascom, 1974). Twelve years later, Edward Goldberg of Scripps Institution of Oceanography concluded that "the waste capacity of US coastal waters is not now fully used" (Goldberg, 1986). This perspective is difficult to comprehend today. Yet individuals, industries, and nations continue to dispose of waste in the ocean, both intentionally and unwittingly. The most recent estimate of microplastic accumulation on the seafloor stands at 14 million tons (Barrett et al., 2020). The gradual sloughing off of microscopic debris from the surfaces of modern life is slow, and at a small enough visual scale to generally go unnoticed. Collective accumulation far removed from our homes and backyards often remains out of sight and out of mind, particularly for those privileged enough, socioeconomically, to contribute the most to this growing environmental tragedy.

This essay discusses how we arrived at a condition whereby the world's oceans have been subtly, insidiously, and invisibly contaminated, at a scale that few people today can appreciate. The factors that have brought us to this place are literary as well as chemical, artistic as well as technological, and cultural as well as industrial. The construction of the ocean as a boundless immensity has informed behaviors that urgently require rethinking. We move between science, literature, and the visual arts to historicize the issues at stake while looking toward the future. Marine pollution is a particularly complex problem, one that begs to be addressed not just in terms of specific behavioral changes – finding alternatives to individual yogurt cups and other plastic products, for instance – but in terms of a deeper network of assumptions and constructions that must be interrogated and transformed. We begin with the deep history of both pollution and the structures of thinking about the ocean that make it particularly vulnerable to silent, invisible forms of damage.

DOI: 10.4324/9781003058151-8

Oceans Without History

We often think of the 19th-century Industrial Revolution as the origin of anthropogenic pollution. In fact, we have been introducing pollutants into our surroundings for many centuries. Evidence from ice cores collected in Greenland, which span across the Greek and Roman, Medieval, and Renaissance eras (500–3,000 years ago), shows that high levels of lead and silver – metals that are naturally present in rock but toxic in larger quantities – appear as the result of mining and smelting. While the use of metals declined somewhat in the next millennium, scientists suggest that excessive lead mining may have contributed to widespread toxicity and could have precipitated the fall of the Roman empire (Hong et al., 1994).

Long before combustion engines were invented in the 19th century, humans began irreversibly altering the planet. The last 150 years, however, have dramatically accelerated anthropogenic environmental impacts. Since 1830, we have witnessed the advent of electricity, the rise of the Industrial Revolution, "better living through chemistry", and an exponential increase in the need for products that enhance the speed and convenience with which we live our lives – all accompanied by an equally rapid increase in the need to find locations to dispose of the refuse from our modern lives. Early in the Industrial Revolution, towns in Europe were so ensconced in soot and smoke produced by factories that the peppered moth, a species with white coloring speckled with black spots, evolved a darker variant that was better camouflaged in England's smoke-stained towns (Hooper, 2002; Mosley, 2013). By the early 1900s, the whiter variant was difficult to find. Later, in the post–World War II era, magazines such as *Life*, *Good Housekeeping*, and *Time* boasted of the ease of living when we rely on advanced chemicals designed by scientists. Products were synthesized to control pests, keep our homes clean, reduce the time spent on housework, and replace all manner of natural or traditional materials. For example, DDT pesticide was added to children's wallpaper glue as part of a larger project to seal the pristine interiors preferred in the post–World War II era from the outside world, and to serve as protection from insect-borne diseases (Epstein, 2014) (Figure 6.1). Around the same time, innumerable products began to be made of the miracle material, plastic, which did not break down, rust, or mold. Plastic was easy to wash – and even easier to toss in the garbage, which in Western countries was conveniently carted off each week to a landfill far from sight.

Chemicals popularized in the 1940s aimed to create cleaner, disease-free environments, but became their own sources of toxicity, ultimately harmful to humans and wildlife. DDT in agricultural runoff poisoned marine estuaries; in 1967, one study found concentrations of DDT as high as 32 pounds per acre in a salt marsh on Long Island (Woodwell et al., 1967). The manufacture of DDT produced its own waste, and scientists estimate that as many as half a million barrels of both DDT and acid sludge byproduct from its production were intentionally disposed and still remain across nearly 3,000 km^2 of the ocean seafloor off Southern California (Kivenson et al., 2019). Even as DDT was being manufactured and sprayed

Figure 6.1 "DDT Is Good for Me-e-e!", Time Magazine, June 30, 1947. Image credit: Science History Institute, Philadelphia. https://digital.sciencehistory.org/works/1831ck18w.

in large quantities in the 1950s, under the guise of protecting public health, a postwar boom in petroleum extraction fueled the plastics industry, setting the stage for the accumulation of macroplastics, as well as microplastic fragments in the oceans – a significant problem whose scale and impacts we are only now coming to appreciate.

Invisible to the naked eye when sunk to the deep ocean floor, reduced to tiny plastic particles difficult to match back to the item from which they came, or slowly accumulating in fish, shorebirds, and the tiny plankton that fuel food webs, these pollutants enter a marine environment that for ages appeared capable of absorbing every kind of human disturbance into its apparently infinite vastness. Almost everyone who has stood by the sea has had the profound experience of its scale, power, and human-minimizing extent. In the Romantic era (c. 1789–1860) many writers turned to the ocean as a supreme example of an environment that was beyond the turbulent nature of human affairs. Romanticism, understood as a specific aesthetic movement rather than a general feeling, describes a period of expression in Western art and literature that celebrated nature both as an alternative to the early Industrial Revolution and as a model for the "natural rights" of individual humans in the age of democratic revolutions against monarchical rule and aristocratic privilege. Thus, the specific forms of pollution that developed in the mid-20th century, with profound impacts on the oceans, actually have a deeper history in ways of thinking about the marine environment that developed much earlier. A powerful example of how the marine environment centered these concerns is afforded by the so-called "Apostrophe to the Ocean" in Lord Byron's long narrative poem, *Childe Harold's Pilgrimage* (Byron, 1816–1817). This poem is about the wandering of a displaced young man, the prototypical Romantic protagonist, who seeks respite from the concerns of the world. In the "Apostrophe to the Ocean" section – an "apostrophe", in literature, refers to writing that addresses or personifies a non-human entity – Harold celebrates how human history and human life have no impact on the sea, which he finds comforting. "Roll on, thou deep and dark blue Ocean", he exclaims, "ten thousand fleets sweep over thee in vain". In these lines, the passage of innumerable navies over the water leaves no trace. Humans, Byron believes, may "mark … the earth with ruin", but our "control stops with the shore". The "Apostrophe to the Ocean" addresses the sea as a place entirely removed from and unimpacted by human society. It is the image of eternity, a mirror of God, rather than life on earth, an environment of invisibility and inscrutability, a "fathomless" depth, and – preeminently and finally – a place separate and "alone".

The view of the ocean presented in *Childe Harold's Pilgrimage* reappears again and again in the history of literature and art, with the consistent feature that the ocean appears large enough to absorb all manner of violence, inequity, and pollution. Nathaniel Hawthorne's poem "The Ocean", published in 1825, likely reflected on the death of his father, a ship's captain, on a voyage. Hawthorne harbored a great deal of guilt about his father's side of the family; his great-great grandfather was a judge at the Salem witchcraft trials, and therefore responsible for the murders of innocent women presumed to be witches, while his father died on a mercantile voyage to Dutch Suriname, a plantation colony where export goods – sugar, cotton, and indigo – were produced by enslaved Africans under a regime of cruelty that was one of the most violent in the colonies. But "guilt" and "care", in the poem, are terrestrial concerns, entirely absorbed under the "fury" of the ocean's waves. For Hawthorne, the ocean constitutes an antidote, an alternative, a place elsewhere: Its "solitudes" are "blest", and "there is purity" and "peaceful sleep". Existing outside of history, the ocean constitutes a timeless realm in which the Romantic self idealistically transcends all that is troubling.

BOX 6.1 Nathaniel Hawthorne, *The Ocean* (1825)

The Ocean has its silent caves,
Deep, quiet, and alone;
Though there be fury on the waves,
Beneath them there is none.
The awful spirits of the deep
Hold their communion there;
And there are those for whom we weep,
The young, the bright, the fair.
Calmly the wearied seamen rest
Beneath their own blue sea.
The ocean solitudes are blest,
For there is purity.
The earth has guilt, the earth has care,
Unquiet are its graves;
But peaceful sleep is ever there,
Beneath the dark blue waves.

The history of art and literature presents hundreds of examples of "ahistorical" oceans – those not polluted by acts of human degradation. Fiction is also replete with examples. Herman Melville's *Moby-Dick* (1851) opens with the novel's narrator – a melancholic, world-weary figure, just as in *Childe Harold's Pilgrimage* – asserting that it is "high time to get to sea" in order to escape the oppressive nature of commercial and industrial life in New York City; he tells us that thousands of his fellow city-dwellers, normally "tied to counters, nailed to benches, [and] clinched to desks", can be found on any given Sabbath day standing on the seashore and indulging their escapist "ocean reveries" (Melville, 1993, p. 3).

It is not a long leap from these examples to those drawn from the history of visual art. The reflective quality of water served many painters as a surface on which the glory of heavens could be revealed. A painting like Frederic Church's *Syria by the Sea* (1873), for instance, powerfully contrasts the ruins of human history with the setting sun, whose reflection on the limpid waters of the bay suggests

Figure 6.2 Frederic Edwin Church, Syria by the Sea, 1873. Oil on canvas. 56″ × 85″. Detroit Institute of Arts, USA. Gift of Mrs. James F. Joy. Image: © Detroit Institute of Arts/Bridgeman Images.

a perennial cycle of days and seasons that outlasts and transcends the rise and fall of civilizations (Figure 6.2).

Even a painting not about a literal ocean can carry such meanings. Casper David Friedrich's *Wanderer Above the Sea of Fog* (c. 1818) presents the clouds below the solitary mountain hiker as an ocean, signified not just by the sea described in the title but by the wave-like forms of the fogs and mists, and the island-like appearance of the peaks and spires. One of the most famous expressions of Romanticism, the *Wanderer* – like *Childe Harold's Pilgrimage* and *Moby-Dick* – presents a solitary male figure interested in transcending or escaping the limitations of everyday life (Figure 6.3). In the line of literary and visual expression represented by this painting, marine environments, whether real or metaphorical, lift human beings (particularly men) out of history and its violence, oppressions, and banalities.

"The Sea Is History": A New Narrative for Marine Environments

With the publication of Rachel Carson's *Silent Spring* in 1962, the influence of human activities on the natural environment burst into view for the general public. Her poetic yet scientifically accurate description of DDT's and other pesticides' impacts captured readers and conveyed an urgency that had not been felt before. Carson owned the moment, using her experience as a scientist to push back against

Figure 6.3 Caspar David Friedrich, Der Wanderer über dem Nebelmeer (The Wanderer Above the Sea of Fog), c. 1818. Oil on canvas. 37.3″ × 29.4″. Hamburger Kunsthalle, Germany. Image courtesy of Hamburger Kunsthalle/Art Resource.

the unregulated production of chemicals for use in agriculture and insect control. *Silent Spring* focused on the ways in which water transports chemicals across

> the chains of life it supports – from the small-as-dust green cells of the drifting plant plankton, through the tiny water fleas to the fishes that strain plankton from the water and are in turn eaten by other fishes or by birds, mink, raccoons – in an endless cyclic transfer of materials from life to life.
> (Carson, 2002, p. 46)

Although the book turned Carson's attention to more terrestrial processes than her prior works, her emphasis on water as the fundamental link in the chain emerged from what she defined in *The Sea Around Us* as the essential relationship between humans and the 70% of the earth that is water. The title is connective, describing the oceans as a scientifically approachable environment "around us" rather than an incomprehensible mystery projected out there. More particularly, the title refers to the ways in which the sea defines and shapes – and is shaped and defined by – solid masses, including the moon with its tidal influence, sediments deposited on the ocean floor, or a volcanic island born from eruptions and sea-level change. Throughout her career, Carson approached water not as the medium for what Melville called the "ungraspable phantom of life" (1993, p.3), but as the medium that connects all forms of life and matter within a network that is simultaneously marine, astronomical, chemical, terrestrial, and human. Points of interface and interconnections always drew her eye, as in her book *The Edge of the Sea* (1955), which opens not by projecting a fathomless, ahistorical profundity but by issuing a very different kind of invitation.

> When we go down to the low tide line … we enter a world that is as old as the earth itself … For us as living creatures, it has special meaning as an area in or near which some entity that could be distinguished as Life first drifted in shallow waters – reproducing, evolving, yielding that endlessly variety stream of living things that has surged through time and space to occupy the earth.
> (Carson, 1998, p. xii)

Carson's emphasis on water and the seashore as the building blocks of history made her acutely aware of how toxic chemicals like DDT migrated through food chains and ecological networks. Hers was not the only voice sounding the alarm. In just a few decades, Western societies had become accustomed to a life of convenience, with little thought as to where the products used daily were being disposed of or whether there were unintended side effects. In the US images were broadcast nationwide of rivers on fire in Pennsylvania and Ohio due to slicks resulting from oil pollution (Figure 6.4).

The environmental movement of the 1960s was ignited by such images and by other realizations that the use of a chemical (DDT) in homes could be contributing to the demise of bald-headed eagles, pelicans, and fish species, such as Atlantic salmon, and that without proper oversight, chemicals used for human benefit could also cause damage to our children and our livelihoods. These developments led to the eventual establishment of the Environmental Protection Agency (EPA) in the early 1970s, and to the strengthening of the role of the Federal Drug Administration (FDA) in the United States. In Europe, the London and Oslo Conventions, both convened in 1972, sought to regulate the deliberate disposal of wastes from vessels, aircraft, and ocean platforms. In 1972, the US Congress passed the Marine Protection, Research, and Sanctuaries Act (MPRSA), effectively regulating the dumping of "any material which would adversely affect human health, welfare, or amenities, or the marine environment,

Figure 6.4 Cuyahoga River Fire, November 1, 1952. (Fire hoses being used to push back flammable oil pollution on the river's surface.) Gelatin silver print. 6.5″ × 8.5″. Courtesy of Cleveland Public Library.

ecological systems, or economic potentialities" (Marine Protection, Research, and Sanctuaries Act, 1972). This included everything from radioactive waste, which was knowingly dumped in the ocean at that time, to petroleum products, construction materials, and raw sewage. Shortly thereafter, the first report of plastic pellets detected in the ocean, off the east coast of North America in the Sargasso Sea, was published (Carpenter & Smith, 1972). The momentum that was building for considering the way unmitigated technological advances were impacting the planet became far stronger.

While the impacts of modern life in homes, backyards, and factories were beginning to be understood, Carson's literary and scientific intervention – to counter the older narrative of a mysterious watery realm immune from impact and meaningful precisely to the degree that it was unknown, unknowable, and otherwise untouchable – required other kinds of innovations to publicize. Some of these innovations were technological. A new era of ocean discovery was initiated in the 1960s with the invention of submarines, such as *Alvin*, one of the first deep ocean submersibles in the world. Previously, it was impossible to thoroughly explore what some have called the "fathomless" depths of the oceans. In 1965 *Alvin* completed a 6,000 foot (1,800 meter) manned dive, and since then it has made nearly 5,000 dives worldwide. *Alvin* and vessels like it have received numerous upgrades over the decades (Twombly et al., 2014), but the original body remains the same – and without this invention we would know very little about deep-sea hydrothermal vents (or, for that matter, about famous shipwrecks, such as the *Titanic*).

Over the past 50 years, oceanographers and writers like Sylvia Earle and Carl Safina have depended on modern technologies – including submersibles, SCUBA gear, and undersea research stations like Aquarius – to bring both scientific and poetic nuance to marine ecosystems (Earle, 2010; Safina, 1997). Remarkable discoveries have captured the intricacy of marine ecosystems – like worms and crabs that can live at over 700°F (Ramirez-Llodra et al., 2007) – even as stunning descriptions of the undersea environment, previously only glimpsed at shallow depths through goggles, have brought technicolor visibility to the seafloor's "unlooked-on bed" (Rossetti, n.d.). Seussian and almost other-worldly images of deep-sea worms and mysterious bioluminescent fish began to appear in magazines, textbooks, films, and popular writings. Ocean profundity – in that root sense of the word associated with the depths – no longer needs to be understood as a projection of fathomlessness. Indeed, one of the most recent descents in a vessel similar to the *Alvin*, operated by Victor Vescovo, reached the bottom of the Mariana Trench, the deepest area of the oceans. He accomplished this deepest dive to date (35,849 feet) in 2019 (Morelle, 2019).

Side-by-side with these technological innovations, writers and artists have challenged the still-seductive narrative of a depthless, all-absorbing, and history-less sea. Adrienne Rich's poem "Diving Into the Wreck" (Rich, 1973), in which SCUBA diving operates as a metaphor, opens with the narrator donning a wetsuit ("the body-armor of black rubber") and her diving gear; before she descends, she "read[s] the book of myths, and load[s] the camera". Once she reaches the wreck, however, things change. The ocean life she encounters dissociates her from her daily life and brings her closer to actual history than the "book of myths" ever could. For Rich, a feminist poet, the anthropomorphized wreck appears as a metaphor for a history – whether that of patriarchal culture itself, or of a prior female self within it – that she will no longer read through received ideas (the "myths") but will directly encounter as a scuttled past. Facing "the wreck and not the story of the wreck", Rich uses the undersea environment not to escape history, but much more powerfully to encounter and de-romanticize it. The ocean environment, made available by SCUBA, appears not *outside* history but instead as the wreck *of* history.

Rich was not thinking about microplastics, DDT barrels, or medical waste when she described the wreck of history sunk in the ocean. But her poem channels the reader away from marine Romanticism and toward concepts that are applicable to concerns about marine pollution.

Other writers offer similar challenges to readers steeped in traditional thinking about the sea. When a poet like Derek Walcott, a Caribbean writer engaged with the relationship between African traditions and colonial forms in his native St. Lucia, stands on the shore, he does not see the "ocean reveries" described in *Moby-Dick*. Rather than boundlessness, Walcott sees the terrific violence of African bondage. In his 1978 poem "The Sea is History", he writes, "the packed cries, / the shit, the moaning" of the thousands of slave ships that transported enslaved people across the Atlantic (Walcott, 2014, p. 253). The luxury of marine escapism and the purity of an all-absorbing ocean does not exist for Walcott, and thus, his poem contains imagery quite unlike that in the works of Byron, Hawthorne, or Friedrich. The surface of Walcott's ocean presents "heaving oil, heavy as chaos". Rivers have "scum" on their "drying lip". The unmet promises of "emancipation" vanish and wither "as the sea's lace dries in the sun" (pp. 253–255). Just as Adrienne Rich dons a mask and SCUBA gear to explore the wreck of history, Walcott invites readers to "strop on these goggles" to look directly at the past (p. 254). "Strop", rather than "strap", is an intentional usage: In the same sense that a razor is stropped (sharpened), goggles are Walcott's incisive edge into history, the tool he offers us for seeing history as it is reflected in the seas – seas that were the scene of the horrors of the transoceanic journey between Africa and the Americas that transported people to the plantations on which they were enslaved. Seeing the ocean as a place of peace is a form of privilege.

Marine history, seen through the lenses offered by Rich and Walcott, involves gendered and racialized forms of violence. Their interventions, restoring history to the scene of sea-gazing, may help us see forms of environmental violence, as well. As new ways of seeing what we are at stake of losing have developed, our awareness of the impacts of modern life on marine ecosystems has increased – as has interest in protecting the ocean. A survey conducted in 2018 of over 30,000 people in 21 different countries found that public perception of threats to the marine environment is at an all-time high, with 70% of respondents believing that the ocean is in danger due to human activities, such as pollution, overfishing, climate change, and biodiversity loss. The availability of technology has benefited us in many ways, increasing our knowledge of ecosystems that until a few decades ago we had no way of exploring, but it also reveals the side effects of progress; one of the many discoveries of Vescovo's deepest dive in 2019 were the plastic items scattered along the ocean's deepest crevices, along with synthetic particles, such as nylon, polyethylene, rayon, and other fibers found in every organism sampled from the Mariana Trench (Jamieson et al., 2019).

Stropping on goggles is more important than ever. We are now more than 20 years into the new century, surrounded by an enormous expanse of salty water teeming with marine creatures that many know relatively little about. Not enough people are aware, for example, that tiny organisms, phytoplankton, produce over half the oxygen necessary to sustain the planet, and that climate change is occurring so rapidly that warming conditions may affect mixing in such a way that fewer nutrients will be available in the surface layer – the level to which sunlight is able to reach to allow photosynthesis to take place in these organisms. Global warming is also increasing the frequency of harmful algal blooms (HABs), which release biotoxins resulting in toxicity to shellfish, fish, and other marine animals

(Gittings et al., 2018). Other effects of climate change, such as ocean acidification and hypoxia (reduced concentration of oxygen in warmer waters), contribute to this altered state of phytoplankton (Turley et al., 2010). While great attention is paid when large oil spills occur (Abbriano et al., 2011) that cause immediate visible damage to ecosystems and sensitive wildlife (Paine et al., 1996), our daily use of smaller amounts of fuel – for cars, airplanes, and industrial purposes – has caused global ocean temperatures to increase 1°F on average over the past 100 years. Nineteen of the warmest years on record have occurred since 2000, and 2020 was tied with 2016 for the warmest year yet (Green & Jacobs, 2021), threatening the tiny oceanic organisms that can be described as the "other lung of planet earth" (McRae, 2020).

As phytoplankton drifting on the surface layers of the ocean have begun to suffer from an early case of emphysema due to climate change, the ocean has been ravaged by more direct forms of pollution during the 20th and 21st centuries. Ocean acidification and warming temperatures can be described as pollutant effects because these phenomena are caused by the use and burning of fossil fuels. Traditionally, however, pollution is thought of as the introduction of synthetic chemicals produced by humans – including pesticides; excess nutrients from agriculture (Howarth, 2008); industrial compounds like plasticizers and fluorinated chemicals that magically make our clothing wrinkle and stain-free; and pharmaceuticals that tend to find their way out of our medicine cabinets and into waterways (Brander, 2022). This chemical cocktail affects wildlife while also compromising human health (Colborn et al., 1996). Much of our knowledge about these environmental impacts depends on the insights made by conservationists and scientists of Carson's era – namely, that products promoting a better way of life for many (protecting humans from disease-bearing insects, such as mosquitoes; promoting the growth of crops; improving health through medication) also have a negative impact on oceanic and inland water bodies, through the chains of aqueous transport described in *Silent Spring* and based on the kinds of interconnectedness detailed in *The Sea Around Us*.

Many contaminants found in runoff (liberated from lands via stormwater runoff, or that enter coastal areas or estuaries in the form of effluents from wastewater treatment plants) are persistent in that they bind to sediments or are harbored in the fatty tissues of animals, including humans. Some are broken down by sunlight and hydrolysis. However, a unique pollutant type that is both highly persistent and capable of traveling long distances – into the depths of the ocean and to remote polar locations – is also a three-dimensional object which other chemicals and living organisms can adsorb or adhere to: Plastics.

The Mongoose and the Bottle: Global Plastics

Plastics originated in the mid-19th century. In the 1840s, Charles Goodyear and Thomas Hancock independently took out patents in the US and the UK, respectively, for "vulcanized" rubber. Western factories began producing tires for bicycles, and later for cars and trucks, launching a plastic revolution that deepened over the next century. Another early plastic, called "Parkesine" (invented by Alexander Parkes), was displayed in 1862 at the international exhibition in London. It was made from cellulose and used to make molds for printing, buttons, and combs. We really began producing plastics *en masse* in the early 1900s. What organics chemists know as "polyoxybenzylmethylenglycolanhydride" was presented to the public as "Bakelite". Developed in 1909, Bakelite depended upon a phenol formaldehyde resin that, when combined with wood flour as a filler, could be used to make radio cases, sunglasses, toys, jewelry, rotary phones, bowling balls, and handles on pots and pans. Soon to follow were many other "thermoset" plastics, Bakelite's sisters polystyrene (1929), polyester (1930), polyvinylchloride (PVC) and polyethylene (1933), and nylon (1935). During World War II, oil companies invested in factories to convert crude oil into plastic, resulting in an overload of polymers once peaceful times resumed. The problem of too much plastic was solved, from the industrial point of view, with the invention of Tupperware (1948). Tupperware products were inexpensive, durable, and versatile. Parents were happy to have toys and cups for their kids that did not break or shatter, and oil companies had another outlet for their products (Gilbert, 2016).

If the romanticization of the oceans, historically speaking, involves their unaffectable vastness and unreachable horizons, a correlative perception fueling marine pollution involves the false projection of an "away" for plastics. Most people are aware that plastics do not biodegrade, but they trust that they are recycled. In truth, recycling rates have always been relatively flat (around 9–10%) and remain so because many plastics are difficult to recycle and because countries lack economic incentives to recycle or reuse most items (Jambeck et al., 2015). Western culture has invented and mass-produced a non-biodegradable material that is far less expensive to make than reuse. Oceanic gyres, including the North Pacific gyre, also known as the Great Pacific Garbage Patch, are a daily reminder of our disposable culture (LeBreton et al., 2018). A plastic Miss Piggy driving a small pink corvette from a Happy Meal™ eaten in 1986 may outlive your great grandchildren.

It is now trendy to limit one's plastic use, and some percentage of the public has finally grasped that the plastic problem cannot be wished away by occasionally remembering to toss one's single-use plastic water bottle into the correctly labeled bin. As of 2020, the US harbors just 4% of the world's population while creating 17% of total global plastic waste (Borrelle et al., 2020). In effect, each American contributes over four times their global share of plastic pollution, an amount comparable to our greenhouse gas footprint. More visible evidence, in the form of "macro" plastics, is accumulating, and many bans on single-use items now exist. "Bioplastics" have also become a popular replacement. Manufactured from corn, soy, or avocado pits, these straws, cups, and lids are advertised as being compostable. But research shows that these biologically derived plastics can be just as harmful to the environment (Robbins, 2020). Composting them successfully requires specialized facilities. Regardless of the impression that some may have and of prominent movements led by environmental groups to reduce reliance on single-use plastics, the production of plastics continues to increase exponentially (Jambeck et al., 2015).

The fact that nearly all US states with bans on single-use plastic are along either coast of North America (National Conference of State Legislatures, 2021) – and that countries with large swaths of coastline are the first to cry foul when plastic begins to wash ashore – is no accident. People seem to be more motivated to protect habitat associated with water than with land (Nichols, 2014), and an image of a turtle with a straw stuck in its nose gets far more press than microplastics being used as mulch on farmland in Kansas. In the same opening passages of *Moby-Dick* that describe the inherent orientation of humans to the oceans, Melville suggests that water represents a "chief element" of visual art and visual appreciation; watercourses and waterways represent what he calls the "magic stream" of art, drawing the eye with a sort of magnetic power (Melville, 1993, p. 3). Even so, and with an increasing number of images helping us see the accumulation of plastics in marine and coastal regions, the gravity of the situation remains difficult to convey. We have reached a point at which scientists are calling for a redefinition of the carbon cycle due to the sheer amount of plastic (which contains carbon) now residing on the ocean floor, floating in the water column, and carried by the wind (Zhu, 2021). By 2050, plastics are predicted to comprise 15–20% of our carbon footprint. Plastics are dangerous to wildlife (Thompson et al., 2009) and are even embedded in human tissues to such a degree that scientists have coined a new term – "plasticenta" – to describe the microplastics contaminating the organ that nourishes unborn babies (Ragusa et al., 2021). Craig Santos Perez, an Indigenous Chamoru poet from Guam, writes of the points where plastics, human birth and life, and ocean devastation intersect. In "The Age of Plastics", he describes the plastic probe of an ultrasound machine visualizing the embryo in "my wife's belly", the "inflatable tub" in which she labors, the plastic chemicals in her placenta, and the plastic pacifier that his daughter sucks (Perez, 2018). Perez connects all those plasticized points along a developmental trajectory of life in which food and plastics become inseparable, as life and plastics, birth and death, become permanently interfused in a complex and painful meditation. Perez moves from the plastic-mediated birth of his daughter to the darkly ironic and tragic vision of how plastic never dies. The title of the poem captures its irony; the "age of plastics" represents the origins of life in a modern content, but the "age of plastics" is ultimately ageless, never-ending, eternally durable and devastating to the "paradise / of the Pacific gyre" in which Perez's homeland is located.

The time for change has arrived, and the past decade has brought with it what marine scientist Jane Lubchenco calls "a new narrative for the ocean" (Lubchenco & Gaines, 2019). The term "narrative", in this call to action, matters; the traditional story of the ocean, told in literature and art, represents a threat inextricably linked to physical pollutants. We know now that the ocean is best understood not as an "ungraspable phantom", to recall Melville's language, but as a real, physical environment subject to the ravages of history and the deposition of its byproducts. Modern life has damaged it over time, perhaps irreversibly; too often, we treat its waters as an unacknowledged waste dump rather than honoring them as the beating heart of our planet. Today, instead of assuming that the ocean is too big to fail, we must acknowledge that it is too big to ignore (Lubchenco & Gaines, 2019). To do so requires a comprehensive understanding of a problem that is simultaneously oriented to the stories we tell and the actions we take.

Scientists have revealed the extent and insidiousness of marine pollution. Literature scholars have revealed the assumptions about oceanic purity made by Romantic-era writers. A comprehensive approach to grappling with the destruction of the marine ecosystems requires that we study not just what we are doing, but the long history of narratives about the oceans that have led us to this point. We must also engage in acts of both rethinking our actions and rewriting our marine stories. Even as scientists have revealed the polluted condition of marine and terrestrial ecosystems under modern systems of production and consumption, literary scholars have challenged the assumption of oceanic timelessness and invincibility by interpreting how contemporary poetry depicts "ocean waste in ways that necessarily encompass [the] histories and agencies of the more-than-human world" (Bloomfield, 2019).

As in many efforts having to do with conservation and climate change, younger activists and writers are particularly adept at translating science into action and awareness. Dozens of poems and lyrical passages about ocean plastics can be found on blogs, websites, and digital classroom projects. Compelling work has also been published in more traditional venues. In "Giving a Mongoose a Plastic Bottle", 16-year-old poet Uma Menon describes a mongoose wrestling a bottle as a metaphor for the more general struggle of animal life against plastic pollution. In the first half of the poem, Menon transitions from a single event in a particular geography (in Menon's "motherland", India, she witnesses the mongoose) to the accumulation of plastic products in a global geography: "Plastic across the oceans, across/ rice fields, across a cardboard/ building where a forest once breathed." Most of the poem's short lines are enjambed, a technique in which a reader is required to proceed to the next line to complete each grammatical phrase. Enjambment produces a feeling of breathlessness in this poem, capturing the suffocating extent of a world in which the mongoose may choke on the bottle and – just as the poetic lines blur and spread into one another – plastics are spread across the lands and the seas of the world. The final line of the poem – "&&&)" – represents a cry of despair at the extent and scope of the problem. The poet has carried the reader from her own motherland to the circulatory systems in which plastics persist in the environment of everyone's motherland, the earth itself. Oceans, rice fields, forests, and deserts come together at the apparent conclusion of this poem – which is not, in fact, a conclusion, but an invitation to the reader to fill in the "&&&" with their own points of origin, places of observation, and observed ecological niches. None of those, Menon tells us, is secure from the danger of suffocation.

BOX 6.2 Uma Menon, *Giving a Mongoose a Plastic Bottle* (2019). (Poem reprinted with permission from Mawenzi House Publishers Ltd).

In my motherland, I see
a mongoose sinking
its warrior teeth into plastic trash.
The sunlight reflects its struggle,
plastic across the oceans, across
rice fields, across a cardboard
building where a forest
once breathed. This land
gave me a life & I've
given it bottle caps. Warning:
Choking Hazard-mongoose
can't read. Immune to snake
venom, susceptible to plastic
bottles. The lake has closed
its eyes-closed its eyes &
swallowed brown water, sprinkled
with plastic bags. Warning:
To avoid danger of suffocation,
keep this plastic bag away
from babies & children (& lakes
& oceans & mongooses
& hornbills & turtles &
peacocks & rice fields
& forests & deserts &
&&&)

We must accept Menon's invitation to adopt a global perspective. As has happened with climate change, the areas of the world most highly impacted by plastic pollution are often the poorest and most vulnerable to exploitation by nations that produce an outsized portion of nonrenewable plastic waste. In the UK, for example, enough plastic to fill two double-decker buses is tossed in the waste bin every 30 seconds. This plastic used to be primarily shipped to China for sorting and supposed recycling, but it was impossible to do this in a mutually beneficial or profitable manner. Plastic is now shipped to countries such as Malaysia in Southeast Asia – which refused to accept large shipments of plastic from the UK and the US in 2020 (Malaysia returns 42 containers of 'illegal' plastic waste to UK, 2020). There is simply a lack of global consensus on where to put our plastic pollution. Ocean currents are driving plastic debris to remote areas of the world, such as the isolated Cocos Keeling Islands in the Indian Ocean (Lavers et al., 2019).

Only a few hundred people occupy the area, but accumulated piles of microplastic debris (<5 mm in size) in the form of degraded toothbrushes, plastic bags, and straws weigh nearly 240 tons as of 2019. The pollution of some areas of the world for the benefit of others may be understood as a form of colonialism – the use of lands outside one country to stockpile unwanted goods or waste products. A particularly insidious example of this colonialist mindset blames the countries forced to accept unusable plastics for financial or political reasons; poorer countries with large swaths of coastline, such as the Philippines, Indonesia, and a number of African nations, have been accused of mismanaging their waste and contributing disproportionately to polluting the ocean (Liboiron, 2018).

In fact, Western habits may be rendering the land and homes of others unusable and uninhabitable. Many people have become aware that atoll nations, such as Kiribati, the Maldives, Tuvalu, and the Marshall Islands, are experiencing sea-level rise that threaten their very existence. A plastics analogue is now being chronicled by filmmakers, photographers, poets, scientists, and journalists. Artists have begun using imagery from the Pacific Ocean, for example from Japan, to contrast the original intent of these paintings with the plastic inundation of modern times (Figure 6.5). Their collective work – research on physical environments with the power of visual and written media to shape how we see the world – may be having an effect. The United Nations has begun work toward a binding global agreement to reduce plastic pollution similar to the 2015 Paris Climate Agreement (Bergmann et al., 2022). Countries such as the US have introduced legislation to reduce plastic production and frame economic incentives

Figure 6.5 Russ White, The Great Trash Wave (after Hokusai), 6-color screenprint, ed. of 25, 18 × 24″, 2014. Image courtesy of the artist.

for recapture and reuse. A resurgence in the visceral concern for marine environmental impacts may be becoming apparent as we enter the third decade of the 21st century. However, a new narrative for the ocean must replace the old to challenge that "book of myths" that Adrienne Rich identifies in her poem. "The thing I came for", writes Rich, is "the wreck itself and not the story of the wreck/ the thing itself and not the myth/... the evidence of damage/ worn by salt and spray into this threadbare beauty".

Questions for Reflection

1. What types of anthropogenic pollution were occurring before the Industrial Revolution began in the mid-1800s? How did that change/increase with the advent of modern technologies, such as electricity and motor vehicles?
2. Can you suggest why must people today believe that the ocean is in danger due to human activities, yet as a larger global society we fail to address the root(s) of the problem? Give three reasons why this might be.
3. What do you consider to be the greatest hurdles to reach a global agreement to reduce marine plastic pollution? Do you believe that effective implementation of the agreement will be easy? Why or why not?

References

Abbriano, R., Carranza, M. M., Hogle, S. L., Levin, R. A., Netburn, A. N., Seto, K. L., Snyder, S. M., & Franks, P. J. S. (2011). Deepwater Horizon oil spill: A review of planktonic response. *Oceanography*, *24*(3), 294–301. https://www.jstor.org/stable/24861323

Barrett, J., Chase, Z., Zhang, J., Banaszak Holl, M. M., Willis, K., Williams, A., Hardesty, B. D., & Wilcox, C. (2020). Microplastic pollution in deep-sea sediments from the Great Australian Bight. *Frontiers in Marine Science*, *7*, 576170. https://doi.org/10.3389/fmars.2020.576170

Bascom, W. (1974). The disposal of waste in the ocean. *Scientific American*, *231*(2), 16–25. https://www.jstor.org/stable/24950139

Bergmann, M., Almroth, B. C., Brander, S. M., Dey, T., Green, D. S., Gundogdu, S., Krieger, A., Wagner, M., & Walker, T. R. (2022). A global plastic treaty must cap production. *Science*, *376*(6592), 469–470. https://doi.org/10.1126/science.abq0082

Bloomfield, M. (2019). Widening gyre: A poetics of ocean plastics. *Configurations*, *27*(4), 501–523. https://doi.org/10.1353/con.2019.0033

Borrelle, S. B., Ringma, J., Law, K. L., Monnahan, C. C., Lebreton, L., McGivern, A., Murphy, E., Jambeck, J., Leonard, G. H., Hilleary, M. A., Eriksen, M., Possingham, H. P., De Frond, H., Gerber, L. R., Polidoro, B., Tahir, A., Bernard, M., Mallos, N., Barnes, M., & Rochman, C. M. (2020). Predicted growth in plastic waste exceeds efforts to mitigate plastic pollution. *Science*, *369*(6510), 1515–1518. https://doi.org/10.1126/science.aba3656

Brander, S. M. (2022). Rethinking our chemical legacy and reclaiming our planet. *One Earth*, *5*(4), 316–319. https://doi.org/10.1016/j.oneear.2022.03.020

Byron, G. G. (1816–1817). Childe Harold's pilgrimage. In *The Works of Lord Byron Including Several Poems now First Collected Together with an Original Biography* (Vol. 1) (pp. 13–29). M. Thomas & J. Maxwell, printer.

Carpenter, E. J., & Smith Jr., K. L. (1972). Plastics on the Sargasso Sea surface. *Science, 175*(4027), 1240–1241. https://doi.org/10.1126/science.175.4027.1240
Carson, R. (1998). *The edge of the sea.* Mariner Books (Original work published 1955).
Carson, R. (2002). *Silent spring.* Houghton Mifflin (Original work published 1962).
Carson, R. (2018). *The sea around us.* Oxford University Press (Original work published 1951).
Colborn, T., Dumanoski, D., & Meyers, J. P. (1996). *Our stolen future: Are we threatening our fertility, intelligence and survival? – a scientific detective story.* Dutton Adult.
Earle, S. A. (2010). *The world is blue: How our fate and the ocean's are one.* National Geographic Books.
Epstein, L. (2014). Fifty years since Silent Spring. *Annual Review of Phytopathology, 52*(1), 377–402. https://doi.org/10.1146/annurev-phyto-102313-045900
Gilbert, M. (2016). *Brydson's plastics materials.* Butterworth-Heinemann.
Gittings, J. A., Raitsos, D. E., Krokos, G., & Hoteit, I. (2018). Impacts of warming on phytoplankton abundance and phenology in a typical tropical marine ecosystem. *Scientific Reports,* 8(1), 2240. https://doi.org/10.1038/s41598-018-20560-5
Goldberg, E. D. (1986). The assimilative capacity of the oceans for wastes. In C. S. Giam & H. J.-M. Dou (Eds.), *Strategies and advanced techniques for marine pollution studies: Mediterranean Sea* (pp. 1–8). NATO ASI Series. Springer.
Green, T., & Jacobs, P. (2021, March 17). *2020 tied for warmest year on record, NASA analysis shows.* NASA. https://www.nasa.gov/press-release/2020-tied-for-warmest-year-on-record-nasa-analysis-shows.
Hawthorne, N. (n.d.). *The Ocean.* Poetry Foundation (Original work published 1825). https://www.poetryfoundation.org/poems/57286/the-ocean
Hong, S., Candelone, J.-P., Patterson, C. C., & Boutron, C. F. (1994). Greenland ice evidence of hemispheric lead pollution two millennia ago by Greek and Roman civilizations. *Science, 265*(5180), 1841–1843. https://doi.org/10.1126/science.265.5180.1841
Hooper, J. (2002). *Of moths and men: An evolutionary tale.* W.W. Norton & Company.
Howarth, R. W. (2008). Coastal nitrogen pollution: A review of sources and trends globally and regionally. *Harmful Algae,* 8(1), 14–20. https://doi.org/10.1016/j.hal.2008.08.015
Jambeck, J. R., Geyer, R., Wilcox, C., Siegler, T. R., Perryman, M., Andrady, A., Narayan, R., & Law, K. L. (2015). Plastic waste inputs from land into the ocean. *Science, 347*(6223), 768–771. https://doi.org/10.1126/science.1260352
Jamieson, A. J., Brooks, L. S. R., Reid, W. D. K., Piertney, S. B., Narayanaswamy, B. E., & Linley, T. D. (2019). Microplastics and synthetic particles ingested by deep-sea amphipods in six of the deepest marine ecosystems on Earth. *Royal Society Open Science,* 6(2), 180667. https://doi.org/10.1098/rsos.180667
Kivenson, V., Lemkau, K. L., Pizarro, O., Yoerger, D. R., Kaiser, C., Nelson, R. K., Carmichael, C., Paul, B. G., Reddy, C. M., & Valentine, D. L. (2019). Ocean dumping of containerized DDT waste was a sloppy process. *Environmental Science & Technology,* 53(6), 2971–2980. https://doi.org/10.1021/acs.est.8b05859
Lavers, J. L., Dicks, L., Dicks, M. R., & Finger, A. (2019). Significant plastic accumulation on the Cocos (Keeling) Islands, Australia. *Scientific Reports,* 9(7102), 1–9. https://doi.org/10.1038/s41598-019-43375-4
Lebreton, L., Slat, B., Ferrari, F., Sainte-Rose, B., Aitken, J., Marthouse, R, Hajbane, S., Cunsolo, S., Schwarz, A., Levivier, A., Nobl, K., Debeljak, P., Maral, H., Schoeneich-Argent, R., Brambini, R., & Reisser, J. (2018). Evidence that the Great Pacific Garbage Patch is rapidly accumulating plastic. *Scientific Reports,* 8, 4666. https://doi.org/10.1038/s41598-018-22939-w

Liboiron, M. (2018, December 21). How plastic is a function of colonialism. *Teen Vogue.* https://www.teenvogue.com/story/how-plastic-is-a-function-of-colonialism

Lubchenco, J., & Gaines, S. D. (2019). A new narrative for the ocean. *Science, 364*(6444), 911. https://doi.org/10.1126/science.aay2241

Malaysia returns 42 containers of 'illegal' plastic waste to UK. (2020, January 20). *BBC News.* Retrieved June 21, 2022, from https://www.bbc.com/news/uk-51176312

Marine Protection, Research, and Sanctuaries Act, Public Law 92–532 (1972). https://www.govinfo.gov/content/pkg/COMPS-1680/pdf/COMPS-1680.pdf

McRae, G. (2020, April 16). Will climate change threaten earth's other lung? *The Revelator.* https://therevelator.org/phytoplankton-climate-change/

Melville, H. (1993). *Moby-Dick; or, the whale.* Wordsworth Classics (Original work published 1851).

Menon, U. (2019, September 30). *Giving a mongoose a plastic bottle.* Skipping Stones 31. https://www.skippingstones.org/wp/2019/09/30/poetry-by-uma-menon-16-florida/

Morelle, R. (2019, May 13). Mariana Trench: Deepest-ever sub dive finds plastic bag. *BBC News.* https://www.bbc.com/news/science-environment-48230157

Mosley, S. (2013). *The chimney of the world: A history of smoke pollution in Victorian and Edwardian Manchester.* Routledge.

National Conference of State Legislatures. (2021, February 2). *State Plastic Bag Legislation.* https://www.ncsl.org/research/environment-and-natural-resources/plastic-bag-legislation.aspx

Nichols, W. J. (2014). *Blue mind: The surprising science that shows how being near, in, on, or under water can make you happier, healthier, more connected, and better at what you do.* Little Brown.

Paine, R. T., Ruesink, J. L., Sun, A., Soulanille, E. L., Wonham, M. J., Harley, C. D. G., Brumbaugh, D. R., & Secord, D. L. (1996). Trouble on oiled waters: Lessons from the *Exxon Valdez* oil spill. *Annual Review of Ecology & Systematics, 27,* 197–235. https://www.jstor.org/stable/2097234

Perez, C. S. (2018, November 1). *The age of plastic.* The Margins: Asian American Writers' Workshop. https://aaww.org/the-age-of-plastic/

Ragusa, A., Svelato, A., Santacroce, C., Catalano, P., Notarstefano, V., Carnevali, O., Papa, F., Rongioletti, M. C. A., Baiocco, F., Draghi, S., D'Amore, E., Rinaldo, D., Matta, M., & Giorgini, E. (2021). Plasticenta: First evidence of microplastics in human placenta. *Environment International, 146,* 106274. https://doi.org/10.1016/j.envint.2020.106274

Ramirez-Llodra, E., Shank, T. M., & German, C. R. (2007). Biodiversity and biogeography of hydrothermal vent species: Thirty years of discovery and investigations. *Oceanography, 20*(1), 30–41. https://www.jstor.org/stable/24859973

Rich, A. (1973). Diving into the wreck. *Diving Into the Wreck: Poems, 1971–1972.* W.W. Norton & Co.

Robbins, J. (2020, August 31). *Why bioplastics will not solve the world's plastics problem.* Yale Environment 360. https://e360.yale.edu/features/why-bioplastics-will-not-solve-the-worlds-plastics-problem

Rossetti, C. G. (n.d.). By the sea. (Original work published 1858). http://famouspoetsandpoems.com/poets/christina_rossetti/poems/16473

Safina, C. (1997). *Song for the blue ocean: Encounters along the world's coasts and beneath the seas.* Holt Paperbacks.

Thompson, R. C., Swan, S. H., Moore, C. J., & vom Saal, F. S. (2009). Our plastic age. *Philosophical Transactions of the Royal Society B: Biological Sciences, 364(1526),* 1973–1976. https://doi.org/10.1098/rstb.2009.0054

Turley, C., Eby, M., Ridgwell, A. J., Schmidt, D. N., Findlay, H. S., Brownlee, C., Riebesell, U., Fabry, V. J., Feely, R. A., & Gattuso, J.-P. (2010). The societal challenge of ocean acidification. *Marine Pollution Bulletin*, *60*(6), 787–792. http://dx.doi.org/10.1016/j.marpolbul.2010.05.006

Twombly, M., Scalamogna, A., & Stegmaier, A. (2014). *The evolution of Alvin*. National Geographic. https://www.nationalgeographic.com/news-features/evolution-of-alvin/?sf3188984=1

Walcott, D. (2014). The sea is history. In G. Maxwell, (Ed.), *The poetry of Derek Walcott, 1948–2013*. Farrar, Straus, and Giroux (Original work published 1978).

Woodwell, G. M., Wurster Jr., C. F., & Isaacson, P. A. (1967). DDT residues in an east coast estuary: A case of biological concentration of a persistent insecticide. *Science*, *156*(3776), 821–824. https://doi.org/10.1126/science.156.3776.821

Zhu, X. (2021). The plastic cycle–an unknown branch of the carbon cycle. *Frontiers in Marine Science*, *7*, 609243. https://doi.org/10.3389/fmars.2020.609243

7 Oceans and the Changing Climate

D. O. Suman

A recent comprehensive study by a high-level international organization, the Intergovernmental Panel on Climate Change (IPCC), presented *The Special Report on the Oceans and Cryosphere in Changing Climate* (Intergovernmental Panel on Climate Change [IPCC], 2019) that examines the latest projections of impacts to the ocean environment depending on different behaviors of the global community during the 21st century. The most optimistic scenario assumes low emissions of greenhouse gases (GHG), high mitigation actions,[1] and a two-thirds chance of limiting global warming to warming to less than 2°C by 2100. The most pessimistic scenario assumes high GHG emissions and an absence of policies to combat climate change.

Atmospheric carbon dioxide (CO_2) concentrations in 2021 were 416 ppm, compared with preindustrial concentrations of 280 ppm. Predicted concentrations in 2100 range between less than 400 ppm and close to 1,000 ppm for the optimistic and pessimistic scenarios, respectively. Increased concentrations of atmospheric CO_2 and other GHG will lead to higher atmospheric and surface seawater temperatures. Compared with the baseline period of 1986–2005, the global mean sea surface temperature is projected to increase between 0.7°C and 3°C by 2100, depending on the responses of the global community.

Human activities causing increases in atmospheric GHG include fossil fuel combustion, biomass burning, and deforestation, among others. Ultimately, our actions are responsible for significant physical and chemical changes to the ocean system that have great implications for people who use coastal and marine living resources and space. The geographical ranges of some species may shift, and others may experience changes in growth and reproduction or become locally extinct. Population changes may impact the food web and, of course, industrial and artisanal fishers. Coastal communities are already experiencing exposure to rising sea levels and more intense storms, causing damage to life and property.

In response to these human-driven changes in the ocean system, governments and resource users must mitigate the cause of the problem and adapt to the changing circumstances. *Mitigation*, as defined in the footnote, refers to actions that reduce stressors such as, in this case, increasing emissions of GHG to the atmosphere. *Adaptation* involves actions that reduce the adverse impacts of the changes. Both mitigation and adaptation actions may be felt differently by various

DOI: 10.4324/9781003058151-9

social sectors or user groups. In other words, the impact of the actions may not be distributed equally across society and may enhance inequalities.

This chapter examines important changes to the ocean system resulting from climate change: (1) rising sea levels, (2) changes in oceanic conditions (sea surface temperatures, currents) that affect the availability of fishery resources, and (3) the increasing acidity of seawater (ocean acidification). These changes threaten human livelihoods and wellbeing. This chapter also illustrates ways people are adapting to these environmental changes, often resulting in winners and losers. Addressing climate change impacts in coastal and marine areas presents decision-makers with both great challenges and opportunities.

Sea Level Rise (SLR)

Higher atmospheric temperatures are melting portions of the polar ice caps, the Greenland ice sheet, and high-altitude glaciers. Ice melt, in addition to thermal expansion of seawater, will lead to higher sea levels and flooding of low-lying coastal areas. The global mean sea surface elevation has already risen some 1.4 mm/yr (0.8–2.0 mm/yr) between 1901 and 1990. Predictions for global sea level rise by 2100 range between 0.43 and 0.84 meters relative to 1986–2005 levels. Sea level rise is site-specific, however, and regions may differ ±30% of the global average depending on local conditions, such as glacial rebound, subsidence, local hydrology, loss of coastal wetlands, and ocean currents. Evidence also suggests that higher surface seawater temperatures lead to increases in the intensity and frequency of tropical hurricanes and typhoons, particularly intense storms of Categories 4 and 5. Storm surges associated with tropical storms, when compounded with SLR and high tides, expose coastal areas to greater flooding risks.

Human Dimensions of SLR

Rising seas resulting from global climate change are already having great impacts on the people who inhabit coastal areas. Properties and critical infrastructure may be damaged or lost to coastal erosion and flooding, leading to high financial losses or even loss of life. Social demographics may change in various ways as some peoples are displaced and search for alternative livelihoods. SLR may also cause loss or reduction of ecosystem services that intact coastal environments provide (storm protection, biodiversity). In this sense, SLR compounds other drivers that threaten ecosystem services (urbanization, coastal sprawl, habitat loss, pollution).

About half of the world's population lives in coastal areas and is potentially at risk from coastal flooding. Large coastal urban areas present major concerns due to population and essential infrastructure and real estate assets. Nicholls et al. (2007) ranked coastal cities most exposed to coastal flooding in 2007 and 2070 in terms of exposed population and assets. Considering exposed populations, 15 of the top 20 cities were located in Asia, 3 in Africa, and 2 in the US (Miami and New York). Miami placed ninth on the list with exposed populations of 2,003,000 (2007) and 4,795,000 (2070). However, when examining assets exposed to coastal flooding, Miami placed first with exposed assets of $416 billion (2007) and $3.5

trillion (2070). The top five cities in exposed assets were Miami (USA), Guangzhou (China), New York-Newark (USA), Kolkata (India), and Shanghai (China), in descending order. Miami has recently been identified as the most economically vulnerable city to SLR in the world, according to the US National Climate Assessment (Melillo et al., 2014). Box 7.1 details the vulnerabilities that Miami faces in light of SLR.

BOX 7.1 The Case of Miami: The Human Dimensions of SLR

Miami is a key example of an urban area highly vulnerable to flooding, resulting from SLR coupled with tropical storms. Miami's geography increases its exposure to sea level rise and flooding. Miami's highest elevations of about nine meters above sea level are found along a fossilized coral ridge. However, over 95% of land in Miami is less than six meters (about 20 feet) above sea level (Weiss et al., 2011). Porous and permeable limestone bedrock underlies the entire region, so rising seas also increase groundwater levels. Most of the coastal ecosystems (mangrove forests and seagrass beds) in the northern half of the county have been degraded or removed during 120 years of urban development, reducing nature-based protections. Moreover, Southeast Florida lies in an extremely active hurricane zone. The average return period for a hurricane strike (Categories 1–5) in Miami Beach is five years, while the return period for a strong hurricane (Category 3–5) is 18 years (Malmstadt, 2019; National Hurricane Center, 2019). Sea levels in South Florida are estimated to have already increased some 20 cm since the 1930s (Zervas, 2009) and, as sea level rise accelerates, high tides and storm surges will be superimposed on higher tidal base levels (Sweet et al., 2017). Miami already experiences flooding of streets during high-tide events or "king tides" ("sunny day flooding"), and these events will become ever more frequent. Using US Army Corps of Engineers (USACE) estimates for SLR, the Union of Concerned Scientists predicted that streets in Miami Beach will flood 380 times per year by 2045 (Union of Concerned Scientists, 2017).

Potential property losses in Miami from flooding are astronomical due to the combination of increased storm intensity (Patricola & Werner, 2018), sea level rise, rapid coastal development, and population increases. A sea level rise of 0.5 meters by 2070 without adaptation measures could threaten assets valued at $3.5 trillion and displace 300,000 persons in Miami (Hauer et al., 2016; Treuer et al., 2018).

Climate gentrification occurs when affluent persons purchase properties at higher elevations currently occupied by lower-income residents. There is evidence that climate gentrification has already begun in Miami (Keenan et al., 2018). Real estate prices have increased faster in higher elevation neighborhoods than those at lower elevations. Real estate located at higher elevations in mainland Miami neighborhoods that feature low-income communities of color (Little Haiti, North Wynwood, Little River) are being targeted by speculators and investors promoting projects for higher-income residents who may wish to escape low-lying coastal areas vulnerable to flooding (Bolstad, 2017).

Miami faces an uncertain future with limited options due to its porous limestone substrate, low elevation, high hurricane risk, limited space to retreat to higher ground, high costs involved in adaptation measures, abundance of climate deniers, and the lack of political will to invest today in protections that future generations will enjoy (Molinaroli et al., 2019).

Besides the high vulnerability of many urban coastal areas, such as Miami, to SLR and flooding, many smaller coastal communities throughout the world are already being displaced due to SLR. Often, residents of these communities are among the most socially vulnerable populations. Several US coastal communities present dramatic examples of vulnerability to impacts from sea level rise and coastal erosion.

Coastal wetlands and barrier islands in southern Louisiana are disappearing due to sea level rise, erosion from hurricane storm surges, and reduced deposition of Mississippi River sediments resulting from USACE flood control and levee projects. Isle de Jean Charles, home to members of the United Houma Nation tribe and the Biloxi-Chitimacha Confederation of Muskogees Inc., once had an extension of 22,000 acres, while only 320 acres remain today. In 2016, the US Department of Housing and Urban Development awarded the State of Louisiana $48.3 million in resettlement funds (Louisiana Office of Community Development, 2020). The resettlement effort has been rocky, however. Although Louisiana purchased land for resettlement, in 2019 the community rejected the offer because the members considered that the state's planning process was culturally insensitive; in addition, many people simply did not want to move (Dermansky, 2019). The situation for "the country's first climate refugees" has yet to be resolved.

Many native Alaskan villages are also experiencing the impacts of climate change (Environmental Protection Agency, 2016). As air and sea temperatures increase, sea ice and permafrost melt and storms intensify, leading to increased coastal erosion. Numerous communities have been forced to retreat and, in some cases, relocate. One Yupík Eskimo community, Newtok, has contemplated relocation for almost two decades as coastal erosion and flooding have caused contamination of drinking water, damage to the wastewater treatment system, and loss of the dock and shoreline structures. Shishmaref, an Iñpuiat community, voted to relocate from their barrier island three times during the past decades. The high costs of relocation and limited financial assistance available from the federal government are constant challenges for these communities (Martin, 2018).

The "climate refugee" issue is perhaps the starkest for local-lying Pacific Island nations and other small island developing states (SIDS). Clearly, there is a need to resolve these issues, as perhaps 143 million persons may be displaced from climate change impacts by 2050 (McDonnell, 2018). Rising sea levels threaten the Marshall Islands that have elevations of less than two meters (about 6.5 feet) above sea level, and tidal flooding occurs monthly. As a trust territory of the US, Marshallese may immigrate freely to the US. Uncertainty remains, however, about a fund allocated by the US Congress to support relocation to the US mainland (Davenport & Haner, 2015). Relocation scenarios are more uncertain for other Pacific Islander climate refugees. New Zealand denied the request for asylum from a Kiribati man based on his claim as a "climate refugee". His appeal to the UN Human Rights Committee was rejected on the basis that he was not in "immediate danger". The Committee did suggest, however, that people fleeing immediate danger could not be forced to return to their homes (Lyons, 2020; People urgently fleeing climate crisis, 2020). This non-binding opinion from the UN may be an important signal for nations' obligations under international law (the 1951 Convention relating to the Status of Refugees and its 1968 Protocol) and may open the door to protection for climate refugees. Flooding, coastal erosion, and sea level rise create "immediate danger" and "serious threats to the lives" of many persons living in low-lying coastal regions throughout the world. However, many questions remain regarding whether countries will recognize climate refugees under international refugee law and who will pay for relocation.

Adaptive Responses to SLR

The range of adaptive responses to SLR is extensive (National Research Council, 2007; Pilkey & Cooper, 2014; Pilkey et al., 2016). The ultimate decision regarding the mix of adaptive strategies depends on numerous factors: geography of the shoreline (elevation, subsidence, groundwater level, wave energy), comparative costs and funding sources, sense of urgency, political will to invest today in measures benefiting future generations, scientific and technological capabilities, social equity, and legislative mandates, among others.

Decisions to remain in place, defend infrastructure, and fight against nature may be costly, exacerbate coastal erosion, cause great environmental impact, and ultimately be unsuccessful. Shoreline armoring via the construction of seawalls, bulkheads, riprap, groins, and breakwaters is a common measure. Beach nourishment, although a softer alternative with added recreational benefits, is expensive, often temporary, and not without its own environmental impacts.

Nature-based solutions, such as the creation, restoration, or protection of mangrove forests and salt marshes, seagrass beds, or oyster beds, may be applicable on some sheltered shores, and may also provide environmental benefits of new habitats and increased marine biodiversity. Nature-based alternatives might be integrated into hardened shoreline structures. Other adaptation measures may include flood-proofing, floating homes, stricter construction standards, elevation of roads and buildings, floodgates, and pumps to remove flood waters.

The ideal strategy is managed retreat away from areas vulnerable to SLR and flooding. Nevertheless, the high costs of retreat, the challenges of proactive planning, the cultural aversion to relocation, and the potential displacement of low-income communities residing in higher elevation locations (climate gentrification) are all obstacles. Box 7.2 describes Miami's initial efforts to adapt to SLR and coastal flooding.

BOX 7.2 The Case of Miami – Adaptation Measures

Local initiatives to address the threat of SLR began in 2010 with the formation of the Southeast Florida Regional Climate Change Compact (SFRCCC) signed by local government leaders from the four-county region. In October 2012, the SFRCCC released its Regional Climate Action Plan with 110 action items that local governments could adopt to promote resilience under the threats presented by climate change. Miami-Dade County and, particularly, the City of Miami Beach have initiated numerous planning measures to respond to climate change – including identification of critical infrastructure (roads, bridges, wastewater treatment plants, stormwater drains, seawalls, emergency services, etc.) and adoption of means to protect them, as well as revisions to building codes and county comprehensive plans. The City of Miami Beach has perhaps been the most proactive in adopting measures that address sea level rise. These include raised roads in some low-lying neighborhoods highly vulnerable to flooding, higher seawalls, revised building standards that mandate higher elevations for buildings' finished floors, and hundreds of one-way stormwater pumps on streets that flood during king tides.

The US federal and state governments have been relatively absent from the flooding threats in Southeast Florida (except for the National Flood Insurance Program and beach nourishment projects). However, in 2020 the USACE released a $4.6 billion proposal (Miami-Dade Back Bay Coastal Storm Risk Management Study) calling for construction of a flood wall through central Miami-Dade County and flood gates that would close three rivers during storm surges. The proposal drew sharp criticism because the flood wall would divide neighborhoods, use eminent domain to purchase hundreds of properties, and fill in more areas of Biscayne Bay (United States Army Corps of Engineers, 2020). Critiques of the Draft Environmental Impact Statement released in 2020 pointed to the need to rely more on nature-based solutions, particularly increased restoration of mangrove forests and seagrass beds. County commissioners formally rejected the proposal in August 2021 and proposed to cooperate in the development of a more acceptable plan.

Changing Climate and Fisheries

Climate change impacts fishery resources on numerous levels – from physiological changes at the cell level to broad ecosystem shifts. All have the potential to significantly affect the fish populations from the community to global levels, and subsequently, the fishing sectors that depend on these resources.

Physiological changes occur at the organism level due to changes in seawater temperature and dissolved oxygen. Individuals may reach an oxygen- and capacity-limited thermal tolerance (OCLT), particularly in their larval or spawning stages; larvae have high metabolic rates and lower energy reserves, while egg tissue requires an ample supply of oxygen. Above certain temperature limits oxygen supply capacity becomes limiting, and hypoxemia develops. This appears to be the case with *Gadus morhua* (Atlantic cod) in the North Sea where larvae and adult spawners demonstrate sensitivity to winter warming. Cod recruitment in the North Sea correlates with cooler seawater temperatures (Pörtner et al., 2010). Furthermore, Muñoz et al. (2015) report evidence of the upper limits on thermal tolerance for Pacific salmon that is thought to be the result of a mismatch between oxygen supply and demand. Tissues require increased amounts of oxygen at higher temperatures, resulting in increased heart rates that reach a maximum limit. Estimates of catastrophic losses of Pacific salmon by 2100 in western North America range from 17% at average projections for increased temperature, to 98% for maximum warming scenarios. Cheung et al. (2013) predict an additional impact of these physiological changes caused by reduced oxygen supply: maximum body weight of fish may shrink between 14 and 24% from 2000 to 2050 under a high-emissions scenario.

Changes may also occur at the ecosystem level due to different physiological responses of organisms at various levels of the food web. The increased mortality of North Sea cod (G. *morhua*) in early life stages may be due to a mismatch in the timing of the thermal window of its larval food. The climate sensitivity of one species may induce changes in food web interactions. For example, a decrease in abundance of copepods (*Calanus* species) with warmer temperatures correlates with the timing of decreased G. *morhua* stocks in the North Sea, suggesting that the reduced availability of juvenile cod's food source results in decreased cod recruitment.

Additional evidence comes from the Northeast US continental shelf, where the abundant cooler water habitat (5–15°C) appears to be contracting, resulting in a decrease of the thermal habitat of two zooplankton species that are prey for Atlantic cod larvae and juveniles in Georges Bank and the eastern Gulf of Maine. Friedland et al. (2013) suggested that the decreased availability of zooplankton prey due to warmer temperatures may affect the feeding and growth of larval and juvenile cod and may influence cod recruitment in those regions.

Temperature appears to be the forcing factor for fish population shifts poleward from their historic ranges. Populations follow shifts in their food sources or move into waters that are metabolically more favorable. Numerous examples of poleward shifts exist, and they are most noticeable at the northern and southern range limits of various populations.

Models developed by Pörtner and Peck (2010) suggest that, by 2050, some pelagic and demersal species may move poleward some 600 km (haddock and Atlantic cod) and 223 km (flatfish). Faraday (2015) also reported that cod stocks found north to Greenland and south to North Carolina appear to be shifting northward in the northwest Atlantic. Rijnsdorp et al. (2009) observed increases in the northern range limits of species common near the Iberian Peninsula (Lusitanian species, such as anchovy, sprat, and horse mackerel). Boreal species (Atlantic cod and plaice) show increased abundance at their northern limit and decreased abundance at their southern limit. These researchers suggest that this shift is due to higher production or survival of pelagic eggs or larval stages or perhaps to changes in the quantity and quality of nursery habitats.

The American lobster (*Homarus americanus*) shows a similar poleward shift. The inshore lobster fishery operates from Maine to New Jersey (USA), while the offshore fishery ranges from Maine to North Carolina. During the past decade the southernmost range has moved 43 km northward, and Gulf of Maine stocks also appear to be moving northward (Faraday, 2015).

Globally, we can expect high levels of species invasions in the Arctic and Southern Ocean regions. Local extinctions will occur in the tropics and semi-enclosed seas. Areas of highest species turnover are the subpolar regions of the Southern Ocean and the Arctic. Models of Cheung et al. (2009) estimate that 83% of fish species will show a poleward shift, with a median poleward displacement of almost 300 km by the timeframe of 2040–2060, compared to ranges during 2001–2005.

Human Dimensions of Fishery Changes

These changes in fish populations and ranges present great implications for fishing businesses and communities. Warmer seawater temperatures will cause the poleward migration of some species and will also result in local extinctions of other populations, particularly in the tropics. These migrations will favor nations and communities whose fisheries are blessed with increased stocks and potential new target species. But nations with decreased net primary production, decreased catch, and loss of target species will experience adverse social and economic impacts. The social impacts for disadvantaged nations and fishing sectors may include increased levels of poverty, loss of cultural traditions and livelihoods, loss of social capital, and emigration away from fishing centers (Ojea et al., 2020). The potential relocation of fishing fleets and processing plants, changing exports and seafood prices, and decreased profits from the fishing sectors all present severe financial challenges to disadvantaged nations and communities. The predictions are dire; climate change may reduce global fisheries revenue by $10 billion/yr (Gaines et al., 2018).

Barange et al. (2014) linked models of physical, biological, and human responses to climate change in 67 Exclusive Economic Zones, noting increased productivity at high latitudes and decreased productivity at low and mid-latitudes. Their models predict increased fisheries biomass in Iceland, Norway, the Kuroshio Current, and the Gulf of Guinea. Meanwhile, they expect decreases in the Canary

Current, the California Current, and the Humboldt Current. Linking productivity to nations' economic dependency on fisheries, they identify areas of greatest concern to be nations that are highly dependent on fisheries and will experience decreased productivity due to climate change. Areas of greatest concern are in South and Southeast Asia, Southwest Africa (Nigeria to Namibia), and Peru. South Asia is the most vulnerable region, given its poor fisheries regulation and high population growth.

Models of Blasiak et al. (2017) examine some 147 nations and consider exposure to high sea surface temperatures, their sensitivity (number of fishers, fish landings, and exports), as well as their adaptive capacities (governance, GDP, government subsidies, social indices) to conclude that seven of the ten most vulnerable nations are SIDS, such as Kiribati, Micronesia, Solomon Islands, Maldives, Vanuatu, Samoa, and Tuvalu.

The poleward displacement of fish stocks has already begun to present fishery management challenges that will only increase. Stocks may move from one fishery management jurisdiction to another within one nation, or into a neighboring nation. Faraday and Bigford (2019) note that the Loligo squid population appears to be shifting northward from New York/New Jersey waters to Rhode Island. That resource, however, is still managed by the Mid-Atlantic Fishery Management Council rather than the New England Fishery Management Council. Thus, fishery management regions or international borders may present barriers to fishers' adaptation.

Adaptive Responses

Mitigation or reduction of GHG emissions is essential for reducing the impacts to fisheries mentioned above. In addition, fisheries management must be enhanced to ensure sustainability and modified to be more flexible and incorporate the climate-induced migrations (Faraday & Bigford, 2019; Gaines et al., 2018). This rarely occurs, as fishery management today generally ignores the dynamic spatial distributions of fish stocks caused by climate change. Rather, allocations are usually based on historical spatial distributions. Fishery legislation must be revised to expressly recognize climate change and require stock assessments that include climate change predictions. For example, in the US, the Magnuson–Stevens Act (MSA) sets out ten National Standards (N.S.) that must guide fishery management plans. Although N.S. 2 requires that fishery management plans be based on the "best scientific information available", the N.S. fail to mention climate change. Some groups recommend the amendment of the MSA to include N.S. 11, which would require fishery management plans to include proactive adaptive measures to account for climate change (Carter, 2019).

Many types of adaptation are possible, depending on the species and their locations. As ranges of stock shift poleward, authorities must reconsider spatial fishery management and, where necessary, formalize new effective agreements to manage transboundary stocks. In some instances, it may be advantageous to create new institutions, while in others, regional fishery authorities must foster flexible management strategies as the stocks migrate. These management authorities must collaborate and develop procedures for joint monitoring of oceanic conditions, assessing stocks, and sharing or transferring management of fish stocks. The critical geographical areas for stocks will be their leading and retreating edges, and these must be managed conservatively to avoid overfishing. Initial research in this area indicates the critical importance of integrating climate change into sustainable fisheries management. Researchers from the US National Marine Fisheries Service and Fisheries and Ocean Canada have developed models to evaluate the international management system for Pacific hake, based on climate-related movements of the stock (Jacobsen et al., 2022). Their models indicate decreased catches and increased catch variability with the incorporation of spatial dynamics. Therefore, sustainable fisheries management requires not only the use of historic catch, but also the incorporation of climate-driven spatial data to avoid overfishing and disruption to the fish stocks.

Decision-making must also consider equity to shareholders or fishers with permits in areas with declining stocks. For example, disadvantaged fishers will have to adapt to new situations, perhaps by being granted flexible fishing licenses that allow them to shift their efforts to new species that are sufficiently abundant for harvest. Perhaps permits could be regional, rather than species-specific. Provisions must also be adopted to grant financial assistance and access to credit for displaced fishers. Attention should also be given to the disproportionate impact on certain fishing sectors; industrial fishing vessels have greater mobility to follow stocks, while small-scale fishers lack this advantage.

In some cases, existing marine protected areas (MPAs) and marine reserves may be relocated to protect new spawning aggregation sites or critical areas. In other cases, designation of new MPAs may be appropriate to minimize conflicts between resource users or protect new migration corridors or leading migratory edges.

Catch diversification, development of value-added products, promotion of fishing cooperatives, increased access to climate information, and the identification of new markets may reduce fishers' vulnerability. However, the existing and new challenges faced by capture fisheries point to the increasingly attractive option of sustainable ocean aquaculture.

Ocean Acidification

The atmospheric concentration of CO_2 in 2021 was over 416 ppm, a level 50% greater than that of preindustrial times. The ocean is believed to be a sink for approximately 25% of total anthropogenic CO_2 (Doney et al., 2020). As CO_2 dissolves in seawater, it alters the carbonate–bicarbonate equilibrium, causing a decrease in the saturation states of aragonite and calcite, forms of $CaCO_3$ that are precipitated from seawater by marine organisms to construct shells and skeletons. The surface ocean pH is estimated to have decreased globally by an average of ~0.1 pH units from preindustrial times, or today about 0.002 units/yr (Doney et al., 2020; Ekstrom et al., 2015; Hall-Spencer & Harvey, 2019), a phenomenon known as ocean acidification (OA). The IPCC report (2019) suggests with high certainty that by 2081–2100, pH values of surface open ocean waters will drop between 0.036–0.042 and 0.287–0.291 pH units compared to 2006–2015, depending on the scenario. Although this appears to be a small decrease, it can have severe impacts on certain marine organisms.

More acidic seawater and reduction in the aragonite saturation state translate into increased difficulty in construction of calcium carbonate shells and skeletons by bivalves (oysters, scallops, mussels, clams, abalone), scleractinian tropical corals (hard corals), and planktonic pteropods, as well as crustose coralline algae that form pavements on rocks in the photic zone. Although species vulnerabilities vary, generally the larval stages of shellfish appear to be more sensitive to OA, thus adversely affecting recruitment. In addition, thermal stress from increasing temperatures is thought to exacerbate impacts of OA.

In the tropics, OA may lead to decreases in ecosystem diversity, species richness, and spatial heterogeneity (Hall-Spencer & Harvey, 2019). OA is a factor that will lead to significant declines in the health of coral reefs. Net calcification of hard corals declines with decreased aragonite accretion and increasing bioerosion. The result will be a decrease in complexity of the coral reef, lowered taxonomic diversity and species richness, and less complex reef structure. Coral diversity is depressed, and macroalgae abundance increasingly limits the recruitment of juvenile corals. OA also leads to increasing dominance of non-calcified species (soft corals and anemones over hard corals), decreases in crustose coralline algae with proliferation of turf algae, and reduction in size and abundance of calcified animals (sea urchins).

Changes in coral community structure associated with OA can lead to indirect effects on reef-associated invertebrates and fish communities, as well as altered trophic interactions. For example, reduction in calcareous herbivores (sea urchins) contributes to the overgrowth of weedy turf algae. It is difficult, however, to isolate coral reef declines due solely to OA because of the cumulative impact of additional stressors, such as ocean warming, coral bleaching, and coral diseases.

Human Dimensions of OA

OA will have increasing significant impacts on the harvest of shellfish and crabs. Of great concern in North America are fisheries and port communities dependent on Dungeness crab, Red King crab, and Pacific oyster. Estimates of global economic losses to shellfish production by 2100 from OA amount to $6 billion ($1 billion Europe; $400 million USA) (Doney et al., 2020). Narita et al. (2012) suggest that global losses in mollusk production due to OA may reach $100 billion by 2100. Because aquaculture produces more than 90% of global shellfish harvest, and cupped oysters (*Crassostrea* spp.) contribute about 30% of the global mollusk production, this activity is particularly vulnerable to OA, as described in Box 7.3 (Food and Agriculture Organization of the United Nations, 2020).

BOX 7.3 Pacific Northwest Oyster Cultivation and OA

In the US, oysters are the most important cultivated marine species by value, accounting for a value of $186 million. In the Pacific Northwest (USA), shellfish farming specializes in culture of the Pacific oyster (*Crassostrea gigas*), stocked by hatcheries that produce between 40 and 60 billion oyster larvae/yr. OA has already impacted the oyster aquaculture industry in that region, causing economic losses of $110 million and directly jeopardizing 3,200 jobs (Ekstrom et al., 2015). Strong upwelling events of seawater undersaturated with aragonite occurred in 2006–2008 with pH values around 7.6 at hatchery intakes. The lowered aragonite saturation state of inflowing water and the survival of larval groups were highly correlated (Barton et al., 2015). Only 25% of the 2007 season's normal production survived, and poor shell integrity of surviving hatchery-reared C. *gigas* larvae increased their vulnerability to parasites and diseases. It is believed that the increased atmospheric CO_2 concentrations exacerbated the intensity and duration of the upwelled acidic waters (CO_2 enriched) and contributed to shoaling of the Pacific Northwest saturation horizon. With increasing concentrations of atmospheric CO_2 in the future, this acidification trend is expected to continue.

Since the 2006–2008 event, Washington shellfish hatcheries have formed partnerships with the University of Washington and NOAA's Ocean Acidification Program (mandated by federal law and established in 2011) to monitor seawater carbonate parameters, implement early warning systems, and optimize water treatment systems that could reduce exposure of oyster larvae to low pH waters.

While the Pacific Northwest and southern Alaska are areas featuring high exposure to OA, as well as high sensitivity of commercially important species to acidic waters, other areas in the US (Gulf of Mexico, mid-Atlantic, and New England) also show high vulnerability to OA because of local amplifiers (coastal eutrophication and high river discharge) that lower pH (Ekstrom et al., 2015). Mid-Atlantic and New England areas are socially vulnerable because of their high economic dependence on shellfish farming, while Gulf of Mexico sites show high vulnerability due to low political engagement, poor utilization of scientific information, and lack of alternative employment opportunities – all factors that limit the ability to adapt.

Tropical coral reefs occupy less than 0.1% of the ocean floor, yet they provide habitat for about 25% of known marine species and have among the highest biodiversity of any marine ecosystem; perhaps 1–10 million species live in or near coral reefs (Hoegh-Guldberg et al., 2017). Coral reefs face multiple threats from pollution, diseases, bleaching caused by increasing seawater temperatures, and OA. Degradation of this important ecosystem has serious impacts for coastal communities and some 500 million people in 90 countries that depend on coral reef ecosystems for fisheries and food security, shoreline protection, and recreation and tourism. Estimates of the economic value of goods and services offered by coral reefs reach $375 billion/yr (Hoegh-Guldberg et al., 2017).

Adaptive Responses

OA calls for implementation of vigorous adaptation strategies that are often site-specific. Adaptation measures for shellfish aquaculture operations include adjusting the chemical composition of inflow water and real-time monitoring of seawater chemical characteristics. Potentially, selective breeding of oyster broodstock may identify lines that are less sensitive to decreased pH. Measures should be adopted to reduce exposure to low pH water and control for factors that amplify impacts of OA and result in more acidic waters, particularly reduction of nutrient loading (eutrophication). Increased diversity of harvested species (particularly seaweed cultivation that may be favored by OA) may reduce social vulnerability. Partnerships between the shellfish industry and research institutions, as illustrated in the Pacific Northwest oyster case, increase access to scientific information and expertise that can help reduce impacts.

Coral reef ecosystems are in a more precarious situation as they face multiple threats from increased seawater temperature, bleaching, diseases, direct anthropogenic physical impacts, and pollution. The rapid rate of environmental change appears to be faster than possible coral evolution and acclimation to new conditions. Reduction of nutrients and other pollutants that stress corals, decrease pH, and reduce their resilience to climate change is essential. Improved fishery management and integrated coastal management initiatives are general recommendations that may help ensure the continued health of reef ecosystems. In addition, current research efforts are focusing on identification of coral genotypes that are best able to withstand increased temperatures, followed by selective breeding and restoration by means of coral gardens that provide stock to restore damaged reefs (Drury & Lirman, 2021; Lirman & Schopmeyer, 2016).

Conclusion

The changing climate is already causing significant impacts on ocean and coastal environments, and also on the people and communities that depend upon these resources and environments for their livelihoods and wellbeing. Even if the global community were to embrace measures to curb emissions of GHG today, SLR and OA impacts will continue into the future because of past emissions of GHG (IPCC, 2022). This chapter has examined the enormous human dimensions of several of these anthropogenic changes and examined some possible avenues that can be taken to adapt to these changes. Successfully meeting these challenges will require great political will, costly investments, integration of the best available science into decision-making, implementation of adaptive management strategies, and incorporation of climate change into all areas of ocean and coastal management. Adaptation to the changing climate is clearly one of the greatest challenges future generations of ocean and coastal managers will face.

Questions for Reflection

1. What are the relationships between exposure and vulnerability? Much of the Netherlands is situated below sea level, yet it may not be as vulnerable to sea level rise as areas of Bangladesh or the Philippines. How could that be?
2. Can you suggest any possible adaptive measures for:
   - The Oyster Aquaculture industry in Oregon and Washington, in light of OA?
   - Groundfish fishers in the Gulf of Maine facing increasing SST?
   - Hapag-Lloyd shipping company's routes from Rotterdam to Yokohama with decreasing Arctic sea ice?
   - The South Florida Water Management District facing sea level rise and increasing precipitation?
   - USFWS concern regarding the future of polar bears in the Arctic?
   - Ocean Wind Farms in Scotland?

Note

1 *Mitigation* means actions that reduce the emissions of carbon dioxide and other greenhouse gases to the atmosphere, such as reduction of deforestation and agricultural burning, phaseout of vehicles that use fossil fuels, and the switch to renewable "clean energy" sources (wind and solar).

References

Barange, M., Merino, G., Blanchard, J. L., Scholtens, J., Harle, J., Allison, E. H., Allen, J. I., Holt, J. & Jennings, S. (2014). Impacts of climate change on marine ecosystem production in societies dependent on fisheries. *Nature Climate Change, 4,* 211–216. https://doi.org/10.1038/nclimate2119

Barton, A., Waldbusser, G. G., Feely, R. A., Weisberg, S. B., Newton, J. A., Hales, B., Cudd, S., Eudeline, B., Langdon, C. J., Jefferds, I., King, T., Suhrbier, A., & McLaughlin, K. (2015). Impacts of coastal acidification on the Pacific Northwest shellfish industry and adaptation strategies implemented in response. *Oceanography, 28*(2), 146–159. https://doi.org/10.5670/oceanog.2015.38

Blasiak, R., Spijkers, J., Tokunaga, K., Pittman, J., Yagi, N., & Österblom, H. (2017). Climate change and marine fisheries: Least developed countries top global index of vulnerability. *PLoS ONE, 12*(6), e0179632. https://doi.org/10.1371/journal.pone.0179632

Bolstad, E. (2017, May 1). *High ground is becoming hot property as sea level rises.* Scientific American ClimateWire. https://www.scientificamerican.com/article/high-ground-is-becoming-hot-property-as-sea-level-rises/

Carter, A. (2019). *A national standard for climate-ready fisheries.* Center for American Progress. https://www.americanprogress.org/article/national-standard-climate-ready-fisheries/

Cheung, W. W. L., Lam, V. W. Y., Sarmiento, J. L., Kearney, K., Watson, R., & Pauly, D. (2009). Projecting global marine biodiversity impacts under climate change scenarios. *Fish and Fisheries, 10,* 235–251. https://doi.org/10.1111/j.1467-2979.2008.00315.x

Cheung, W. W. L., Sarmiento, J. L., Dunne, J., Frölicher, T. L., Lam, V. W. Y., Palomares, M. L. D., Watson, R., & Pauly, D. (2013). Shrinking of fishes exacerbates impacts of global ocean changes on marine ecosystems. *Nature Climate Change*, 3, 254–258. https://doi.org/10.1038/nclimate1691

Davenport, C., & Haner, J. (2015, December 1). The Marshall Islands are disappearing. *New York Times*. https://www.nytimes.com/interactive/2015/12/02/world/The-Marshall-Islands-Are-Disappearing.html?searchResultPosition=1

Dermansky, J. (2019, January 11). *Isle de Jean Charles Tribe turns down funds to relocate first US "climate refugees" as Louisiana buys land anyway*. DeSmog. https://www.desmogblog.com/2019/01/11/isle-de-jean-charles-tribe-turns-down-funds-relocate-climate-refugees-louisiana.

Doney, S. C., Busch, D. S., Cooley, S. R., & Kroeker, K. J. (2020). The impacts of ocean acidification on marine ecosystems and reliant human communities. *Annual Review of Environment and Resources*, 45, 83–112. https://doi.org/10.1146/annurev-environ-012320-083019

Drury, C., & Lirman, D. (2021). Genotype by environment interactions in coral bleaching. *Proc. Royal Soc. B.*, 288(1946), 20210177. https://doi.org/10.1098/rspb.2021.0177

Ekstrom, J. A., Suatoni, L., Cooley, S. R., Pendleton, L. H., Waldbusser, G. G., Cinner, J. E., Ritter, J., Langdon, C., van Hooidonk, R., Gledhill, D., Wellman, K., Beck, M. W., Brander, L. M., Rittschof, D., Doherty, C., Edwards, P. E. T., & Portela, R. (2015). Vulnerability and adaptation of US shellfisheries to ocean acidification. *Nature Climate Change*, 5, 207–214. https://doi.org/10.1038/NCLIMATE2508

Environmental Protection Agency. (2016). *Adapting to climate change: Alaska*. https://www.epa.gov/sites/production/files/2016-07/documents/alaska_fact_sheet.pdf

Faraday, S. E. (2015). Moving targets: fisheries management in New England in the midst of climate change. In R. S. Abate (Ed.), *Climate change impacts on ocean and coastal law: US and international perspectives* (pp. 73–90). Oxford University Press. https://doi.org/10.1093/acprof:oso/9780199368747.001.0001

Faraday, S. E., & Bigford, T. E. (2019). Fisheries and climate change: Legal and management implications. *Fisheries*, 44(6), 270–275. https://doi.org/10.1002/fsh.10263

Food and Agriculture Organization of the United Nations. (2020). *The state of world fisheries and aquaculture 2020. Sustainability in action*. United Nations. https://doi.org/10.4060/ca9229en.

Friedland, K. D., Kane, J., Hare, J. A., Lough, R. G., Fratantoni, P. S., Fogarty, M. J., & Nye, J. A. (2013). Thermal habitat constraints on zooplankton species associated with Atlantic cod (*Gadus morhua*) on the US Northeast Continental Shelf. *Progress in Oceanography*, *116*, 1–13. https://doi.org/10.1016/j.pocean.2013.05.011

Gaines, S. D., Costello, C., Owashi, B., Mangin, T., Bone, J., Molinos, J. G., Burden, M., Dennis, H., Halpern, B. S., Kappel, C. V., Kleisner, K. M., & Ovando, D. (2018). Improved fisheries management could offset many negative effects of climate change. *Science Advances*, 4(8), eaao1378. https://doi.org/10.1126/sciadv.aao1378

Hall-Spencer, J. M., & Harvey, B. P. (2019). Ocean acidification impacts on coastal ecosystem services due to habitat degradation. *Emerging Topics in Life Sciences*, 3, 197–206. https://doi.org/10.1042/ETLS20180117

Hauer, M. E., Evans, J. M., & Mishra, D. R. (2016). Millions projected to be at risk from sea-level rise in the continental United States. *Nature Climate Change*, 6, 691–695. https://doi.org/10.1038/NCLIMATE2961

Hoegh-Guldberg, O., Poloczanska, E. S., Skirving, W., & Dove, S. (2017). Coral reef ecosystems under climate change and ocean acidification. *Frontiers in Marine Science*, 4, 158. https://doi.org/10.3389/fmars.2017.00158

Intergovernmental Panel on Climate Change. (2019). Summary for Policymakers. In: *IPCC Special Report on the Ocean and Cryosphere in a Changing Climate.* [H. -O. Pörtner, D. C. Roberts, V. Masson-Delmotte, P. Zhai, M. Tignor, E. Poloczanska, K. Mintenbeck, A. Alegría, M. Nicolai, A. Okem, J. Petzold, B. Rama, N. M. Weyer (Eds.)]. Geneva, Switzerland. https://doi.org/10.1017/9781009157964.

Intergovernmental Panel on Climate Change. (2022). *Synthesis report of the IPCC Sixth Assessment Report (AR6): Summary for policymakers.* United Nations. https://www.ipcc.ch/report/sixth-assessment-report-cycle/

Jacobsen, N. S., Marshall, K. N., Berger, A. M., Grandin, C., & Taylor, I. G. (2022). Climate-mediated stock redistribution causes increased risk and challenges for fisheries management. *ICES Journal of Marine Science, 79*(4), 1120–1132. https://doi.org/10.1093/icesjms/fsac029

Keenan, J. M., Hills, T., & Gumber, A. (2018). Climate gentrification, from theory to empiricism in Miami-Dade County, Florida. *Environmental Research Letters, 13*(5). https://iopscience.iop.org/article/10.1088/1748-9326/aabb32/pdf.

Lirman, D., & Schopmeyer, S. (2016). Ecological solutions to reef degradation: Optimizing coral reef restoration in the Caribbean and Western Atlantic. *PeerJ, 4*, e2597. https://doi.org/10.7717/peerj.2597

Louisiana Office of Community Development. (2020). *Resettlement of Isle de Jean Charles: Background and overview.* US Department of Housing and Urban Development. http://isledejeancharles.la.gov/sites/default/files/public/IDJC-Background-and-Overview-6-20_web.pdf

Lyons, K. (2020, January 20). Climate refugees can't be returned home, says landmark UN human rights ruling. *The Guardian.* https://www.theguardian.com/world/2020/jan/20/climate-refugees-cant-be-returned-home-says-landmark-un-human-rights-ruling

Malmstadt, J. C., Elsner, J. B., & Jagger, T. H. (2019). Risk of strong hurricane winds to Florida cities. *J. Applied Meteorology and Climatology, 49*(10), 2121–2132. https://doi.org/10.1175/2010JAMC2420.1

Martin, A. (2018, October 18). *An Alaskan village is falling into the sea. Washington is looking the other way.* The World: Public Radio International. https://www.pri.org/stories/2018-10-22/alaskan-village-falling-sea-washington-looking-other-way

McDonnell, T. (2018, June 20). *The refugees the world barely pays attention to.* National Public Radio. https://www.npr.org/sections/goatsandsoda/2018/06/20/621782275/the-refugees-that-the-world-barely-pays-attention-to

Melillo, J. M., Richmond, T. T. C., & Yohe, G. W. (Eds.) (2014). *Climate change impacts in the United States: The third national climate assessment.* US Global Change Research Program. https://doi.org/10.7930/J0Z31WJ2

Molinaroli, E., Guerzoni, S., & Suman, D. (2019). Do the adaptations of Venice and Miami to Sea Level Rise offer lessons for other vulnerable coastal cities? *Environmental Management, 64*(4), 391–415. https:doi.org/10.1007/s00267-019-01198-z

Muñoz, N. J., Farrell, A. P., Heath, J. W., & Neff, B. D. (2015). Adaptive potential of a Pacific salmon challenged by climate change. *Nature Climate Change, 5*, 163–166. https://doi.org/10.1038/nclimate2473

Narita, D., Rehdanz, K., & Tol, R. S. J. (2012). Economic costs of ocean acidification: A look into the impacts on global shellfish production. *Climatic Change, 113*, 1049–1063. https://doi.org/10.1007/s10584-011-0383-3

National Hurricane Center. (2019). *Estimated return period in years for hurricanes passing within 50 nautical miles of various locations on the US coast.* https://www.nhc.noaa.gov/climo/#cp100.

National Research Council. (2007). *Mitigating shore erosion along sheltered coasts.* National Academies Press.

Nicholls, R. J., Hanson, S., Herweijer, C., Patmore, N., Hallegatte, S., Corfee-Morlot, J., Chateau, J., & Muir-Wood, R. (2007). *Ranking of the world's cities most exposed to coastal flooding today and in the future* (OECD Environment Working Paper No. 1). https://doi.org/10.1787/011766488208

Ojea, E., Lester, S. E, & Salgueiro-Otero, D. (2020). Adaptation of fishing communities to climate-driven shifts in target species. *One Earth, 2*(6), 544–556. https://doi.org/10.1016/j.oneear.2020.05.012

Patricola, C. M., & Wehner, M. F. (2018). Anthropogenic influences on major tropical cyclone events. *Nature, 563*, 339–346. https://doi.org/10.1038/s41586-018-0673-2

People urgently fleeing climate crisis cannot be sent home, UN rules. (2020, January 20). *BBC News*. Retrieved May 6, 2022, from https://www.bbc.com/news/world-asia-51179931

Pilkey, O. H., & Cooper, A. G. (2014). *The last beach.* Duke University Press.

Pilkey, O. H., Pilkey-Jarvis, L., & Pilkey, K. C. (2016). *Retreat from a rising sea.* Columbia University Press.

Pörtner, H. O., & Peck, M. A. (2010). Climate change effects on fish and fisheries: towards a cause-and-effect understanding. *Journal of Fish Biology, 77*(8), 1745–1779. https://doi.org/10.1111/j.1095-8649.2010.02783.x

Rijnsdorp, A. D., Peck, M. A., Engelhard, G. H., Möllmann, C., & Pinnegar, J. K. (2009). Resolving the effect of climate change on fish populations. *ICES Journal of Marine Science*, 66(7), 1570–1583. https://doi.org/10.1093/icesjms/fsp056

Sweet, W., Kopp, R., Weaver, C., Obeysekera, J., Horton, R., Thieler, E. R., & Zervas, C. (2017). *Global and regional sea level rise scenarios for the United States* (NOAA Technical Report NOS CO-OPS 083). NOAA/NOS Center for Operational Oceanographic Products and Services. https://tidesandcurrents.noaa.gov/publications/techrpt83_Global_and_Regional_SLR_Scenarios_for_the_US_final.pdf

Treuer, G., Broad, K., & Meter, R. (2018). Using simulations to forecast homeowner response to sea level rise in South Florida: Will they stay or will they go? *Global Climate Change*, 48, 108–118. https://doi.org/10.1016/j.gloenvcha.2017.10.008

Union of Concerned Scientists. (2017). *Encroaching tides in Miami-Dade County, Florida.* UCS Fact Sheet, pp. 1–10. https://www.ucsusa.org/sites/default/files/attach/2016/04/miamidade-sea-level-rise-tidal-flooding-fact-sheet.pdf

United States Army Corps of Engineers (USACE). (2020). *Miami-Dade back bay coastal storm risk management draft integrated feasibility report and programmatic environmental impact statement.* https://usace.contentdm.oclc.org/utils/getfile/collection/p16021coll7/id/14453

Weiss J. L., Overpeck, J. T., & Strauss, B. (2011). Implications of recent sea level rise science for low-elevation areas of coastal cities of the conterminous USA. *Climate Change, 105*, 635–645. https://doi.org/10.1007/s10584-011-0024-x

Zervas, C. (2009). *Sea level variations of the United States 1854–2006* (NOAA Technical Report NOS CO-OPS 053). NOAA. https://tidesandcurrents.noaa.gov/publications/Tech_rpt_53.pdf

8 Marine Renewable Energy

Policy, People, and Prospects

B. J. Wickizer, D. Brandt, B. Robertson, and H. S. Boudet

The ocean's ability to provide a myriad of resources has made it fundamental to human societies across history. Beyond sustenance from fishing, chief among these resources during the past century has been energy. Before 1990, marine energy resources were almost exclusively fossil fuel based, namely, offshore oil and natural gas extraction. Even today, more than a quarter of global oil and gas production occurs offshore (US Energy Information Administration, 2016). Over the next 20 years, worldwide offshore oil extraction is projected to remain steady, while offshore natural gas extraction is expected to grow by 50% (International Energy Agency [IEA], 2018).

Yet, this offshore oil and gas production has not come without costs. Offshore oil and gas activities have resulted in substantial accidents and spills. The United States alone has experienced 45 significant oil spills (more than 10,000 barrels) in the past 50 years (National Oceanic and Atmospheric Administration, 2017), with several (e.g., Santa Barbara oil spill, Exxon Valdez, BP Horizon) leading to significant environmental impacts and associated policy changes. Moreover, the growing climate crisis – and contributions from fossil fuel combustion – has led us to seek alternative renewable energy sources, both onshore and offshore.

Beginning in the early 1990s, marine renewable energy resources – offshore winds, waves, tides, and currents – began to emerge as viable technology alternatives that not only presented fewer risks to the ocean environment and its assets, but also produced negligible greenhouse gas emissions (see Box 8.1). Marine renewables may also enhance electricity grid resilience and complement existing intermittent renewable sources. In addition, they can provide more cost-effective, sustainable electricity to isolated coastal communities. Although it may seem counterintuitive, offshore fossil fuel operations are poised to benefit from marine renewables, and vice versa. For example, marine renewables are capable of supplying power to electrify oil and gas rig operations (IEA, 2018), while natural gas generators' quick ramping ability can complement variable marine renewable generation (Lee et al., 2012). Moreover, oil and gas firms have ventured into the marine renewable energy space to diversify their portfolios, hedge against fossil fuel price volatility, and adapt to changing market and policy landscapes.

DOI: 10.4324/9781003058151-10

BOX 8.1 Orsted Case Study

Perhaps the most notable example of an offshore oil and gas firm entering the offshore wind space is the Dutch firm Dansk Olie og Naturgas (DONG). DONG was founded in 1972 and managed oil and gas resources in the North Sea. In the early 2000s, DONG became involved in electricity markets and offshore wind. In 2017, it divested itself from all oil and gas holdings to focus exclusively on renewable projects, renaming itself Orsted. Orsted now has more operational offshore wind capacity globally than any other company. It has taken a keen interest in the fledgling US market, and as of 2020, it was responsible for 30% of the planned or operational offshore wind projects on the US Atlantic Coast (Musial et al., 2020), including the United States' only operational wind farm.

The experience that oil and gas firms possess operating in the marine context could be a boon for offshore wind. As wind technology continues to advance, supply chains become more mature, the regulatory environment assumes more structure, and costs continue to decline, more oil and gas firms are likely to enter the offshore wind arena. The level of interest of oil and gas firms may also hinge on how aggressively decarbonization policies are pursued. In the case that decarbonization is prioritized, offshore oil extraction is predicted to decline by 2030 and beyond (IEA, 2018), making offshore wind potentially even more attractive.

In this chapter, we review the development and proliferation of marine renewable energy, detail the function and status of various technologies, highlight related public perceptions, and examine relevant policies.

Marine Renewable Energy

The wealth of energy resources in the ocean is immense and is, quite possibly, the linchpin energy resource for future sustainable economic growth. Estimates suggest that the worldwide offshore wind resource could theoretically provide up to 420,000 terawatt-hours (TWh) of energy annually (IEA, 2019), while worldwide wave and tidal resources could theoretically generate 29,500 TWh (Huckerby et al., 2017) and 25,880 TWh (Neill et al., 2018) annually, respectively. For perspective, worldwide energy consumption in 2019 amounted to 158,839 TWh (Global Change Data Lab, n.d.); thus, the theoretical energy potential of offshore wind, wave, and tidal resources is 264%, 19%, and 16% of present energy consumption, respectively. As of 2019, offshore wind had reached 27 gigawatts (GW) of installed capacity globally (Musial et al., 2020), and marine hydrokinetic energy (i.e., wave, tidal, and current) had reached less than 1 GW of global installed capacity (IEA, 2018). Here we differentiate between traditional energy stocks (reserves of oil and gas) and renewable energy flows (wind, waves, currents, etc.). Another way to think of this differentiation is as two different energy "currencies" – one is a currency of fuel, while the other is electricity.

Offshore Wind

While considered relatively new within the marine energy technology realm, the development and growth of the offshore wind industry is staggering, and it is considered the most mature marine renewable technology. The first offshore wind farm, Vindeby Offshore Wind, was developed off the coast of Denmark in 1991. It featured eleven 450-kilowatt (kW) turbines, or ~5 megawatts (MW) of installed capacity.

Compared to onshore wind, offshore wind generally enjoys greater consistency and higher wind speeds, making it more productive. Additionally, prime offshore wind areas are often proximate to areas of high electricity demand, unlike many onshore wind areas, thus reducing necessary grid infrastructure investments and transmission energy losses. Fewer restrictions on footprint area and turbine height also generally exist for offshore wind (IEA, 2018).

Although offshore wind installations do not utilize valuable terrestrial land, the ocean has a significant number of stakeholders competing for space, necessitating collaborative development approaches. Offshore wind development has potential implications for fishing, marine protected areas, shipping, and coastal aesthetics, among other uses, so coordination is essential. The most notable direct environmental impacts related to offshore wind are mortality and aversion behavior in birds and bats, as well as auditory disruption and displacement of marine mammals and sea turtles from construction operations (Bailey et al., 2014). Impacts and conflicts can be reduced if appropriate siting and mitigation measures are pursued, although this requires substantial upfront research and outreach.

Offshore wind farms are predominantly seafloor-mounted and feature a three-bladed horizontal axis turbine. Initial offshore wind installations utilized modified terrestrial wind components. Yet, as turbines moved offshore over the past 30 years, the opportunity to increase the scale of the tower, blades, and rated capacity expanded greatly (Figure 8.1). GE is currently testing the Haliade-X 12 MW turbine – a single turbine that is 260 m tall and boasts an impressive capacity factor of 63%. The capacity factor compares how much energy was generated to the maximum that could have been generated over a specified period if the system ran at full electrical output capacity or power.

The next development domain for offshore wind is in deeper water – at depths beyond the ability to mount the turbine to the seafloor directly – utilizing floating platforms. The benefits of deepwater offshore wind are significant. Offshore winds are stronger and more consistent, leading to increased system capacity factors and allowing for increased turbine capacities (i.e., fewer physical turbines for the same energy output). Floating offshore wind farms are nascent, but first-of-a-kind projects have been developed in Scotland (Equinor, n.d.) and Portugal (Energias de Portugal, 2018). The world's first floating offshore wind project, HyWind in Scotland, features five 6 MW turbines, all using single, very deep, spar-type subsurface

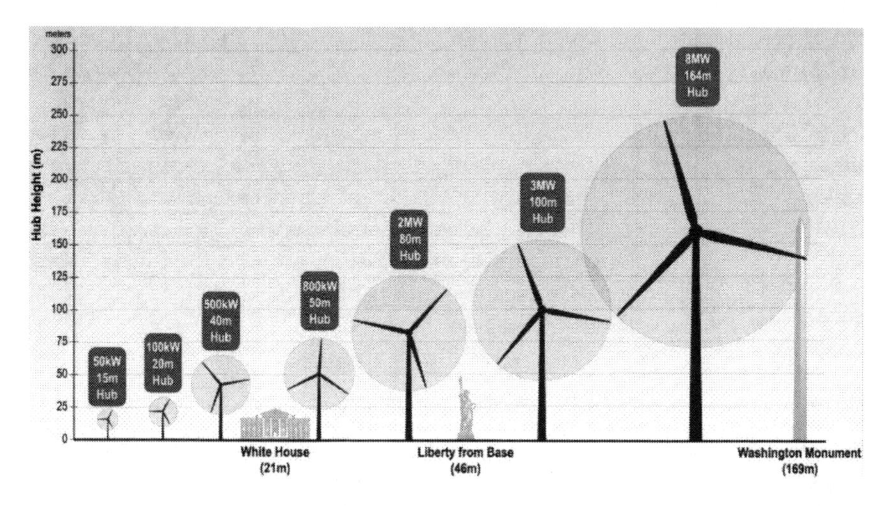

Figure 8.1 Wind turbine height and energy output. Credit: Image by Josh Bauer, NREL.

floating platforms. In Portugal, the Windfloat project features three 8.4 MW turbines but relies on a tri-pontoon, semi-submersible platform design.

Europe contains the vast majority of presently installed offshore wind capacity. The UK has the strongest offshore wind market, comprising 36% of the world's installed offshore capacity alone (Global Wind Energy Council, n.d.). Germany and China have the second and third most installed offshore wind, with 29% and 15% of installed capacity, respectively. Currently, the United States has only one commercial offshore wind farm – Deepwater Wind, off Block Island, Rhode Island – that has a modest capacity of 30 MW. However, as of 2020, 14 projects were in the permitting phase and slated to be operational in the next five years (Musial et al., 2020). Once realized, these projects will provide approximately 6,500 MW of additional capacity.

Offshore Wind Growth and Costs

Despite offshore wind's presently limited footprint, some estimates suggest that its installed capacity will grow to 90 GW in the next decade – a more than three-fold increase from current capacity (IEA, 2018). Offshore wind prices presently exceed those of select other renewables, with prices in 2016, on average, 150% more than onshore wind and 50% more than solar photovoltaic, with respect to levelized cost of energy (LCOE) (IEA, 2018). LCOE measures the lifetime costs of a generation source relative to the energy it will produce during its lifespan. However, advancements in turbine technology, access to areas further offshore, larger wind farms and economies of scale, more advantageous financing, more mature logistics and supply chains, industry experience, and market competition are all driving offshore wind costs down (Energy Sector Management Assistance

Program, 2019). In 2018, offshore wind's global weighted average LCOE was $0.127 per kilowatt-hour, which represented more than a 20% decline from 2010 (International Renewable Energy Agency, 2019). In fact, in the UK in 2019 the lowest historic offshore wind contract price ($0.049/kWh) was secured, representing a 30% cost decline from the next most recent auction in 2017 (Energy Sector Management Assistance Program, 2019). Offshore wind prices are projected to continue declining, with the global LCOE potentially decreasing 30%–60% by 2030 and 45%–75% by 2050 (International Renewable Energy Agency, 2019). Offshore wind in parts of Europe is already price-competitive with fossil fuel sources, and if price reduction trends are realized, this parity will be solidified on a global scale in the coming decades.

Emerging Marine Renewable Energy Technologies

Energy from waves, currents, and tides – referred to collectively as marine hydrokinetic energy – represents a promising, but largely untapped, resource. Tidal technology is in the early deployment phase, while wave energy is closer to the development phase (see Box 8.2). Other marine energy sources also exist, such as ocean thermal energy conversion and marine salinity gradients, which utilize differences in temperature and in salt concentrations, respectively. However, these technologies are even more immature than marine hydrokinetics.

BOX 8.2 Wave Energy Test Sites Case Study

While offshore wind and tidal energy generation projects are in commercial or final near-commercial stages, the commercial wave energy sector is still nascent. Rather than focusing efforts on individual project development, the wave energy sector has focused on the development of dedicated technology testing sites. These pre-permitted, well resourced, and centralized locations provide technology developers with a low-risk opportunity to test technologies and further research and development to ensure cost reductions, reliability increases, and improved understanding. The European Marine Energy Center, the Wave Energy Test Site, and PacWave are the dominant global testing sites.

The European Marine Energy Center (EMEC) in Scotland's Orkney Islands has been testing wave energy technologies since 2003 and is the location of more full-scale and scaled wave energy converters than any other site globally. EMEC operates two testing locations north of Scotland: a full energetic, open ocean site at Billia Croo and a more sheltered testing site at Scapa Flow. The Billia Croo site consists of five cabled berths, between 50 and 70 m, each with up to 11 kilovolts of electrical capacity.

The Wave Energy Test Site (WETS) sits on the Marine Corps Base Hawaii in Oahu. The WETS was first developed in 2003 to test point-absorber devices designed by Ocean Power Technologies. Given the warm climate, clear ocean, and lower natural wave energy resources, the WETS is often touted as the ideal location for first sea trials of any wave energy conversion technology – before graduating to fully energetic locations like EMEC or PacWave. Since 2003, WETS has greatly expanded to include three berths at varying depths and exposure to open ocean conditions.

PacWave is sited 10 km off the coast of Newport, Oregon, and is being developed by Oregon State University on behalf of the US Department of Energy. The PacWave site is extremely energetic during winter months, allowing for survivability and resiliency testing, and has very benign summer months, allowing for shorter initial deployment tests. PacWave will be commissioned in 2023 and features two locations – a fully pre-permitted, grid-connected South site and a non-grid-connected North Site. The PacWave South site features five berths with a rated capacity of 4 MW per site (20 MW total) and sits in approximately 60 m of water. PacWave will provide WEC technology developers with the final system tests prior to development of commercial-scale projects.

Wave Energy Converters

Wave energy is broadly classified as the kinetic and potential energy associated with the movement of ocean waves – primarily based on the height and wave period (time difference between two subsequent wave crests) of the waves. In contrast to other renewable energy technologies, wave energy conversion (WEC) technology concepts vary widely in their form and generation mechanism. This is driven by the multiple potential and kinetic energy resources available in waves. The energy content of the wave energy resource depends on two parameters – significant wave height and wave energy period. This is in direct contrast to other renewable resources, which are categorized by a single parameter: wind (wind speed), solar (irradiation), tidal (water velocity), etc. As a result of the dual dependency, the opportunities and variables to consider when building WEC technologies are inherently larger and more complex.

In the broadest sense, WECs can be classed as oscillating water columns, terminators, attenuators, or point absorbers. Oscillating water columns (shown in Figure 8.2d) with the air turbine utilize the surge (forward/backward) and heave (up/down) motions of the water surface, within an enclosed air volume, to compress air and drive an air turbine. The Pico OWC and the Islay Limpet OWC – both of which are now decommissioned – harnessed wave energy and generated electricity this way for over a decade. Both Pico and Limpet were shore-mounted, yet newer projects, such as the Ocean Energy Limited OE35 buoy deployed off Hawaii in 2020, are floating systems.

Terminators are identified by their long wave crest parallel dimensions, such as the Pelamis-type system shown in Figure 8.2b. In general, these systems are

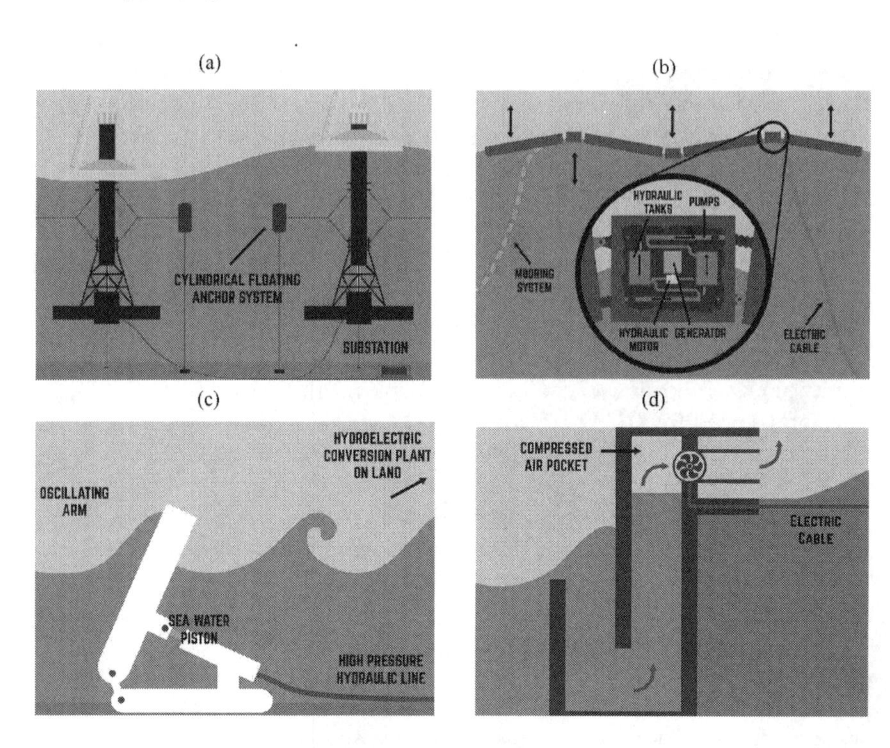

Figure 8.2 (top left 2a; top right 2b; bottom left 2c; bottom right 2d): Variety of Wave Energy Conversion (WEC) designs. Credit: Images courtesy of the Pacific Institute for Climate Solutions' Wave Energy: A Primer for British Columbia report.

sufficiently large scale so that portions of the WEC are in different parts of the wave profile (peak vs. tough) and can harness the kinetic energy from the associated out-of-phase motions. Attenuators feature long wave crest perpendicular dimensions and focus on the kinetic and potential energy associated with the wave particle velocities and the forward–backward surge of these water particles in shallower water regions, as shown by the "flap-type" system in Figure 8.2c. Finally, point absorbers are small compared to the wavelength and harness the kinetic and potential energy associated with relative displacement between a surface float, driven by water elevation changes, and a relatively static subsurface heave plate or the seafloor; this concept is visually shown in Figure 8.2a.

The future development of wave energy will be driven by several social, technical, and economic factors; however, the scale and locational value aspects of the resource are unique and provide a different "service" to the grid. Wave energy benefits from being available in all the world's oceans – which cover 70% of the planet – and, as a result, these resources are nearby growing coastal urban centers. As the demand for electricity grows in urban coastal cities, wave energy's proximity is incredibly valuable, offering distinctive promise in comparison to other renewables that require long transmission lines, with associated power losses, to transmit the electricity to customers from remote terrestrial locations.

As with all energy extraction or conversion technologies, environmental risks and impacts are inherent. The International Energy Association's Ocean Energy Systems Environmental State of the Science report (Copping & Hemery, 2020) provides a living database and an up-to-date review of global environmentally focused marine energy research to help ensure responsible development of these resources, and, thus far, it indicates very limited impacts. As a nascent sector, wave energy requires more deployment time to fully characterize the environmental risks and full-life-cycle costs and benefits – all necessary precursors to large-scale development.

Tidal and In-Stream Energy Converters

Tidal and In-Stream Energy Converters (TISEC) harness the kinetic energy in the ocean's moving tidal currents, which are driven by the gravitational forces of astronomical bodies. As the celestial bodies rotate around the world, gravitational attraction "pulls" the global oceans and creates temporal variation in water depths. As this "bulge" moves around the world, there is a tremendous movement of water from one region to another and a significant, energy-dense renewable energy flow.

Various archetypes of tidal energy devices are being designed, developed, and tested. Broadly speaking, tidal technologies can be categorized as barrages, lagoons, or in-stream. Tidal barrages are the oldest of tidal technologies and involve forcing all the tidally fluctuating water through a "barrage", or wall, of turbines. As the tide ebbs and flows, the barrage generates electricity. Tidal barrages exist in Korea, Canada, and Russia, while the Rance Tidal Station in France has been operating on this principle since 1966. Tidal lagoons involve the construction of large retaining walls across a bay or coastal inlet and "gates" to hold the water in/out until the maximum water head (elevation) differential is achieved. Water is then directed through low-head turbines and electricity is generated. Both barrages and lagoons require very specific location geography and conditions for project development.

The dominant tidal converter archetypes deployed are in-stream technologies, which build on knowledge from both the ship propulsion and the wind industries. Most feature a horizontal axis nacelle with three or four active blades. Understanding that energy associated with moving water is tied to the density of the fluid, tidal energy has ~1000× more energy per unit volume than air and, as such, tidal energy technologies feature significantly shorter blades to account for the increased loading. In-stream devices are further categorized as either seafloor mounted or floating. The MeyGen project in Scotland is currently the world's largest tidal energy plant, featuring four seafloor-mounted 1.5 MW three-bladed horizontal axis turbines (see Box 8.3). In the floating arena, the Orbital Marine Power 02 features two 1 MW turbines deployed off a floating central spar – allowing for ease of installation, removal, and maintenance of the turbine. A new and interesting concept, diverging from the horizontal axis theme and focusing on lower speed water velocities, is an underwater kite being developed by Minesto, which utilizes the lift around the water foils to drive a motor and generator system.

BOX 8.3 MeyGen Case Study

The MeyGen tidal energy project, located two kilometers offshore between Scotland's northernmost coast and the island of Stroma, completed its first phase of development in April 2018, with the installation of four 1.5 MW tidal turbines in some of the fastest flowing waters in the UK. As of June 2019, it had exported 17 GWh to the grid (Short, 2020). Future phases include installation of another 51 turbines and a subsea power hub that would allow multiple turbines to connect to a single transmission cable, limiting disruption to the marine environment and infrastructure costs. The first phase of development included environmental monitoring equipment to assess the turbine interactions with the marine environment. It is the largest global planned tidal stream project with a permit to produce up to 398 MW and a current grid capacity of 252 MW. After the first stage of completion, it is meeting the electricity needs of 2,600 homes.

One of the only commercial multi-turbine array tidal energy projects, the project aims to prove the commercial viability of tidal energy and supports the Scottish government's desire to increase its use of renewable sources to reduce carbon emissions. Scotland, known for its strong wave and tidal resources, is also host to the European Marine Energy Centre, offering momentum to Scotland's wave and tidal energy development. The limited distribution infrastructure in Scotland presents a current challenge to wider-spread use of tidal energy.

Tidal energy has several valuable attributes. The electricity generated from tidal power is almost perfectly predictable – unlike the vast majority of renewable energy flows – which has significant benefit and value to the electricity grid. Tidal resources are also often proximate to urban demand centers and can provide valuable local generation. However, tidal resources are generally more localized compared to wave resources and not broadly available to all countries with coastlines. As with wave energy, the impact of tidal devices on the environment is still an active research area. The limited deployments undertaken at global testing sites have shown minimal environmental impacts, providing confidence for future deployments (Copping & Hemery, 2020).

Marine Hydrokinetic Growth and Costs

Marine hydrokinetic energy installed capacity is projected to grow to 5 GW by 2030 and 21 GW by 2040 (IEA, 2018). Thus, marine hydrokinetics' projected capacity in 2040 will be roughly equal to offshore wind's installed capacity in 2020. The global LCOE for tidal and wave energy is presently $0.44 kWh and $0.50 kWh, respectively. Hence, marine hydrokinetics overall are not currently cost-competitive with other energy technologies, although select marine hydrokinetic projects do exist that provide comparatively lower cost energy because of the uniquely favorable conditions associated with their locations. Cost declines similar to those of offshore wind can be expected as the technology matures, economies of scale become more favorable, the technology becomes more proven, better financing is available, supply chains become solidified, and firms gain experience.

Public Perceptions of Marine Renewable Energy Development

Public support is critical to the success of energy development, particularly in democratic settings (Ansolabehere & Konisky, 2014; Boudet, 2019; Carley et al., 2020). Examples abound of entire industries (Renn & Marshall, 2016) and specific projects (Firestone et al., 2012a) that have been stymied due to public opposition. Thus, it is important to continually assess and understand public perceptions, as well as to engage affected stakeholders in the development process (Conway et al., 2010).

The public generally reports limited familiarity with energy technologies (Spence et al., 2015; Thomas et al., 2017), including marine renewable energy technologies (Wiersma & Devine-Wright, 2014). Early studies of public perceptions thus linked opposition to a lack of information; educational efforts were stressed to build public acceptance (Slovic, 1987). Scholars and practitioners focused on overcoming not-in-my-backyard (NIMBY) attitudes, or the idea that people oppose an energy project due to its proximity for reasons of selfishness, ignorance, or irrationality (Dear, 1992; Schively, 2007). Similarly, studies on renewable energy technologies highlighted the "social gap" between generalized public support for renewables and opposition to specific projects (Bell et al., 2005, 2013; Walker, 1995). Indeed, concern about public opposition to onshore projects was one driver of offshore proposals, which were viewed as "a problem-free alternative to siting onshore" (Haggett, 2011, p. 503).

More recent research, however, has largely debunked the idea of NIMBYism (Rand & Hoen, 2017; Wüstenhagen et al., 2007). Scholars have increasingly focused on justice in facility siting, in terms of both the distribution of risks and benefits *and* the procedures for public engagement. A growing recognition that people, due to limited time and resources, often use mental shortcuts (Ho et al., 2019) – such as ideological predispositions, environmental and altruistic values (Jacquet, 2012), media coverage (Vasi et al., 2015), and elite cues (Clarke et al., 2015) – to filter information and form opinions about new energy technologies has led scholars to abandon suggestions of purely educational efforts and instead encourage more meaningful public engagement and involvement in decision-making (Arnstein, 1969; Batel, 2020; Devine-Wright, 2005).

Moreover, offshore siting of marine renewable energy projects has not proven problem-free (Boudet & Ortolano, 2010; Firestone & Kempton, 2007; Haggett, 2011; Wiersma & Devine-Wright, 2014). Visual impacts, ignorance of local context and place attachments among affected stakeholders, disconnects between local impacts and global benefits, distant and multinational ownership, and inadequate public engagement processes can doom offshore projects, too (Haggett,

2011). Perhaps the best example of this is the Cape Wind project off the coast of Cape Cod, Massachusetts, which was initiated in 2001 and dissolved in 2015. Ultimately, a small but powerful group of local citizens was able to stymie the project through permit contestation and litigation. Moreover, local officials, Native American tribes, fishermen, and other stakeholders levied numerous concerns. Cape Wind illustrates the importance of prioritizing stakeholder preferences early within project development. Additionally, technological advances since Cape Wind have facilitated superior siting opportunities by allowing projects to be located in deeper water, potentially reducing aesthetic concerns and other conflicts. In fact, the Vineyard Wind project, scheduled to be operational by 2023 after beginning construction in late 2021, is slated to be located in the same general area as Cape Wind was, though further offshore and with larger, but fewer, turbines. Drawing on a framework developed by Boudet (2019), we review some of the main drivers of public perceptions of energy, namely, aspects of the technology proposed, the people affected, the place in which a project is proposed, and the process by which decisions are made.

Technology

Perceptions of the risks and benefits associated with a technology or project are important drivers of attitudes and actions (Slovic, 1987; Stoutenborough et al., 2015). Slovic (1987) highlighted the role of two different factors in shaping risk perceptions: (1) "Dread" risk – relating to a technology's uncontrollability, catastrophic potential, fatal consequences, involuntariness, high risk to future generations and inequitable distribution of costs and benefits – and (2) "unknown" risk – relating to how a technology's risks may be unobservable, new, delayed, and unknown to science. The more dread inspired by a technology and the more unknown its risks, the higher public risk perceptions will be. Other aspects shaping public perceptions of offshore energy development include its costs, its effect on energy prices, the scale of its footprint, and its ecological and aesthetic impacts.

Looking specifically at marine energy, offshore oil and gas development began in the Gulf of Mexico in the 1930s. Its relatively easy transition from onshore to offshore in the Gulf benefited from an established onshore industry, for which the risks and benefits were known (Freudenburg & Gramling, 1993). Despite this relatively easy transition, it was public outcry about an oil spill from offshore development in Santa Barbara, California, in 1969 that is widely viewed as one of the main triggers of the modern environmental movement, highlighting the role of focusing events, like accidents, in shaping public perceptions of energy development (Freudenburg & Gramling, 1994; Molotch, 1970). Though offshore production continues in the United States, it has been limited in California, and the Trump administration's attempts to open other offshore areas have largely been rebuffed, even in Republican-leaning states (Konisky & Woods, 2018).

The offshore wind sector has not enjoyed such a swift transition from land to sea. The risk of a spoiled viewshed, lack of tangible benefits for proximate communities, and concerns about relative cost compared to other energy resources emerged early as public concerns about offshore wind (Haggett, 2011). Despite surveys showing high support for wind energy among the US public (Rand & Hoen, 2017), as of this writing, only one functioning 30 MW offshore installation exists at Block Island, Rhode Island, though recent research indicates that resident support for the project has only increased since its launch (Russell et al., 2020). In contrast, Europe has been much more aggressive in its pursuit of offshore wind development, reaching over 20 GW of offshore capacity recently, and driving innovation and cost declines in the sector (IEA, 2019), which may in turn contribute to increased public support for the technology.

In contrast to oil and wind, no onshore energy equivalent exists for wave and tidal energy. Surveys indicate positive opinions of these technologies, especially compared to fossil fuel sources (Hazboun & Boudet, 2020), despite limited familiarity and concerns about impacts to viewsheds and marine life (Boudet et al., 2020; Conway et al., 2010; Dreyer et al., 2017; Wiersma & Devine-Wright, 2014). Perceived benefits include carbon mitigation, as well as potential contributions to energy independence, security, and local jobs and revenue creation (Boudet et al., 2020; Conway et al., 2010; Dreyer et al., 2017; Wiersma & Devine-Wright, 2014).

People

Sociodemographic factors (age, gender, ethnicity, income, education, and political ideology) have been routinely tested in terms of how they shape public perceptions of energy technologies. Reviewing the literature on offshore renewables, Wiersma and Devine-Wright (2014) found that younger people, those with higher income or education, and women expressed more positive attitudes, though some types of ocean users (e.g., frequent beach goers, fishermen, and tourist-focused business owners) expressed concerns. Environmental values also play a role (Kerr et al., 2014), particularly views on climate change, as does trust in developers and regulators (Simas et al., 2012). Compared to renewables, oil and gas development is more politically polarizing, with conservatives generally more supportive and liberals more opposed (Hazboun & Boudet, 2020).

Beyond the individual, community-level factors can also play a role. For example, the embeddedness of the oil industry in the livelihoods of Louisianans helped facilitate its offshore transition (Freudenburg & Gramling, 1994). Elite cues (e.g., from elected officials, business leaders) and perceptions of what other people think also shape attitudes, both positively and negatively (Crowe, 2020; Sokoloski et al., 2018).

Place

The place in which a project is proposed – the landscape, physical infrastructure, existing economy, and institutions – also shapes public perceptions and attitudes. The landscape itself can shape visual impacts from offshore development (Freudenburg & Gramling, 1994). Moreover, projects proposed in places that are important to local populations due to natural amenities or commercial activity will often face opposition (Giordono et al., 2018; Hall & Lazarus, 2015). At an individual level, positive emotional connections to a specific location can create "place attachment". If this location is then threatened by a proposed energy development, those attached to it are likely to take action to protect it (Devine-Wright & Howes, 2010; McLachlan, 2009). In addition, past experiences with energy industries, particularly extractive industries, and compatibility with existing industries and regulations can shape views of new proposals (Fornahl et al., 2012; Olson-Hazboun et al., 2018). In other words, how a project is viewed depends on where it is proposed.

Process

Aspects of the process by which decisions are made about a development – in terms of public engagement, transparency, economic involvement, and fairness – are also key in determining public perceptions (Firestone et al., 2018, 2012b; Haggett, 2008). Building public trust in developers and regulators through a fair, open, and transparent process can be particularly important when developers come from outside the local community, which is highly likely given the scale and expense of offshore development (Firestone et al., 2012b; Wüstenhagen et al., 2007). At the same time, meaningful engagement and participation requires developers and government officials to give up some level of authority, control, and power – which can prove difficult.

Marine Renewable Energy Policy

Marine renewable technologies are subject to a variety of policy and governance regimes. Establishing a clear, streamlined regulatory environment that provides reliability and efficiency, while also balancing stakeholder concerns and providing effective oversight, is critical for the success of marine renewable energy. In fact, regulatory inexperience and disorganization have been implicated in certain early project failures, and policymakers and regulators have endeavored to make policy changes based on these early lessons.

Law of the Sea

The bedrock policy for marine activities and sovereignty is the United Nations 1982 Law of the Sea Convention (LOSC) that governs the rights of sovereign states to control the waters adjacent to their coasts, including commercially developing marine energy. The two most significant dictates for marine renewables that the LOSC provides are related to territorial seas and exclusive economic zones.

The LOSC outlines that coastal nation-states have essentially exclusive rights to control activities, apply laws, and use resources within the territorial sea that extends 12 nautical miles from its low-water line. The LOSC also stipulates that coastal nations may claim an exclusive economic zone extending 200 nautical miles into the sea from the low-water line, in which they have the right to conserve, exploit, and manage the living and non-living resources present in the water, as well as the seabed and subsurface. Though outside of the scope of the LOSC, countries whose governance involves shared authority between the federal and regional levels may extend marine renewable energy development and leasing jurisdiction to subnational governments. Additionally, even in the case that a project is located within federal jurisdictional waters, approval or comment from subnational authorities may be necessary in relation to the laying of undersea cable to connect the project to the grid.

Zoning and Leasing

A fundamental policy issue for marine renewable energy is determining what areas within a nation's jurisdictional waters are available for development, and how that development will be sanctioned. The development of zoning regimes and structured leases allows regulators to better control and plan for marine renewable development and reduce uncertainty. Countries generally have a single agency that is responsible for overseeing marine renewable leases; however, other permits and approvals are often required from non-leasing agencies. Leasing agencies may be responsible for coordination between other relevant permitting agencies, though the amount of coordination undertaken by leasing agencies varies. The development of leasing policy structures has been an iterative process in many countries, and leasing authority has often been re-designated to different agencies.

Marine renewable leases are often allotted through competitive auctions. One consideration for auctions is whether they are structured around providing leases to the highest bidders or structured around providing leases to companies offering the lowest bids to provide power. The former prioritizes maximizing revenue royalties, while the latter prioritizes wind power competitiveness and electricity cost savings. In addition to issuing leases, leasing authorities may be responsible for overseeing revenue collection related to energy production royalties.

Perhaps the most important aspect of leasing is determining what areas should be leased through planning and zoning. Marine renewable energy project zoning policies vary widely: from open-door policies that allow project developers to apply for a lease in a desired location, to leasing authorities designating broad zones determined to be generally suitable for development, to leasing authorities identifying specific project locations and dictating precise project requirements (Fitch-Roy, 2016). Open-door policies were common early on with offshore wind development; however, the current predominant approach is designating large areas zoned for development. The reason for this movement toward zoned management is likely due to difficulties with early offshore wind projects that offered the lesson that putting in more diligence upfront and identifying lease areas that avoided major conflicts or stakeholder opposition was likely to render better outcomes in the long run.

Balancing Uses

The ocean and coastal environments are associated with a myriad of uses and values. In some cases, these uses and values can coexist without issue; however, in other cases, conflicts may arise. Marine Spatial Planning (MSP) is an approach to navigating uses and values in an attempt to manage the marine environment in pursuit of specific objectives, often determined through political processes, and preserve its assets for various stakeholders and beneficiaries (O'Hagan, 2020).

Although still young in comparison to land use planning, approximately 70 countries presently employ MSP approaches, and marine renewable energy projects are one of the features of MSP efforts. Ideally, MSP allows decision makers to see the forest through the trees and foster a more holistic, rather than piecemeal, approach to determining the appropriateness and compatibility of ocean uses, taking into consideration not only what uses are occurring in the present, but also considering what may occur in the future. Stakeholder participation and transparency are essential characteristics of MSP approaches, although participation can assume different forms, some of which may be more genuine or meaningful than others. MSP is only as effective as the criteria used to inform it, and MSPs often have limited scopes that are focused on only one or two uses or sectors (Jones et al., 2016). If certain stakeholders are not included early on or certain uses or impacts are not meaningfully considered, then the management decisions informed by MSP efforts may not be as effectual or legitimate. MSP has been used in conjunction with marine renewable energy zonal management to delineate areas for development. Two of the primary potential benefits of MSP for marine renewable energy are the streamlining of the regulatory approval process and avoidance of contentious siting (Ryan et al., 2019).

Requirements for environmental review and permitting are a ubiquitous feature of marine renewable energy project approval. Most countries require environmental impact assessments to evaluate potential impacts of planned projects (United Nations Environment Programme, 2018). Generally, these assessments are necessary when significant environmental damage could be associated with a project. However, in many cases, particularly in the European Union and the United States, project developers and regulators can perform more cursory reviews prior to performing a full environmental impacts statement, potentially avoiding a full review if a project meets certain criteria for minimal environmental impacts. Environmental impact statement statutes primarily provide transparency. Even in the case that environmental impacts are anticipated, permitting agencies generally retain discretion to approve projects. Most statutes also stipulate requirements for some degree of public participation in the review process, although the forms this takes vary. As mentioned above, it behooves project developers and regulators to consider various marine uses and potential stakeholder opposition in selecting project locations, as it is likely to ease the environmental review process.

Policy and Marine Renewable Energy Growth

Offshore wind occupies a steadfast position as a competitor in the energy market at this point. However, its growth, as well as the level of development and deployment of marine hydrokinetic energy, is tied to future energy policies. The EU's legally binding goal of a 20% renewable energy share of total energy consumption and its cap-and-trade regime have created favorable conditions for marine renewable development in Europe. Furthermore, in 2018 the European Union agreed upon a goal of 32% renewable energy share. Europe also provided renewable sources with priority in connecting to the grid. These types of favorable policies are an important piece of why offshore wind has experienced outsized success in Europe.

The United States does not have a federal renewable energy standard. Moreover, the Obama administration's Clean Power Plan, which was expected to foster additional renewable energy growth, was replaced with the less robust Affordable Clean Energy Rule by the Trump administration. The United States is not the only nation to experience energy policy backsliding. For example, though Australia implemented carbon pricing in 2011, this policy was short-lived and repealed in 2014. However, 29 US states have some form of renewable portfolio standard (RPS), providing binding targets for renewable energy share. Of the states with RPSs, 16 are coastal, and seven of these states have an eventual goal of 100% renewable energy or zero emissions from the electricity sector. The Biden administration has signaled an interest in catalyzing renewable energy growth, including articulating a goal of deploying 30 GW of offshore wind energy by 2030 (Exec. Order No. 14008, 2021). Thus, although there are many encouraging policy instruments supporting marine renewable energy development in a variety of places, the long-term existence of these policies can be tenuous, particularly as political winds shift between administrations, and policy reversals are likely to have consequences for the trajectory of future marine energy deployment.

Another question related to marine renewable energy's future growth is the ability to access the grid, particularly as project locations move further offshore, as well as achieve necessary grid upgrades. Grid infrastructure inadequacies have already proven to be a limiting factor in maximizing the benefits of offshore wind in certain locations. Some countries have undertaken strategic planning efforts to coordinate grid upgrades with project development. Determining who pays for grid upgrades and if costs will be distributed across all who benefit is an important aspect of grid management, with implications for marine renewable development. Ensuring grid access and upgrades is not only important for maximizing marine renewable energy's possibilities, but also for reducing perceived risks, potentially catalyzing greater investment and development.

Conclusion

The world's present energy paradigm has been characterized as a hybrid regime, occupying a space between our historic fossil-fuel-based conventional energy paradigm and a future sustainable energy paradigm based on renewable energy and energy efficiency. In the coming decades, the growth of marine renewable energy will be bound more broadly to energy policies and the movement toward a sustainable energy regime. If decarbonization is prioritized and additional policy commitments foster renewable energy procurement, it will accelerate marine renewable deployment. Although offshore wind and marine hydrokinetics are slated to grow by 2030 and beyond under current policies, if meaningful global decarbonization policies are pursued, global offshore wind capacity is predicted to grow an additional 81% by 2030 and an additional 117% by 2040; global marine hydrokinetic capacity is predicted to grow an additional 40% by 2030 and an additional 62% by 2040 (IEA, 2018). Thus, the timeline for realizing marine renewable energy growth and competitiveness, and bringing a sustainable blue economy to fruition, depends largely on the vigor with which decarbonization policies are pursued more generally.

Questions for Reflection

1. How have technological innovations affected marine renewable energy development?
2. Which marine renewable energy technology is the most mature? Why might this be?
3. Why is siting important in the development of marine renewable energy projects?
4. What role can policy play in accelerating marine renewable energy development?
5. Could the increasing development of renewable ocean energy be a factor affecting the 2022 increase in gasoline prices?

References

Ansolabehere, S., & Konisky, D. M. (2014). *Cheap and clean: How Americans think about energy in the age of global warming*. MIT Press.

Arnstein, S. R. (1969). A ladder of citizen participation. *Journal of the American Institute of Planners*, *35*(4), 216–224. https://doi.org/10.1080/01944366908977225

Bailey, H., Brookes, K. L., & Thompson, P. M. (2014). Assessing environmental impacts of offshore wind farms: Lessons learned and recommendations for the future. *Aquatic Biosystems*, *10*(1), 1–13. Retrieved July 1, 2022 from https://aquaticbiosystems.biomedcentral.com/articles/10.1186/2046-9063-10-8

Batel, S. (2020). Research on the social acceptance of renewable energy technologies: Past, present and future. *Energy Research & Social Science*, 68, 101544. https://doi.org.10.1016/j.erss.2020.101544

Bell, D., Gray, T., & Haggett, C. (2005). The 'social gap' in wind farm siting decisions: Explanations and policy responses. *Environmental Politics*, *14*(4), 460–477. https://doi.org/10.1080/09644010500175833

Bell, D., Gray, T., Haggett, C., & Swaffield, J. (2013). Re-visiting the 'social gap': Public opinion and relations of power in the local politics of wind energy. *Environmental Politics*, *22*(1), 115–135. https://doi.org/10.1080/09644016.2013.755793

Boudet, H. S. (2019). Public perceptions of and responses to new energy technologies. *Nature Energy*, *4*(6), 446–455. https://doi.org/10.1038/s41560-019-0399-x

Boudet, H. S., Brandt, D., Stelmach, G., & Hazboun, S. (2020). *West Coast perceptions of wave energy*. Pacific Marine Energy Center. https://ir.library.oregonstate.edu/concern/technical_reports/pr76f9588

Boudet, H. S., & Ortolano, L. (2010). A tale of two sitings: Contentious politics in liquefied natural gas facility siting in California. *Journal of Planning Education and Research*, *30*(1), 5–21. https://doi.org/10.1177/0739456X10373079

Carley, S., Konisky, D. M., Atiq, Z., & Land, N. (2020). Energy infrastructure, NIMBYism, and public opinion: A systematic literature review of three decades of empirical survey literature. *Environmental Research Letters*, *15*(9), 093007. https://doi.org/10.1088/1748-9326/ab875d

Clarke, C. E., Hart, P. E., Schuldt, J. P., Evensen, D. T., Boudet, H. S., Jacquet, J. B., & Stedman, R. C. (2015). Public opinion on energy development: The interplay of issue framing, top-of-mind associations, and political ideology. *Energy Policy*, *81*, 131–140. https://doi.org./10.1016/j.enpol.2015.02.019

Conway, F., Stevenson, J., Hunter, D., Stefanovich, M., Campbell, H., Covell, Z., & Yin, Y. (2010). Ocean space, ocean place: The human dimensions of wave energy in Oregon. *Oceanography*, *23*(2), 82–91. https://www.jstor.org/stable/24860714

Copping, A. E., & Hemery, L. G. (Eds.). (2020). *OES-environmental 2020 state of the science report: Environmental effects of marine renewable energy development around the world*. Ocean Energy Systems. Retrieved July 1, 2022 from https://www.ocean-energy-systems.org/news/oes-environmental-2020-state-of-the-science-report

Crowe, J. A. (2020). The effect of partisan cues on support for solar and wind energy in the United States. *Social Science Quarterly*, *101*(4), 1461–1474. https://doi.org/10.1111/ssqu.12799

Dear, M. (1992). Understanding and overcoming the NIMBY syndrome. *Journal of the American Planning Association*, 58(3), 288–300. https://doi.org/10.1080/01944369208975808

Devine-Wright, P. (2005). Beyond NIMBYism: Towards an integrated framework for understanding public perceptions of wind energy. *Wind Energy*, 8(2), 125–139. https://doi.org/10.1002/we.124

Devine-Wright, P., & Howes, Y. (2010). Disruption to place attachment and the protection of restorative environments: A wind energy case study. *Journal of Environmental Psychology*, 30(3), 271–280. https://doi.org/10.1016/j.jenvp.2010.01.008

Dreyer, S. J., Polis, H. J., & Jenkins, L. D. (2017). Changing tides: Acceptability, support, and perceptions of tidal energy in the United States. *Energy Research & Social Science*, *29*, 72–83. https://doi.org/10.1016/j.erss.2017.04.013

Energias de Portugal. (2018). *Windfloat Atlantic*. https://www.edp.com/en/innovation/windfloat

Energy Sector Management Assistance Program. (2019). *Going global: Expanding offshore wind to emerging markets*. https://openknowledge.worldbank.org/bitstream/handle/10986/32801/Going-Global-Expanding-Offshore-Wind-To-Emerging-Markets.pdf?sequence=5&isAllowed=y

Equinor. (n.d.). *The future of offshore wind is afloat*. https://www.equinor.com/en/what-we-do/floating-wind.html

Firestone, J., Hoen, B., Rand, J., Elliott, D., Hübner, G., & Pohl, J. (2018). Reconsidering barriers to wind power projects: Community engagement, developer transparency and place. *Journal of Environmental Policy & Planning*, 20(3), 370–386. https://doi.org/10.1080/1523908X.2017.1418656

Firestone, J., & Kempton, W. (2007). Public opinion about large offshore wind power: Underlying factors. *Energy Policy, 35*(3), 1584–1598. https://doi.org/10.1016/j.enpol.2006.04.010

Firestone, J., Kempton, W., Lilley, M. B., & Samoteskul, K. (2012a). Public acceptance of offshore wind power across regions and through time. *Journal of Environmental Planning and Management, 55*(10), 1369–1386. https://doi.org/10.1080/09640568.2012.682782

Firestone, J., Kempton, W., Lilley, M. B., & Samoteskul, K. (2012b). Public acceptance of offshore wind power: Does perceived fairness of process matter? *Journal of Environmental Planning and Management, 55*(10), 1387–1402. https://doi.org/10.1080/09640568.2012.688658

Fitch-Roy, O. (2016). An offshore wind union? Diversity and convergence in European offshore wind governance. *Climate Policy, 16*(5), 586–605. https://doi.org/10.1080/14693062.2015.1117958

Fornahl, D., Hassink, R., Klaerding, C., Mossig, I., & Schröder, H. (2012). From the old path of shipbuilding onto the new path of offshore wind energy? The case of northern Germany. *European Planning Studies, 20*(5), 835–855. http://hdl.handle.net/10419/120052

Freudenburg, W. R., & Gramling, R. (1993). Socioenvironmental factors and development policy: Understanding opposition and support for offshore oil. *Sociological Forum, 8*(3), 341–364. https://doi.org/10.1007/BF01115049

Freudenburg, W. R., & Gramling, R. (1994). *Oil in troubled waters: Perceptions, politics, and the battle over offshore drilling.* SUNY Press.

Giordono, L. S., Boudet, H. S., Karmazina, A., Taylor, C. L., & Steel, B. S. (2018). Opposition "overblown"? Community response to wind energy siting in the western United States. *Energy Research & Social Science, 43*, 119–131. https://doi.org/10.1016/j.erss.2018.05.016

Global Change Data Lab. (n.d.) *Global direct primary energy consumption.* Our world in data. https://ourworldindata.org/grapher/global-primary-energy?time=earliest.latest

Global Wind Energy Council. (n.d.) *Offshore Wind Power.* https://gwec.net/global-figures/global-offshore/

Haggett, C. (2008). Over the sea and far away? A consideration of the planning, politics and public perception of offshore wind farms. *Journal of Environmental Policy & Planning, 10*(3), 289–306. https://doi.org/10.1080/15239080802242787

Haggett, C. (2011). Understanding public responses to offshore wind power. *Energy Policy, 39*(2), 503–510. https://doi.org/10.1016/j.enpol.2010.10.014

Hall, D. M., & Lazarus, E. D. (2015). Deep waters: Lessons from community meetings about offshore wind resource development in the US. *Marine Policy, 57*, 9–17. https://doi.org/10.1016/j.marpol.2015.03.004

Hazboun, S. O., & Boudet, H. S. (2020). Public preferences in a shifting energy future: Comparing public views of eight energy sources in North America's Pacific Northwest. *Energies, 13*(8), 1940. https://doi.org/10.3390/en13081940

Ho, S. S., Leong, A. D., Looi, J., Chen, L., Pang, N., & Tandoc Jr., E. (2019). Science literacy or value predisposition? A meta-analysis of factors predicting public perceptions of benefits, risks, and acceptance of nuclear energy. *Environmental Communication, 13*(4), 457–471. https://doi.org/10.1080/17524032.2017.1394891

Huckerby, J., Jeffrey, H., de Andres, A., & Finlay, L. (2017). *An international vision for ocean energy.* Ocean Energy Systems. https://www.ocean-energy-systems.org/publications/oes-vision-strategy/

International Energy Agency. (2018). *Offshore Energy Outlook 2018.* https://www.iea.org/reports/offshore-energy-outlook-2018

International Energy Agency. (2019). *Offshore Wind Outlook 2019.* https://www.iea.org/reports/offshore-wind-outlook-2019

International Renewable Energy Agency. (2019). *Future of Wind: Deployment, Investment, Technology, Grid Integration and Socio-Economic Aspects*. https://www.irena.org/-/media/Files/IRENA/Agency/Publication/2019/Oct/IRENA_Future_of_wind_2019.pdf

Jacquet, J. B. (2012). Landowner attitudes toward natural gas and wind farm development in northern Pennsylvania. *Energy Policy*, *50*, 677–688. https://doi.org/10.1016/j.enpol.2012.08.011

Jones, P. J., Lieberknecht, L. M., & Qiu, W. (2016). Marine spatial planning in reality: Introduction to case studies and discussion of findings. *Marine Policy*, *71*, 256–264. https://doi.org/10.1016/j.marpol.2016.04.026

Kerr, S., Watts, L., Colton, J., Conway, F., Hull, A., Johnson, K., Jude, S., Kannen, A., MacDougall, S., McLachlan, C., Potts, T., & Vegunst, J. (2014). Establishing an agenda for social studies research in marine renewable energy. *Energy Policy*, *67*, 694–702. https://doi.org/10.1016/j.enpol.2013.11.063

Konisky, D. M., & Woods, N. D. (2018). Environmental federalism and the Trump presidency: A preliminary assessment. *Publius: The Journal of Federalism*, *48*(3), 345–371. https://doi.org/10.1093/publius/pjy009

Lee, A., Zinaman, O., & Logan, J. (2012). *Opportunities for synergy between natural gas and renewable energy in the electric power and transportation sectors* (Report No. NREL/TP-6A50–56324). National Renewable Energy Laboratory. https://www.nrel.gov/docs/fy13osti/56324.pdf

McLachlan, C. (2009). 'You don't do a chemistry experiment in your best china': Symbolic interpretations of place and technology in a wave energy case. *Energy Policy*, *37*(12), 5342–5350. https://doi.org/10.1016/j.enpol.2009.07.057

Molotch, H. (1970). Oil in Santa Barbara and power in America. *Sociological Inquiry*, *40*(1), 131–144. https://doi.org/10.1111/j.1475-682X.1970.tb00990.x

Musial, W., Beiter, P., Spitsen, P., Nunemaker, J., Gevorgian, V., Cooperman, A., Hammond, R., & Shields, M. (2020). *2019 offshore wind technology data update* (Report No. NREL/ TP-5000–77411). National Renewable Energy Laboratory. https://www.nrel.gov/docs/fy21osti/77411.pdf

National Oceanic and Atmospheric Administration. (2017, April 5). *Largest oil spills affecting US waters since 1969*. https://response.restoration.noaa.gov/oil-and-chemical-spills/oil-spills/largest-oil-spills-affecting-us-waters-1969.html

Neill, S. P, Angeloudis, A., Robins, P. E., Walkington, I., Ward, S. L., Masters, I., Lewis, M. J., Piano, M., Avdis, A., Piggott, M. D., Aggidis, G., Evans, P., Adcock, T. A. A., Židonis, A., Ahmadian, R., & Falconer, R. (2018). Tidal range energy resource and optimization – Past perspectives and future challenges. *Renewable Energy*, *127*, 763–778. https://doi.org/10.1016/j.renene.2018.05.007

O'Hagan, A. M. (2020). Marine spatial planning and marine renewable energy. In A.E. Copping & L.G. Hemery (Eds.), *OES-Environmental 2020 state of the science report: Environmental effects of marine renewable energy development around the world* (pp. 214–241). Ocean Energy Systems. https://doi.org/10.2172/1633204

Olson-Hazboun, S. K., Howe, P. D., & Leiserowitz, A. (2018). The influence of extractive activities on public support for renewable energy policy. *Energy Policy*, *123*, 117–126. https://doi.org/10.1016/j.enpol.2018.08.044

Rand, J., & Hoen, B. (2017). Thirty years of North American wind energy acceptance research: What have we learned? *Energy Research & Social Science*, *29*, 135–148. https://doi.org/10.1016/j.erss.2017.05.019

Renn, O., & Marshall, J. P. (2016). Coal, nuclear and renewable energy policies in Germany: From the 1950s to the "Energiewende". *Energy Policy*, *99*, 224–232. https://doi.org/10.1016/j.enpol.2016.05.004

Russell, A., Firestone, J., Bidwell, D., & Gardner, M. (2020). Place meaning and consistency with offshore wind: An island and coastal tale. *Renewable and Sustainable Energy Reviews*, *132*, 110044. https://doi.org/10.1016/j.rser.2020.110044

Ryan, K., Bates, A., Gopnik, M., Danylchuk, A., & Jordaan, A. (2019). Stakeholder perspectives on the value of marine spatial planning towards advancing offshore wind in the US. *Coastal Management*, *47*(3), 269–291. https://doi.org/10.1080/08920753.2019.1596675

Schively, C. (2007). Understanding the NIMBY and LULU phenomena: Reassessing our knowledge base and informing future research. *Journal of Planning Literature*, *21*(3), 255–266. https://doi.org/10.1177/0885412206295845

Short, A. (2020). *Industrial policy support for tidal technology in the UK: An international comparison*. University of Exeter.

Simas, T., Muñoz-Arjona, E., Huertas-Olivares, C., de Groot, J., Stokes, C., Bailey, I., Magagna, D., Conley, D., Greaves, D., Marina, D., Torre-Enciso, Y., Sundberg, J., O'Hagan, A., & Holmes, B. (2012, October 17). *Understanding the role of stakeholders in the wave energy consenting process: Engagement and sensitivities* [Paper presentation]. International Conference on Ocean Energy, Dublin, Ireland. https://tethys.pnnl.gov/publications/understanding-role-stakeholders-wave-energy-consenting-process-engagement

Slovic, P. (1987). Perception of risk. *Science*, *236*(4799), 280–285. https://doi.org/10.1126/science.3563507

Sokoloski, R., Markowitz, E. M., & Bidwell, D. (2018). Public estimates of support for offshore wind energy: False consensus, pluralistic ignorance, and partisan effects. *Energy Policy*, *112*, 45–55. https://doi.org/10.1016/j.enpol.2017.10.005

Spence, A., Demski, C., Butler, C., Parkhill, K., & Pidgeon, N. (2015). Public perceptions of demand-side management and a smarter energy future. *Nature Climate Change*, 5(6), 550–554. https://doi.org/10.1038/nclimate2610

Stoutenborough, J. W., Vedlitz, A., & Liu, X. (2015). The influence of specific risk perceptions on public policy support: An examination of energy policy. *The ANNALS of the American Academy of Political and Social Science*, 658(1), 102–120. https://doi.org/10.1177/0002716214556472

Thomas, M., Pidgeon, N., Evensen, D., Partridge, T., Hasell, A., Enders, C., Herr Harthorn, B., & Bradshaw, M. (2017). Public perceptions of hydraulic fracturing for shale gas and oil in the United States and Canada. *Wiley Interdisciplinary Reviews: Climate Change*, 8(3), e450. https://doi.org/10.1002/wcc.450

United Nations Environment Programme. (2018). *Assessing environmental impacts – A global review of legislation*. UNEP.

US Energy Information Administration. (2016, October 25). *Offshore production nearly 30% of global crude oil output in 2015*. https://www.eia.gov/todayinenergy/detail.php?id=28492#

Vasi, I. B., Walker, E. T., Johnson, J. S., & Tan, H. F. (2015). "No fracking way!" Documentary film, discursive opportunity, and local opposition against hydraulic fracturing in the United States, 2010 to 2013. *American Sociological Review*, 80(5), 934–959. https://doi.org/10.1177/0003122415598534

Walker, G. (1995). Renewable energy and the public. *Land Use Policy*, *12*(1), 49–59. https://doi.org/10.1016/0264-8377(95)90074-C

Wiersma, B., & Devine-Wright, P. (2014). Public engagement with offshore renewable energy: A critical review. *Wiley Interdisciplinary Reviews: Climate Change*, 5(4), 493–507. https://doi.org/10.1002/wcc.282

Wüstenhagen, R., Wolsink, M., & Bürer, M. J. (2007). Social acceptance of renewable energy innovation: An introduction to the concept. *Energy policy*, 35(5), 2683–2691. https://doi.org/10.1016/j.enpol.2006.12.001

9 Marine Areas beyond National Jurisdiction

H. Harden-Davies

Introduction

More than two-thirds of the ocean lies in marine areas beyond national jurisdiction (ABNJ). These deep and remote spaces of global ocean commons are shared by all but owned by no one. They are also connected to humans, directly or indirectly, in several ways. Marine environments in ABNJ span the sea surface of the open ocean, through the water column to the deep seabed. They include diverse habitats (such as the ocean twilight zone, deep-sea trenches, abyssal plains, hydrothermal vents, and coral reefs) that support rich assemblages of marine life, as yet unexplored by science (Ramirez-Llodra et al., 2010, 2011). Myriad ecosystems of ABNJ underpin vital planetary support functions, such as climate regulation and nutrient recycling, and are connected to coastal areas through oceanic currents (Rogers et al., 2014) and the movements of marine species (Popova et al., 2019). Several marine species traverse ocean ABNJ and coastal areas, representing a source of scientific, economic, and cultural importance for humans (Mulalap et al., 2020; Popova et al., 2019; Vierros et al., 2020). Cultural activities or artifacts of human navigation in marine ABNJ underpin diverse cultural connections between humans worldwide and this vast open space.

Goods are traded and transported via shipping on the sea surface (11 billion tons of maritime trade in 2018) (United Nations Conference on Trade and Development, 2019), while information flows through submarine cables that zigzag the seafloor. Seabed minerals, including polymetallic nodules, cobalt crust, and seafloor metallic sulfides, are being explored (Levin et al., 2020a). Fisheries are a source of revenue for many nation-states (hereafter "States") (Bell et al., 2021; Haas et al., 2021), and genetic resources are a topic of growing interest for scientific research and development (Blasiak et al., 2020; Rogers et al., 2021). Yet questions remain regarding the sustainability of these activities, who benefits from these resources, and how the responsibilities associated with their use are implemented in practice. These questions are especially pertinent at a time of increasing interest in the blue economy worldwide and growing scrutiny on the equitability of resource use (Cisneros-Montemayor et al., 2021).

Marine environments in ABNJ, and the biodiversity that they support, are threatened by human activities that cause them harm. Examples include

DOI: 10.4324/9781003058151-11

destructive, unsustainable, and illegal fisheries practices, as well as pollution and dumping or discard from vessels (Wright et al., 2018). Climate change–related impacts of ocean acidification, warming, and deoxygenation represent confounding impacts that contribute to the uncertain future of marine biodiversity in ABNJ (Levin & Le Bris, 2015). The exploration for seabed minerals now underway raises concerns of the environmental impacts of seabed mining (Levin et al., 2020a, 2020b). Potential future activities, such as offshore aquaculture, geoengineering, or the development of "floating cities", represent further concerns of potential environmental damage. Addressing the cumulative impacts of these activities remains a major challenge for ocean governance (Hassanali, 2021).

The international ocean governance framework divides marine ABNJ into two maritime zones: (i) the water column, or "high seas", and (ii) the international seabed or "the Area". The water column and seabed are connected ecologically and physically, yet there are differences in the legal frameworks. For example, high seas governance is strongly influenced by the principle of freedom of the high seas, whereas the international seabed Area and its mineral resources are the common heritage of mankind. These principles have different historical and ideological roots: the principle of freedom of the high seas emerged hundreds of years ago and is linked to the primacy of mercantile and military interests, while the common heritage of mankind principle emerged during the mid-1960s in support of sharing the benefits from the global ocean commons (Jaeckel, 2020; Vadrot et al., 2022).[1] The practical application of these principles, and the drawbacks of having different legal regimes for an interconnected ocean environment, are ongoing topics of lively international debate.

The limitations of the existing international ocean governance framework, in relation to the conservation and sustainable use of marine biodiversity in ABNJ, are well known (Ardron et al., 2014; Warner, 2014; Wright et al., 2018). There are several sectoral and regional organizations, yet coordination remains challenging. Gaps also exist for marine protected areas, environmental impact assessments, and marine genetic resources in ABNJ, as well as capacity-building and the transfer of marine technology (Humphries & Harden-Davies, 2020). Consequently, international concern regarding marine biodiversity of ABNJ has prompted the United Nations to develop a new treaty: "a new international legally binding instrument for the conservation and sustainable use of marine biological diversity of areas beyond national jurisdiction under the United Nations Convention on the Law of the Sea (UNCLOS)" (United Nations [UN], 2019).

Equity considerations are increasingly prominent in the literature and international debates concerning marine ABNJ (Österblom et al., 2020). All States have the right to access marine ABNJ, and the resources of marine ABNJ are, in principle, open to and shared by all – conditional on responsibilities laid down by UNCLOS (Section 2). Yet, in practice, few have the capacity to access marine ABNJ and fulfill associated responsibilities to protect and preserve the marine environment. For example, the high technological requirements to explore ocean ABNJ can be met by just a few States (Blasiak et al., 2020; Rogers et al., 2021). Consequently, the benefits and burdens are not shared by all. This conundrum raises questions about how the benefits and burdens of conserving and sustainably using marine resources from ABNJ are shared – and how the balance between rights and responsibilities in the global ocean commons plays out in practice.

Understanding these issues requires expertise from several disciplinary areas, including, for example, the natural sciences, to understand the marine environments in ABNJ and inform biodiversity conservation and sustainable use measures; and the social sciences and humanities, to understand the legal frameworks, the historical development, the power dynamics at play, as well as the complex social and cultural dimensions that underlie human connections with ABNJ. It is not possible to provide a comprehensive discussion of all such perspectives in this chapter. Rather, the aim is to provide a high-level introduction to marine ABNJ in a way that is accessible to diverse disciplinary backgrounds and stimulates further study into the human dimensions of this vast global ocean commons.

This chapter explores the relationship between humans and marine ABNJ, introducing the governance of marine ABNJ, outlining perspectives on human connections with marine ABNJ, and introducing the development of the new treaty for marine biodiversity beyond national jurisdiction (BBNJ agreement). Finally, the conclusion offers reflections on emerging and future issues for the governance of marine ABNJ and identifies research gaps relating to human dimensions of marine ABNJ.

Marine Areas beyond National Jurisdiction

Marine ABNJ cover almost half of the Earth's surface. Marine ABNJ do not include other areas beyond national jurisdiction, such as outer space and Antarctica, which have separate international legal frameworks (Berkman, 2010). "High seas" refers to the water column beyond national jurisdiction, measured as 200 nautical miles from a State's territorial sea baseline, i.e., beyond the exclusive economic zone.[2] "The Area" refers to the seabed and subsoil beyond the limits of national jurisdiction,[3] i.e., beyond the continental shelf (normally 200 nautical miles from the territorial sea baseline). International agreement on the establishment of these maritime zones required decades of negotiation, following centuries of differing views on the limits of State control over the ocean. Understanding the historical development provides important context to contemporary challenges facing the governance of marine ABNJ. In this section, the development of the international regimes for the high seas and the Area is briefly outlined, before turning to examine key elements of the governance framework for marine ABNJ today.

Deliberations on the limits of State control of marine spaces and the level of freedom in shared ocean spaces can be traced back several hundred years and are anchored in the context of maritime powers of the 15th, 16th, and 17th centuries exerting influence and promoting mercantile and military interests (Treves, 2015a, 2015b). Some of the earliest triggers for the scholarly debate concerning the freedom of the high seas included a *Papal bull Inter Caetera* issued in 1492 by Pope Alexander VI, and a treaty concluded at Tordesillas the following year that divided the Atlantic between Portugal (to the east) and Spain (to the west). The opposition to this idea stimulated foundational scholarship on the idea of freedom of the seas, including *Mare Liberum* ("free seas"), published in 1609 by the Dutch author Hugo Grotius, arguing the notion of "freedom of the seas" on the basis that the sea could not be occupied or defended by humans; and the conflicting *Mare Clausum* ("closed seas"), published in 1635 by English author John Selden, disputing this argument on the basis that the sea could be controlled by States as illustrated by navigation and fishing activities (Treves, 2015a). A pragmatic compromise known as the "cannon shot rule" became commonplace throughout the 18th century, predicated on the understanding that, regardless of whether marine areas could be possessed, a coastal State could nevertheless exert power and claim possession as far as a cannon would carry (the distance depended on available technology).

Consequently, the idea of dividing the seas into State-controlled "territorial seas" and shared "high seas" became commonplace throughout the 19th and early 20th centuries. At the same time, the notion of freedom of the high seas, particularly for navigation, was reinforced by State practice in relation to shipping and trade. However, recognized limits of the territorial sea varied from 3 to 200 nautical miles. Several attempts were made to formalize the limits of national jurisdiction throughout the 20th century, including during the first (24 February–27 April 1958), second (16 March–26 April 1960), and third (1973–1982) sessions of the United Nations Conference on the Law of the Sea, during which time UNCLOS was negotiated. Yet it was not until the adoption of UNCLOS in 1982, following almost three decades of negotiation, that international agreement was reached on the limits of State jurisdiction, and the high seas and the Area were formally designated.

During the UNCLOS negotiations, the principle of "common heritage of mankind" emerged. It is widely credited to Maltese Ambassador Arvid Pardo tabling the idea during a speech at the United Nations in 1967. In many ways the antithesis of "freedom of the high seas", the common heritage of mankind asserted that all States should benefit from the mineral resources of the seabed (which were expected to generate wealth); activities in the international seabed Area should be for peaceful purposes; and the interests of future generations should be considered (Jaeckel, 2020).

UNCLOS also formally established freedoms of the high seas for navigation, overflight, laying submarine cables, constructing artificial islands, fishing, and scientific research.[4] These freedoms are not absolute but conditional on responsibilities (Freestone, 2009), including to protect and preserve the marine environment and conserve and manage the living resources of the high seas.[5]

In considering the factors that played into the adoption of UNCLOS – including the eventual agreement on the extent of State jurisdiction and the decision to designate the seabed and the water column as separate maritime zones – the geopolitical context was crucial (Vadrot et al., 2022). The UNCLOS negotiations occurred in the aftermath of the Second World War, at a time of significant geopolitical change with several newly independent States seeking to establish control over their territories and resources. The emergence of the common heritage of mankind principle is an example of a change in approach for the global ocean; it is emblematic of the resistance to the prevailing practice of behaving solely in national interests, and it favors the prioritization of equity considerations for both current and future generations. In practice, however, the evolution and implementation of the common heritage of mankind principle remains a lively topic of debate (Jaeckel, 2020; Vadrot et al., 2022).

Although UNCLOS is regarded as the "constitution" for the ocean, the development of the international framework for the governance of ocean areas beyond national jurisdiction did not end with the adoption of UNCLOS in 1982. Already, two implementing agreements have been developed under UNCLOS – both of which are directly relevant to marine ABNJ. The 1994 Agreement relating to the Implementation of Part XI of UNCLOS focused on governance arrangements for mineral resources in the Area, including softening the provisions concerning associated marine technology transfer. It was only with this agreement that several developed States signed UNCLOS, making it possible for UNCLOS to enter into force in 1994. The 1995 UNCLOS Implementing Agreement relating to the Conservation and Management of Straddling Fish Stocks and Highly Migratory Fish Stocks concerned fisheries and became foundational to high-seas fisheries management (Gjerde et al., 2019). A third agreement (the BBNJ agreement) is now being negotiated to further implement UNCLOS obligations for marine ABNJ, including but not limited to the protection of the marine environment and the transfer of marine technology, as well as to fill gaps left by UNCLOS, including for marine genetic resources (Section 4).

In addition to these agreements, there are several other legal instruments, frameworks, and organizations that play a role in managing activities in ABNJ. Some are sectoral and global, others are regional; some have regulatory power, others do not (Wright et al., 2018). It is not within the scope of this chapter to explore them all, but some illustrative examples are provided in Table 9.1.

Yet there are persisting questions surrounding the sustainability of human activities in ABNJ, and who benefits from them. Environmental impacts of shipping include greenhouse gas emissions, air and ocean noise pollution, invasive species introduced by ship ballast water, and chemical pollution. There are concerns

Table 9.1 Examples of international organizations with authority in ABNJ

Sector	*Organization*
Shipping	International Maritime Organization (IMO)
Seabed minerals	International Seabed Authority (ISA)
Marine scientific research	Intergovernmental Oceanographic Commission of UNESCO (IOC)
Fisheries	UN Food and Agriculture Organization; Regional Fisheries Management Organizations
Cetaceans	International Whaling Commission
Environmental Protection	UN Regional Seas Programs

that seabed mining activities could damage the marine environment, including through the impact of sediment plumes in the water column, and persisting questions relating to how benefits are shared and with whom (Jaeckel, 2020; Levin et al., 2020a). Overfishing in the high seas and the use of destructive fishing practices has caused declines in the diversity and abundance of marine species (Dunn et al., 2018). Transparency issues remain a concern in high-seas fishing and supply chains, heightened by growing interest in developing new fisheries in the deep ocean (Wright et al., 2020) and persisting problems of illegal, unreported, and unregulated fishing (Carmine et al., 2020).

This brief summary is intended to demonstrate that the governance of marine ABNJ has a long and complex history that is intricately linked to the role of the ocean in trade, colonial history, and conflict. UNCLOS was not envisioned to be static, nor to operate in isolation, but rather to provide a framework for implementation. The balance between national and common interests in marine ABNJ remains a topic of discussion internationally and can be seen in several contemporary ocean law and policy developments, including a new BBNJ agreement (Section 4).

Human Connections with Marine ABNJ

The high seas are often referred to as "out of sight, out of mind". Some people might see the high seas from an airplane window, catching a glimpse of the seemingly endless ocean, while others might experience the high seas from the deck of a ship or sailboat. Far fewer will travel beneath the waves into the depths of the high seas. Yet despite the remote nature of the high seas, humans are connected to marine ABNJ in several ways. The following discussion offers illustrative examples of the various cultural, ecological, economic, social, and spiritual connections between humans and the high seas.

Ecological connections are based on the movements of oceans and oceanic species. Ocean currents act as great conveyor belts, moving nutrients and oxygen and distributing temperature around the planet. These currents help to regulate climate and sequester carbon, and they form the basis of complex webs of ocean life that underpin ecosystem resilience and services (Popova et al., 2019). Species of marine organisms move and migrate around the ocean, connecting the high seas with coastal areas and their human inhabitants. Several marine species that spend part of their life cycles in the high seas are important for human societies. An estimated 900 highly migratory species of fish, turtles, cetaceans, and seabirds are of cultural, social, spiritual, and/or economic significance to local communities and Indigenous Peoples (Mulalap et al., 2020; Vierros et al., 2020). Some species are important sources of revenue for local communities reliant on wildlife tourism (Spiteri et al., 2021).

Partly due to recognition of the high ecological importance of marine ABNJ, there are growing calls to increase protection for high seas ecosystems (Gjerde et al., 2016). Badman et al. (2016) identify illustrative examples of places of possible "Outstanding Universal Value" in marine ABNJ, such as: The Lost City

Hydrothermal Field on the Mid-Atlantic Ridge, in which hydrothermal vents 800 m below the surface rise up from the seabed to heights of 60 m; the Sargasso Sea in the western Atlantic, the only pelagic ecosystem bordered by floating algae; and the Costa Rica Thermal Dome, a wind-driven upwelling system in the eastern tropical Pacific Ocean that supports high biological productivity as well as spawning, feeding, and migration of endangered and commercially valuable marine species.

There are also several cultural connections between humans and marine ABNJ. Cultural heritage can be considered tangible (e.g., objects, sunken vessels, human remains)[6] and/or intangible (e.g., customs, traditions, art).[7] Historical voyages across the open ocean give the high seas a place in many human cultures, inspiring art, literature, and music. Some aspects of human cultural heritage are celebrated and kept alive through contemporary marine activities, such as offshore sailing and traditional navigation. Others are commemorated, as in memorializing the loss of life at sea. Recognizing and respecting the significance of such human connections with marine ABNJ is emerging as an issue within ocean governance. For example, the strong cultural connections of Pacific Island peoples to the open ocean ABNJ, anchored in ancestral voyaging history and passed down through traditional knowledge practices, have been highlighted at the United Nations in the context of developing the BBNJ agreement (Mulalap et al., 2020; Vierros et al., 2020).

Preserving cultural heritage through the protection of archaeologically and historically significant objects at sea, including in the Area, is part of the framework established by UNCLOS for the protection and preservation of the marine environment and is of recognized importance for both present and future generations (UNCLOS Articles 303(1) and 149). One example is the shipwreck of the *Titanic*, which has been afforded protections as a site of underwater cultural heritage. Yet, there are gaps. For example, a proposal has been made by Turner et al. (2020) to memorialize the victims of the slave trade across the Atlantic "Middle Passage" and to ensure that activities, such as seabed mining, are sensitive to the historic and contemporary cultural significance of this area as the final resting place of approximately 1.8 million victims of slavery (Turner et al., 2020). They propose such measures as memorial ribbons on UN maps of the region, to show respect and recognize cultural significance. Improving protection for intangible and tangible aspects of cultural heritage, as well as preserving natural ecosystems in marine ABNJ, is an area for further work (Badman et al., 2016).

In addition to the cultural, ecological, spiritual, and social connections illustrated above, there are also economic connections through the exploitation of living resources (such as fisheries) and non-living ones (such as minerals). Although ocean ABNJ are, in principle, open to all, in practice just a few powerful actors have the capacity to access ABNJ and exploit the resources. Few States have the scientific capacity to access the high seas or utilize marine genetic resources for such activities as developing or commercializing new biotechnologies or pharmaceutical products (Rabone et al., 2019; Rogers et al., 2021). Similar concerns surround the equitability of benefit-sharing from seabed mining (Jaeckel,

2020). Furthermore, there are questions pertaining to who benefits from high seas fishing, partly because it is heavily reliant on subsidies, and also because high seas primarily supply high-end markets (Carmine et al., 2020), rather than meet basic nutritional needs of coastal communities or contribute to global food security (Schiller et al., 2018). Worse, there are serious concerns of human rights abuses at sea, including human trafficking and slavery aboard fishing vessels (Tickler et al., 2018). In-depth examination of these issues is beyond the scope of this chapter.

Given that inequitable exploitation of marine resources exacerbates inequalities through either the absence of benefits flowing to marginalized groups, or unfair shouldering of the burdens of resource use (Spiteri et al., 2021), the equitable sharing of benefits from the high seas is a topic of growing importance. The complex interconnections between marine ABNJ and the lives, livelihoods, and wellbeing of humans, especially ocean-dependent communities and marginalized groups, is an area in need of further research.

A New Treaty for Biodiversity beyond National Jurisdiction

The development of a new international, legally binding instrument for the conservation and sustainable use of marine biodiversity of ABNJ under UNCLOS (BBNJ agreement) has emerged within the international community as a response to the challenges facing the high seas. While the BBNJ agreement is not the only way to address these challenges, it is widely expected to fill gaps in the governance framework for marine ABNJ and strengthen the implementation of UNCLOS – particularly in relation to the protection and preservation of the marine environment in marine ABNJ, as well as capacity building, and technology transfer. According to The Preamble of President's draft text of the BBNJ agreement (UN, 2019), States:

- Recognize the "need for the comprehensive global regime to better address the conservation and sustainable use of BBNJ".
- Desire to promote sustainable development.
- Desire to act as stewards of the ocean in ABNJ on behalf of present and future generations.

The development of the BBNJ agreement has taken almost two decades thus far and is still in progress at the time of writing (which was February 2021). The long timeframe reflects the complexity and challenges posed by these international negotiations (Humphries & Harden-Davies, 2020; Tiller et al., 2019; Vadrot et al., 2022). The following sections briefly summarize the historical development of the BBNJ agreement before introducing the key issues that the negotiations are addressing.

Brief History of BBNJ Process

In 2004 the United Nations General Assembly (UNGA) established the Ad Hoc Open-ended Informal Working Group to study issues related to the conservation and sustainable use of marine biological diversity beyond areas of national jurisdiction ("BBNJ Working Group"). Nine meetings of the BBNJ Working Group were held between 2006 and 2015, with additional intersessional workshops. Early discussions focused on identifying gaps in the international ocean governance framework, including marine genetic resources, marine protected areas, governance principles such as the ecosystem and precautionary approach, and weaknesses in cooperation and coordination between existing international organizations and frameworks (Wright et al., 2018). The 2011 meeting was a breakthrough. A compromise position was reached, which subsequently led to the establishment of an intergovernmental negotiating process for a "Package Deal" of elements to address the conservation and sustainable use of marine biodiversity in ABNJ, "in particular, together and as a whole":

- Marine genetic resources, including questions on the sharing of benefits
- Area-based management tools, including marine protected areas
- Environmental impact assessments
- Capacity building and the transfer of marine technology

Momentum to establish a negotiation process for a new treaty began to build during the meetings and intersessional workshops of 2013 and 2014.

In 2015, the Working Group recommended the establishment of a Preparatory Committee (PrepCom) to negotiate an international legally binding instrument. Reaching this decision required overcoming obstacles, such as disagreement on the mandate of the PrepCom and opposition to a legally binding instrument (Wright et al., 2018). The UNGA adopted Resolution 69/292[8] establishing the Preparatory Committee (PrepCom) that held four sessions in 2016 and 2017. The outcome of these sessions was substantive recommendations on the elements of the Draft Text of a BBNJ agreement. In 2017 the UNGA adopted Resolution 72/249 convening an intergovernmental conference "to elaborate the text of an [ILBI] on the conservation and sustainable use of marine biological diversity of [ABNJ], with a view to developing the instrument as soon as possible".[9]

Three sessions of the intergovernmental conference were held in 2018 and 2019,[10] for formal negotiations to take place (Tiller et al., 2019). The fourth (and supposedly final) session of the intergovernmental conference was scheduled for March 2020 but was postponed to the earliest possible date to be decided by UNGA, due to the COVID-19 pandemic.[11] At the time of writing, the final session was scheduled for 2022.[12] Intersessional work occurred between the third and fourth sessions of the intergovernmental conference, which included online discussion fora where delegations exchanged views.

As the negotiations are taking place within the arena of the United Nations General Assembly, participation is open to all Member States of the United Nations, members of the specialized agencies, and parties to UNCLOS, in accordance with UNGA Resolution 72/249. The highly technical nature of the issues involved in the negotiations is particularly challenging for smaller delegations. State groupings are one of the ways that delegations who are like-minded on certain issues can work together. (Examples include the African Group, the Alliance of Small Island States, and the Pacific Small Island Developing States.) Some observer organizations are also permitted to participate. The widespread interest in the BBNJ negotiations is illustrated by the emergence of publications on this topic from natural scientists, lawyers, international relations scholars, and political scientists (Tessnow-von Wysocki & Vadrot, 2020).

Overview of the BBNJ Agreement

The BBNJ agreement covers four key elements in addition to some cross-cutting issues. The revised draft text of the agreement (UN, 2019) contains 70 draft articles and two Annexes. The following discussion briefly introduces each of the elements in the BBNJ agreement and key issues in the negotiations.

Marine Genetic Resources, Including Questions on Benefit Sharing

Marine genetic resources have been one of the more contentious issues of the package of elements in the BBNJ agreement (Harden-Davies & Gjerde, 2019). Predicated on the understanding that the genetic libraries of marine organisms could be utilized in scientific research and development to yield new knowledge, products, and processes, concern over the equitable sharing of benefits from marine genetic resources of ABNJ has grown. Because few countries possess the scientific and technological capacity to access the high seas and sample marine life, much of the debate has focused on increasing scientific and technological capacity in developing nations, as well as accessing the outcomes of scientific research. Equally prominent in the debate have been questions of sharing in the outcomes of commercialization, such as financial revenues and other benefits known broadly as "monetary" benefits. Yet the debate has encountered challenges in defining marine genetic resources and the benefits to be derived, as well as the applicable principles and the mechanisms through which benefits might be shared. Against this backdrop, "marine genetic resources, including questions on the sharing of benefits" have been a difficult topic in the BBNJ negotiations.

The revised draft text of the BBNJ agreement (UN, 2019) addresses this issue in Part II. The objectives include contributing to conservation and sustainable use of biodiversity and promoting both scientific research and innovation, as well as capacity building and technology transfer. This Part includes draft provisions relating to scope of application, activities involving marine genetic resources, traditional knowledge, fair and equitable benefit sharing, intellectual property rights, and monitoring.

Area-Based Management Tools, Including Marine Protected Areas

Area-based management tools are important measures for the conservation and sustainable use of marine biodiversity. Processes already exist to establish sectoral measures, such as fisheries closures, sites of environmental interest in the international seabed Area, and Particularly Sensitive Sea Areas in the context of shipping. Yet at present there is no mechanism to establish marine protected areas in the high seas. The revised draft text of the BBNJ agreement (UN, 2019) addresses this issue in Part III. The objectives broadly include enhancing cooperation and coordination in the use of area-based management tools and establishing a comprehensive system of area-based management tools, including marine protected areas, among others. This Part currently includes draft provisions relating to international cooperation and coordination, identification of areas requiring protection, procedural aspects of proposals for the establishment of area-based management tools, consultation on and assessment of proposals, decision-making, implementation, and monitoring and review.

Environmental Impact Assessments

Although UNCLOS (Article 206) establishes an obligation for States to assess the effects of activities that may cause significant and harmful changes to the marine environment, the implementation of this obligation faces challenges. Gaps include absences of coherent principles and approaches to undertake such assessments, mechanisms to consider cumulative impacts, and frameworks to monitor compliance. In the revised draft text of the BBNJ agreement (UN, 2019) this issue is addressed in Part IV and includes draft provisions relating to the obligation to conduct environmental impact assessments: the relationship with other relevant legal instruments, frameworks and bodies; thresholds and criteria for environmental impact assessments; cumulative impacts; transboundary impacts; areas identified as ecologically or biologically significant or vulnerable; strategic environmental assessments; lists of activities requiring (or not requiring) environmental impact assessments; screening of activities; the scope of environmental impact assessments; the assessment and evaluation of environmental impacts; the mitigation, prevention, and management of potential adverse effects; public notification and consultation; preparation and content of environmental impact assessment reports; publication, consideration, and review of assessment reports; and decision-making, monitoring, reporting, and review.

Capacity Building and Technology Transfer

Capacity building and the transfer of marine technology are important to ensure that all States are able to fully participate in the implementation of the BBNJ agreement – to realize their rights, as well as fulfill their responsibilities under the agreement (Vierros & Harden-Davies, 2020). This could include scientific capacity to investigate and monitor BBNJ, to benefit from marine genetic resources, to designate area-based management tools, or to undertake environmental impact assessments. The concept also extends to technologies to monitor the human activities in marine ABNJ, including the monitoring, control, and surveillance of marine protected areas. It could also encompass legal capacity for developing national policy, as well as legal and regulatory frameworks for the implementation of the BBNJ agreement. Part V of the revised draft text of the BBNJ agreement (UN, 2019) addresses this issue and includes provisions related to cooperation, modalities, types of capacity building, monitoring, and review.

Conclusion

This chapter provides a general introduction to marine ABNJ and outlines the key issues involved in international negotiations for marine biodiversity. While impossible to examine in detail the governance of specific resources, such as fisheries or mineral resources in marine ABNJ (a helpful introduction to these topics can be found in Wright et al., 2018), it has sought to highlight that humans are connected to marine ABNJ in several ways. Marine ABNJ include biologically diverse ecosystems that are important for humanity as sources of services, resources, knowledge, and cultural heritage. The limitations of the existing governance framework to address the interconnected nature of marine ABNJ and to conserve and sustainably use marine biodiversity have prompted the UN to develop a new treaty, the BBNJ agreement.

The development of the BBNJ agreement is the latest in a long succession of international deliberations concerning marine ABNJ and the balance between rights and responsibilities of States in the global ocean. The establishment of UNCLOS spanned almost three decades and was partially influenced by one principle that had existed for hundreds of years (the freedom of the high seas) and one that emerged in the mid-1960s (the common heritage of mankind). The historical context behind these principles is an important foundation to understand their role in ocean governance and why, for example, the idea of common heritage is considered applicable to the exploration of mineral resources of marine ABNJ, but not to other resources, such as fisheries. That UNCLOS did not enter into force until the adoption of the 1994 Implementing Agreement is illustrative of the pivotal role of marine ABNJ in international ocean governance. Yet even now, questions of equity and fairness persist regarding the benefits and burdens of marine ABNJ. Although all States, in principle, have the right to access and use the resources of ABNJ, in practice, not all States have the capacity to do so.

This chapter suggests that several perspectives are useful to fully understand the governance of marine ABNJ, including the physical and ecological characteristics of ocean ecosystems; the geopolitical, economic, and philosophical issues that influence international law and policy over time; and the cultural and sociological factors that shape human connections with the ocean. A rich literature already exists on marine ABNJ, yet much remains to be explored in a variety of research disciplines in ocean studies. Areas that would benefit from further interdisciplinary research include social and cultural values of marine ABNJ; intangible and tangible aspects of cultural heritage in ABNJ; balancing rights and responsibilities in marine ABNJ; and voice and representation in decision-making, including the interests of current and future generations.

Questions for Reflection

1. In what ways are humans connected with marine areas beyond national jurisdiction?

2. What are the challenges facing marine areas beyond national jurisdiction?
3. How might a new international legal instrument address the challenges facing marine areas beyond national jurisdiction?
4. To what extent are the benefits from marine areas beyond national jurisdiction shared?
5. What is the relationship between the future BBNJ international agreement and the UN Convention on the Law of the Sea?
6. What is the applicability to the ABNJ of the legal regimes of "freedom of the seas" and "common heritage of mankind"?

Notes

1 *UNCLOS* Article 136. *UNCLOS* Article 1(3) "activities in the Area" means all activities of exploration for, and exploitation of, the resources of the Area; Article 133(a) "resources" means all solid, liquid, or gaseous mineral resources *in situ* in the Area at or beneath the seabed, including polymetallic nodules; Article 133(b). Resources, when recovered from the Area, are referred to as "minerals".
2 *UNCLOS* Articles 57 and 87; *Convention on the High Seas* opened for signature 29 April 1958, 450 UNTS 11 (entered into force 30 September 1962).
3 *UNCLOS* Article 1(1).
4 UNCLOS Article 87.
5 UNCLOS Articles 116–119, 192.
6 See 1972 World Heritage Convention, and 2001 UNESCO Convention. Underwater cultural heritage is defined in the *2001 United Nations Educational, Scientific, and Cultural Organization (UNESCO) Convention on the Protection of the Underwater Cultural Heritage* as "all traces of human existence having a cultural, historical or archaeological character which have been partially or totally under water, periodically or continuously, for at least 100 years". This includes "(i) sites, structures, buildings, artefacts and human remains, together with their archaeological and natural context; (ii) vessels, aircraft, other vehicles or any part thereof, their cargo or other contents, together with their archaeological and natural context; and (iii) objects of prehistoric character" (Article 1(a)).
7 See, for example, the 2003 UNESCO Convention for the Safeguarding of the Intangible Cultural Heritage.
8 UNGA Resolution 69/292 of 19 June 2015 on Development of an international legally binding instrument under the United Nations Convention on the Law of the Sea on the conservation and sustainable use of marine biological diversity of areas beyond national jurisdiction, A/RES/69/292.
9 UNGA Resolution 72/249, International legally binding instrument under the United Nations Convention on the Law of the Sea on the conservation and sustainable use of marine biological diversity of areas beyond national jurisdiction.
10 The first session was convened from 4 to 17 September 2018; the second session, from 25 March to 5 April 2019; and the third session, from 19 to 30 August 2019.
11 United Nations General Assembly Decision A/74/L.41 Draft Decision Submitted by the President of the General Assembly Intergovernmental Conference on an International Legally Binding Instrument under the United Nations Convention on the Law of the Sea on the Conservation and Sustainable Use of Marine Biological Diversity of Areas beyond National Jurisdiction. 74th Session agenda item 74(a) (9 March 2020) UN Doc. A/74/L.41.
12 The official source of information on the intergovernmental conference is https://www.un.org/bbnj/.

References

Ardron, J. A., Rayfuse, R., Gjerde, K., & Warner, R. (2014). The sustainable use and conservation of biodiversity in ABNJ: What can be achieved using existing international agreements? *Marine Policy*, *49*, 98–108. https://doi.org/10.1016/j.marpol.2014.02.011

Badman, T., Douvere, F., Freestone, D., & Laffoley, D. (2016). *World heritage in the high seas: An idea whose time has come*. UNESCO.

Bell, J. D., Senina, I., Adams, T., Aumont, O., Calmettes, B., Clark, S., Dessert, M, Gehlen, M., Gorgues, T., Hampton, J., Hanich, Q., Harden-Davies, H., Hare, S. R., Holmes, G., Lehodey, P., Lengaigne, M., Mansfield, W., Menkes, C., Nicol, S. ... Williams, P. (2021). Pathways to sustaining tuna-dependent Pacific Island economies during climate change. *Nature Sustainability*, *4*(10), 900–910. https://doi.org/10.1038/s41893-021-00745-z

Berkman, P. A. (2010). Biodiversity stewardship in international spaces. *Systematics & Biodiversity*, *8*(3), 311–320. https://doi.org/10.1080/14772000.2010.512623

Blasiak, R., Wynberg, R., Grorud-Colvert, K., Thambisetty, S., Bandarra, N. M., Canário, A. V. M., da Silva, J., Duarte, C. M., Jaspers, M., Rogers, A., Sink. K., & Wabnitz, C. C. (2020). The ocean genome and future prospects for conservation and equity. *Nature Sustainability*, *3*(8), 588–596. https://doi.org/10.1038/s41893-020-0522-9

Carmine, G., Mayorga, J., Miller, N. A., Park, J., Halpin, P. N., Ortuño Crespo, G., Österblom, H., Sala, E., & Jacquet, J. (2020). Who is the high seas fishing industry? *One Earth*, *3*(6), 730–738. https://doi.org/10.1016/j.oneear.2020.11.017

Cisneros-Montemayor, A. M., Moreno-Báez, M., Reygondeau, G., Cheung, W. W., Crosman, K. M., González-Espinosa, P. C., Lam, V. W., Oyinlola, M. A., Singh, G. G., Swartz, W., & Zheng, C. W. (2021). Enabling conditions for an equitable and sustainable blue economy. *Nature*, *591*(7850), 396–401. https://doi.org/10.1038/s41586-021-03327-3

Dunn, D. C., Jablonicky, C., Crespo, G. O., McCauley, D. J., Kroodsma, D. A., Boerder, K., Gjerde, K. M., & Halpin, P. N. (2018). Empowering high seas governance with satellite vessel tracking data. *Fish and Fisheries*, *19*(4), 729–739. https://doi.org/10.1111/faf.12285

Freestone, D. (2009). Modern principles of high seas governance: The legal underpinnings. *Environmental Policy and Law*, *39*(1), 44–49.

Gjerde, K. M., Clark, N. A., & Harden-Davies, H. R. (2019). Building a platform for the future: The relationship of the expected new agreement for marine biodiversity in areas beyond national jurisdiction and the UN Convention on the Law of the Sea. *Ocean Yearbook Online*, *33*(1), 1–44. https://doi.org/10.1163/9789004395633_002

Gjerde, K. M., Reeve, L. L. N., Harden-Davies, H., Ardron, J., Dolan, R., Durussel, C., Earle, S., Jimenez, J. A., Kalas, P., Laffoley, D., Oral, N., Page, R., Ribeiro, M. C., Rochette, J., Spadone, A., Thiele, T., Thomas, H. L., Wagner, D., Warner, R. M. ... Wright, G. (2016). Protecting Earth's last conservation frontier: Scientific, management and legal priorities for MPAs beyond national boundaries. *Aquatic Conservation: Marine and Freshwater Ecosystems*, *26*, 45–60. https://doi.org/10.1002/aqc.2646

Haas, B., Haward, M., McGee, J., & Fleming, A. (2021). Regional fisheries management organizations and the new biodiversity agreement: Challenge or opportunity? *Fish and Fisheries*, *22*(1), 226–231. https://doi.org/10.1111/faf.12511

Harden-Davies, H. R., & Gjerde, K. M. (2019). Building scientific and technological capacity: A role for benefit-sharing in the conservation and sustainable use of marine biodiversity beyond national jurisdiction. *Ocean Yearbook Online*, *33*(1), 377–400. https://doi.org/10.1163/9789004395633_015

Hassanali, K. (2021). Internationalization of EIA in a new marine biodiversity agreement under the Law of the Sea Convention: A proposal for a tiered approach to review and decision-making. *Environmental Impact Assessment Review*, 87, 106554. https://doi.org/10.1016/j.eiar.2021.106554

Humphries, F., & Harden-Davies, H. (2020). Practical policy solutions for the final stage of BBNJ treaty negotiations. *Marine Policy*, *122*, 104214. https://doi.org/10.1016/j.marpol.2020.104214

Jaeckel, A. (2020). Benefitting from the common heritage of humankind: From expectation to reality. *The International Journal of Marine and Coastal Law*, *35*(4), 660–681. https://doi.org/10.1163/15718085-BJA10032

Levin, L. A., Amon, D. J., & Lily, H. (2020a). Challenges to the sustainability of deep-seabed mining. *Nature Sustainability*, 3(10), 784–794. https://doi.org/10.1038/s41893-020-0558-x

Levin, L. A., & Le Bris, N. (2015). The deep ocean under climate change. *Science*, *350*(6262), 766–768. https://doi.org/10.1126/science.aad0126

Levin, L. A., Wei, C. L., Dunn, D. C., Amon, D. J., Ashford, O. S., Cheung, W. W., Colaço, A., Dominguez-Carrió, C., Escobar, E. G., Harden-Davies, H. R., Drazen, J. C., Ismail, K., Jones, D. O., Johnson, D. E., Le, J. T., Lejzerowicz, F., Mitarai, S., Morato, T., Mulsow, S. … Yasuhara, M. (2020b). Climate change considerations are fundamental to management of deep-sea resource extraction. *Global Change Biology*, *26*(9), 4664–4678. https://doi.org/10.1111/gcb.15223

Mulalap, C. Y., Frere, T., Huffer, E., Hviding, E., Paul, K., Smith, A., & Vierros, M. K. (2020). Traditional knowledge and the BBNJ instrument. *Marine Policy*, *122*, 104103. https://doi.org/10.1016/j.marpol.2020.104103

Österblom, H., Wabnitz, C. C., Tladi, D., Allison, E., Arnaud-Haond, S., Bebbington, J., Bennett, N., Blasiak, R., Boonstra, W. J., Choudhury, A., Cisneros-Montemayor, A. M., Daw, T., Fabinyi, M., Franz, N., Harden-Davies, H., Kleiber, D. L., Lopes, P., McDougall, C., Resosudarmo, B. P., & Selim, S. A. (2020). *Towards ocean equity*. World Resources Institute. https://www.oceanpanel.org/blue-papers/towards-ocean-equity.

Popova, E., Vousden, D., Sauer, W. H. H., Mohammed, E. Y., Allain, V., Downey-Breedt, N., Fletcher, R., Gjerde, K. M., Halpin, P. N., Kelly, S., Obura, D., Pecl, G., Roberts, M., Raitsos, D. E., Rogers, A., Samoilys, M., Sumaila, U. R., Tracey, S., & Yool, A. (2019). Ecological connectivity between the areas beyond national jurisdiction and coastal waters: Safeguarding interests of coastal communities in developing countries. *Marine Policy*, *104*, 90–102. https://doi.org/10.1016/j.marpol.2019.02.050

Rabone, M., Horton, T., Harden-Davies, H., Zajderman, S., Appeltans, W., Droege, G., Brandt, A., Pardo-Lopez, L., Dahlgren, T. G., Glover, A. G., & Horton, T. (2019). Access to marine genetic resources (MGR): Raising awareness of best practice through a new agreement for biodiversity beyond national jurisdiction (BBNJ). *Frontiers in Marine Science*, 6, 520. https://doi.org/10.3389/fmars.2019.00520

Ramirez-Llodra, E., Brandt, A., Danovaro, R., De Mol, B., Escobar, E., German, C. R., Levin, L. A., Martinez Arbizu, P., Menot, L., Buhl-Mortensen, P., Narayanaswamy, B. E., Smith, C. R., Tittensor, D. P., Tyler, P. A., Vanreusel, A., & Vecchione, M. (2010). Deep, diverse and definitely different: Unique attributes of the world's largest ecosystem. *Biogeosciences*, *7*(9), 2851–2899. https://doi.org/10.5194/bg-7-2851-2010

Ramirez-Llodra, E., Tyler, P. A., Baker, M. C., Bergstad, O. A., Clark, M. R., Escobar, E., Levin, L. A., Menot, L., Rowden, A. A., Smith, C. R., & Van Dover, C. L. (2011). Man and the last great wilderness: Human impact on the deep sea. *PLoS ONE*, 6(8), e22588. https://doi.org/10.1371/journal.pone.0022588

Rogers, A., Sumaila, U., Hussain, S., & Baulcomb, C. (2014). *The high seas and us: Understanding the value of high seas ecosystems*. Global Ocean Commission.

Rogers, A. D., Baco, A., Escobar-Briones, E., Currie, D., Gjerde, K., Gobin, J., Jaspars, M., Levin, L., Linse, K., Rabone, M., Ramirez-Llodra, E., Sellanes, J., Shank, T. M., Sink, K., Snelgrove, P. V., Taylor, M. L., Wagner, D., & Harden-Davies, H. (2021). Marine genetic resources in areas beyond national jurisdiction: Promoting marine scientific research and enabling equitable benefit sharing. *Frontiers in Marine Science, 8*, 844. https://doi.org/10.3389/fmars.2021.667274

Schiller, L., Bailey, M., Jacquet, J., & Sala, E. (2018). High seas fisheries play a negligible role in addressing global food security. *Science Advances, 4*(8), eaat8351. https://doi.org/10.1126/sciadv.aat8351

Spiteri C., Senechal, T., Hazin, C., Hampton, S., Greyling, L., & Boteler, B. (2021). *Study on the socio-economic importance of areas beyond national jurisdiction in the Southeast Atlantic Region*. STRONG High Seas Project. https://publications.iass-potsdam.de/rest/items/item_6001046_4/component/file_6001047/content

Tessnow-von Wysocki, I., & Vadrot, A. B. (2020). The voice of science on marine biodiversity negotiations: A systematic literature review. *Frontiers in Marine Science, 7*(1044), 614282. https://doi.org/10.3389/fmars.2020.614282

Tickler, D., Meeuwig, J. J., Bryant, K., David, F., Forrest, J. A., Gordon, E., Larsen, J. J., Oh, B., Pauly, D., Sumaila, U. R., & Zeller, D. (2018). Modern slavery and the race to fish. *Nature Communications*, 9(1), 4643. https://doi.org/10.1038/s41467-018-07118-9

Tiller, R., De Santo, E., Mendenhall, E., & Nyman, E. (2019). The once and future treaty: Towards a new regime for biodiversity in areas beyond national jurisdiction. *Marine Policy*, 99, 239–242. https://doi.org/10.1016/j.marpol.2018.10.046

Treves, T. (2015a). Historical development of the law of the sea. In D. R. Rothwell, A. G. Oude Elferink, K. N. Scott, & T. Stephens (Eds.), *The oxford handbook of the law of the sea* (pp. 1–24). Oxford University Press.

Treves, T. (2015b). Law and science in the jurisprudence of the International Tribunal for the Law of the Sea. In H. N. Scheiber, J. Kraska, & M-S. Kwon (Eds.), *Science, technology, and new challenges to ocean law: A law of the sea institute publication* (pp. 13–26). Brill/Nijhoff. https://doi.org/10.1163/9789004299610_003

Turner, P. J., Cannon, S., DeLand, S., Delgado, J. P., Eltis, D., Halpin, P. N., Kanu, M. I., Sussman, C. S., Varmer, O., & Van Dover, C. L. (2020). Memorializing the middle passage on the Atlantic seabed in areas beyond national jurisdiction. *Marine Policy, 122*, 104254. https://doi.org/10.1016/j.marpol.2020.104254

United Nations. (1982, December 10). *United Nations convention on the law of the sea*. https://www.un.org/depts/los/convention_agreements/texts/unclos/unclos_e.pdf

United Nations. (2019, November 18). *Revised draft text of an agreement under the United Nations Convention on the Law of the Sea on the conservation and sustainable use of marine biological diversity of areas beyond national jurisdiction. Intergovernmental conference on an international legally binding instrument under the United Nations Convention on the Law of the Sea on the conservation and sustainable use of marine biological diversity of areas beyond national jurisdiction (fourth session, New York, 23 March – 3 April 2020)* (UN doc A/CONF.232/2020/3). UNGA. https://undocs.org/en/a/conf.232/2020/3

United Nations Conference on Trade and Development. (2019). *Review of maritime transport* (UNCTAD/RMT/2019/Corr.1). United Nations. https://unece.org/fileadmin/DAM/cefact/cf_forums/2019_UK/PPT_L_L-UNCTAD-RMT.pdf

Vadrot, A. B. M., Langlet, A., & Tessnow-von Wysocki, I. (2022). Who owns marine biodiversity? Contesting the world order through the 'common heritage of humankind' principle. *Environmental Politics, 31*(2), 226–250. https://doi.org/10.1080/09644016.2021.1911442

Vierros, M. K., & Harden-Davies, H. (2020). Capacity building and technology transfer for improving governance of marine areas both beyond and within national jurisdiction. *Marine Policy*, *122*, 104158. https://doi.org/10.1016/j.marpol.2020.104158

Vierros, M. K., Harrison, A. L., Sloat, M. R., Crespo, G. O., Moore, J. W., Dunn, D. C., Ota, Y., Cisneros-Montemayor, A. M., Shillinger, G. L., Watson, T. K., & Govan, H. (2020). Considering indigenous peoples and local communities in governance of the global ocean commons. *Marine Policy*, *119*, 104039. https://doi.org/10.1016/j.marpol.2020.104039

Warner, R. M. (2014). Conserving marine biodiversity in the global marine commons: Co-evolution and interaction with the Law of the Sea. *Frontiers in Marine Science*, *1*. https://doi.org/10.3389/fmars.2014.00006

Wright, G., Gjerde, K., Finkelstein, A., & Currie, D. (2020). *Fishing in the twilight zone: Illuminating governance challenges at the next fisheries frontier*. Institute for Sustainable Development and International Relations (IDDRI).

Wright, G., Rochette, J., Gjerde, K. M., & Seeger, I. (2018). *The long and winding road: Negotiating a treaty for the conservation and sustainable use of marine biodiversity in areas beyond national jurisdiction*. Institute for Sustainable Development and International Relations (IDDRI).

Part III

Approaches to the Grand Challenges

Part III includes examples of applied approaches to addressing the "Grand Challenges" described in Part II. These approaches are interdisciplinary and showcase professional pathways for the next generation of marine studies professionals.

DOI: 10.4324/9781003058151-12

10 Ocean Governance

Definitions, Framework, and Implementation

I. Scott and A. K. Spalding

Introduction

In July 2020, Ecuador raised public concerns about an enormous fleet of foreign fishing vessels gathered in international waters close to the Galapagos Islands, a region prized for its rich biodiversity and significant population of threatened species, including sharks and manta rays (Solano & Torchia, 2020). Just a few miles away, the islands and waters around the Galapagos Islands have been deemed a UNESCO World Heritage Site, which means they are subject to special protection and conservation, as agreed to by hundreds of countries. Yet, according to the governor of the Galapagos Archipelago, overfishing in the area bordering protected waters has led to decreasing populations of vulnerable species year upon year. What can be done about the species that cross this invisible border into unprotected waters, or the impact to protected stocks from illegal fishing?

Meanwhile, also in 2020, in a region of Northern Siberia, an oil tank owned by a Russian mining company spilled over 150,000 barrels of oil into a lake near the Arctic Ocean (Kramer, 2020). The oil tank was damaged by melting permafrost, caused by the steadily warming climate in the far north, where rates of warming are estimated to be up to three times the global average (Khurshudyan & Freedman, 2020). Observers have suggested that oil has made its way into a river that flows into the Arctic, though the company that owns the oil tank has denied such accounts (Osborn & Balmforth, 2020). The Arctic is uniquely vulnerable to impacts from oil spills, as little is known about the ecosystem in general, and many countries, including the United States, have only limited infrastructure available in the region to respond to an oil spill in a timely fashion (Nunez, 2014). Can the Arctic Ocean be protected in the face of these rapid changing conditions?

These real-life events illustrate some key concepts that we discuss in this chapter and the importance of studying ocean governance. The ocean operates in multiple dimensions, many of which often require international cooperation. The ocean, its resources, and the human communities that depend on it are also facing unprecedented threats from global climate change and unsustainable fishing practices, both of which span international boundaries, in addition to other growing demands on ocean spaces and resources such as offshore aquaculture, marine renewable energy, deep sea mining, and oil and gas development. The

DOI: 10.4324/9781003058151-13

extent and nature of the ocean suggests that rules of use designed for terrestrial areas may not necessarily apply. As with other so-called "wicked problems",[1] ocean governance is challenging in part because of the scale of the ocean and its related economic and social systems, the difficulty of finding an effective solution to governance issues, and the interconnected and simultaneous nature of problems impacting the ocean such as climate change, poverty, and development (Head & Alford, 2013).

In this context, key elements of ocean governance we will discuss in this chapter include legal frameworks, approaches to the implementation of these frameworks, and related actors and institutions. While an in-depth socio-cultural perspective on the relationships between people and the ocean is beyond the scope of this chapter, we feel it is important to emphasize that current systems of ocean governance are inextricably linked to culture, society, economics, and politics. With more than one-third of the world's population residing within 100 kilometers from the coast, and often depending on coastal resources for survival, it is critical to keep this in mind as you read the rest of this chapter.

So far, we have explained why creating a system of rules and regulations around the relationships between people and the ocean is important, and we have hinted at the interacting social and environmental factors that influence this system. In the remaining sections we provide a definition of the terms used in this chapter, differentiating between *policy*, *management*, *law*, and *governance*, with an emphasis on legal frameworks and governance processes. We introduce the United Nations Convention on the Law of the Sea (UNCLOS) and other key principles of international law, which represent the primary legal framework for global ocean governance. And we discuss the governance approaches, processes, and institutions that exist to implement UNCLOS and associated policies at the international and national levels. The chapter ends with insights to the types of professional pathways you may want to pursue in order to develop a future career in ocean governance.

Defining the Field: Policy, Management, Law, and Governance of the Ocean

Before diving into the legal framework for ocean governance and the approaches and actors involved in its implementation, we point out some key differences between the terms *policy*, *management*, *law*, and *governance*, which are often used interchangeably to refer to the set of rules and regulations that determine the use and distribution of, and access to, ocean resources and spaces. In an applied context, understanding the differences between these terms is crucial to their effective design and implementation, and may also guide you toward appropriate training and practical experiences that might facilitate a future career in ocean governance.

We define domestic or national *policy* as the set of guiding principles that outline the intentions of a national or regional government, or similar organization, toward a given issue. Policy may also include the set of rules that determine how organizations (public or private) are run, with the intention of structuring behaviors, internal actions, and interactions beyond the organization. In a government setting, it is usually referred to as public policy, which includes statutes, laws, regulations, executive decisions, and government programs. Policy can, therefore, be understood as the blueprint for action that informs and guides behavior, as it relates to an institution's approach to an issue. Zacharias (2014) further describes policy as the "motivation behind the law" or a way to set "strategic direction, intent, and objectives to address problems that are often complex and controversial" (p. 90).

Resource *management* extends policies into the realm of implementation and action. In a government setting, management decisions are often made at high administrative levels, ideally through processes of public consultation and careful consideration of scientific input, while implementation and enforcement on the ground is led by technical staff and authorized enforcement officers. Management is often perceived to be guided mostly by the natural and physical sciences. However, in the sense that it is informed by public policies that should themselves be informed by the public interest, resource management is increasingly recognized as a process that is, instead, guided by human needs and perceptions and driven by cultural and political processes (Levine et al., 2015). In other words, management actions are informed by science, influenced by policies and politics, enforced through laws, and typically represent a critical piece of the broader governance system.

Laws are similar to policies, especially in the international context, because they represent government intentions for how to address an issue. Laws, however, are generally more specific in that they describe rules for behavior that can be regulated and enforced by officials, who also have the authority to establish and issue fines and punishments for violations. Legal systems also typically establish procedures for settling disputes between stakeholders, who may include public entities, private entities (nongovernmental organizations or businesses), and individual citizens. Depending on the legal and political system of a country, power may be divided among different branches of government with regard to the creation, execution, and interpretation of law, and authority for creating and enforcing laws may rest more or less with a central (versus regional or local) government. In the United States, for example, the central (federal) government shares power with state governments, each of which can separately pass laws within their areas of authority. In China, on the other hand, only the central government is authorized to make laws, while local or regional governments may only create administrative rules or regulations (Zhang, 2014). In the context of the ocean, the law to be applied in any given situation may include national law (the law of an individual nation, applied to matters over which it has jurisdiction) or a mix of national and international law (which we discuss in more detail below).

Lastly, as understood by political scientists, *governance* is often closely linked to governments and what they do (Kjær, 2004). Ocean governance can be broadly defined as the collection of policies and associated norms and institutional arrangements, management actions, and laws linked to the global ocean (Miles, 1999). However, it is important to keep in mind that governance is NOT limited to the government! Indeed, while governance outcomes are often determined by the actions and behaviors of those in power, which we often perceive to be the government, we must also keep in mind that governments are influenced by the economy, constituent needs and interests, international relations, and the best available science. Beyond the purview of governments, governance actions and outcomes can also be moderated by market-based approaches, self-regulation, moral imperatives, culture and tradition, and other influences around a given issue (Spalding & deYcaza 2020).

The Law of the Sea

Sovereign nations or countries use the ocean for a variety of purposes, including travel, food harvest, religious and cultural traditions, recreation, and shipping. Due to connectivity in the ocean, what happens in one corner of the ocean impacts waters thousands of miles away. Plastics and waste move across national borders, as do fish, whales, and free-floating plants. Minerals and other resources in and below the sea floor cannot be extracted by everyone yet may be outside territorial boundaries that would define national ownership. The cultural tradition of one nation, such as whaling, may violate the laws or moral tradition of another. Awareness of these potential conflicts made it essential for sovereign nations to find a way to create common expectations for how they use, share, and ensure access to the ocean and ocean resources. Importantly, by definition, sovereign nations are not subject to the control of another nation. Although there are bodies that *adjudicate* disputes under international agreements, like the International Court of Justice, there is no higher authority, court, or police that can *enforce* those decisions. Ultimately, ocean governance relies heavily on voluntary cooperation and compliance with international law.

Sources of Law in the Sea

International law derives from three primary sources: *conventions* (treaties), *customs*, and generally accepted *principles of law*. These principles are applied by international tribunals, and may also form the basis of informal dispute resolution or diplomatic negotiations between nations. Principles of international law also provide a foundation for agreements between private parties, including commercial contracts.

Conventions are agreements among nations, while *protocols* are more specific agreements under the umbrella of a *convention*. They may be bilateral (between two parties) or multilateral (more than two parties). Conventions are entered into voluntarily, but they may include a mechanism for enforcement of the terms of the agreement. The International Court of Justice (ICJ), for example, is the United Nations' (UN) tribunal for settling disputes among member nations and parties to UN conventions. Importantly, the ICJ only has authority to settle a dispute if both parties to the dispute have consented to its jurisdiction (International Court of Justice, 2020).

The 1982 United Nations Law of the Sea Convention (UNCLOS) is the most important written agreement addressing the governance of the oceans. It is sometimes referred to as a "constitution for the oceans". To be clear, UNCLOS was built on a number of preceding events and agreements (Sohn et al., 2010). In 1956, based on growing concerns over competing national claims for jurisdiction over a variety of oceanic resources, the first United Nations Conference on the Law of the Sea (UNCLOS I) was held in Geneva, Switzerland. That conference resulted in a series of conventions regarding the shared use of and rules for national jurisdiction over certain portions of the oceans. Subsequent conferences (UNCLOS II and III) were held from 1958 until 1982, culminating in the 1982 agreement we now generally refer to as UNCLOS. While the United States ratified the 1958 conventions and participated in the negotiations of the 1982 convention, it did not sign the convention and has not ratified it (Song, 2005).

Customary law derives from customs, practices, and norms established over time, particularly those practices that states have demonstrated, through prior actions and decisions that they believe have the force of law. Customary law may or may not be formally added to a nation's official legal codes and may not exist in a single written form. A number of customary international legal principles known as the *law of the sea* existed prior to UNCLOS and formed the basis of many of the written provisions of the convention. The core of this body of customary international law is the concept of the "freedom of the seas", a doctrine dated to the 17th century that provides that all nations have the right to freely use the oceans – other than a limited strip of water adjacent to the shore, which might be claimed by a coastal nation – for transportation, fishing, and travel (United Nations, 2020).

Customary law is more difficult to determine and enforce than rules that derive from conventions. Under the freedom of the seas doctrine, the extent to which a country can claim jurisdiction over the ocean beyond its coastline could be disputed. Historically, these so-called *territorial waters* were limited to the distance a cannonball could travel from land – approximately three miles (Baty, 1928) – but not all nations agreed to this limitation. Under UNCLOS, jurisdictional rights are separately and specifically defined for the territorial sea (12 nautical miles from the baseline), the contiguous zone (24 nautical miles from the baseline, or 12 nautical miles from the territorial sea), and the Exclusive Economic Zone (EEZ) (200 nautical miles from the baseline); and nations exercise different rights within each zone (see next section for further description of these zones). While customary law may have evolved from unwritten norms and rules, it may also include specific codified principles. Most notably, key provisions of UNCLOS have become accepted as customary law, including by the United States. For this reason, though the United States has not ratified UNCLOS, it does accept as binding most of its provisions.[2]

The final source of international law is *principles of law* that are generally recognized and accepted by nations around the world (Grant, 2010). These principles of law include notions such as fairness and justice, or good faith – difficult concepts to narrow down to a black-and-white rule – but which do inform the way that disputes are settled between parties.

United Nations Convention on the Law of the Sea (UNCLOS)

UNCLOS is a broad, multilateral treaty that includes over 300 articles and nine annexes. Delving into the specifics of these hundreds of provisions is clearly beyond the scope of this chapter. Instead, we provide here a few of the key concepts of UNCLOS, in order to illustrate some of the general rules for ocean governance.

Central to UNCLOS are provisions determining who controls or has *jurisdiction* over particular ocean space and the seabed, and what rights non-jurisdictional countries have to use or move across jurisdictional waters. As noted above, under the freedom of the seas rule, territorial waters – those waters over which the coastal country had jurisdiction and control – extended approximately three miles from shore. Under UNCLOS, these territorial waters extend for 12 nautical miles from the coastal *baseline*. The baseline is the "low-water line along the coast as marked on large-scale charts officially recognized by the coastal state" (UNCLOS, Art. 5). Within its territorial seas, a nation has the sovereign right over the water and the seabed and can enforce its national laws and control the use of resources. However, foreign vessels must be allowed "innocent passage"[3] across territorial waters.

A second key concept of UNCLOS is the Exclusive Economic Zone, or EEZ, which extends 200 nautical miles from a nation's baseline (Tuerk, 2012). Within the EEZ, a nation has the exclusive right to use, exploit, and sustainably manage resources in the water and in the seabed and subsoil. A nation may further extend the right to exploit the resources of the seabed to the limit of the continental shelf, though not beyond 350 nautical miles. The determination of the limits of the continental shelf is neither simple nor uncontroversial; in the Arctic, overlapping claims to the continental shelf are an essential component of the burgeoning dispute in the region over rights to mineral and other seabed and subsoil resources.

Beyond the EEZ, the waters of the ocean are considered "high seas" and therefore not "owned" by any nation; these areas and resources are often considered a "global commons" or "common resources of mankind". UNCLOS also provides that the seabed and subsoil beyond the jurisdiction of any nation – identified as "the Area" in UNCLOS – are the "common heritage of mankind" (UNCLOS, Art. 136). While this principle is not specifically defined, a variety of other provisions of UNCLOS with regard to the Area suggest that resources in the Area are to be shared equitably and the needs of landlocked nations are to be considered (UNCLOS, Art. 140–145). In addition, the environment in the Area is to be preserved and protected (UNCLOS, Art. 145). In 2018, for the first time since its ratification, delegates initiated negotiations for a new legally binding instrument under UNCLOS for the conservation and sustainable use of biodiversity in areas beyond national jurisdiction (ABNJ).

UNCLOS has a number of other important provisions, including the right to fly aircraft over the high seas, special rules for determining baselines of archipelagic states, and methods for dispute resolution. UNCLOS also addresses pollution and dumping, both from land-based sources and at sea, and creates special responsibilities for signatories of the convention to work together for the preservation and conservation of marine resources.

Climate change poses a number of unique challenges to the operation and administration of UNCLOS. Perhaps the most straightforward is the impact of rising sea levels, which are increasing faster and higher than previously predicted, and threaten to inundate low-lying areas and render uninhabitable land now occupied by millions (Intergovernmental Panel on Climate Change, 2019; Kulp & Straus, 2019). Because the jurisdictional provisions of UNCLOS depend on the determination of a nation's baseline, changes in low-water lines may result in a significant loss of jurisdictional territory for coastal nations (Telesetsky, 2019). How shall we address the situation of a nation that is entirely submerged by rising seas? Can baselines be "fixed", or must they accurately reflect changing coastlines?

While UNCLOS is the overarching mechanism in international legal governance related specifically to the oceans, a variety of other agreements also play significant roles in setting expectations and objectives for the shared use of the oceans and related activities (see Table 10.1).

Table 10.1 International agreements for the shared use of the ocean and related activities

CBD	*Convention on Biological Diversity; framework to preserve, share, and sustain biodiversity.*
UNFCCC	United Nations Framework Convention on Climate Change; created a framework for stabilizing greenhouse gas emissions to limit damaging climate change.
MARPOL	The International Convention for the Prevention of Pollution from Ships; addresses prevention of pollution (accidental or purposeful) from ships.
UNEP Regional Seas	United Nations Environmental Program's Regional Seas Program is a collection of regional agreements and action plans for conservation of marine and coastal environments.
RFMOs	Regional management organizations; international organization of nations with shared interest in regional management of fish stocks. Some focus on a specific stock, others look more broadly at the ecosystem.
SIDS/AOSIS	Small Island Developing States are a group within the United Nations of small and low-lying island nations facing unique environmental and economic risks; the Association of Small Island States is a member organization that pools resources and negotiates on behalf of members.
Convention on Wetlands (Ramsar)	Convention on Wetlands; creates a framework for the conservation and sustainable use of wetlands.

Case Study: International Whaling

Few issues illustrate the challenges of international law better than the case of the commercial hunting of whales. The International Commission on Whaling (ICW), established as part of the International Convention for the Regulation of Whaling (ICRW), passed a moratorium on commercial whaling in 1986 (Smith, 2014). Several members of the ICW formally objected to the moratorium, including Japan and Norway (Kobayashi, 2006). While it formally ceased commercial whaling, Japan continued to take whales for what it called "scientific research" in the Southern Ocean near Antarctica. In 2014, the International Court of Justice found Japan's hunting did not qualify as research, and instead constituted a violation of the ICRW (Wold, 2020). In 2018, Japan announced its intention to withdraw from the ICRW, and in 2019, it resumed commercial whaling in its EEZ. Although no longer a party to the ICRW, Japan is still a party to a number of international agreements, including UNCLOS and CITES, which require it to manage whale populations for conservation, and restrict hunting of certain endangered species of whales (Sand, 2008). Yet Japan appears to have no intention of complying with these treaties, as it has not consulted with the ICW regarding its resumption of commercial whaling, prepared environmental assessments of the impact of its whaling program on threatened whale species, or shared information about its program with its international partners (Wold, 2020). Meanwhile, Norway, which has not left the IWC, resumed commercial whaling in 1993 based on its formal objection to the moratorium, and even increased its catch in 2020. However, Norway has received only muted criticism of its whaling program, even while it exports some of its catch to Japan.

Implementing the Law of the Sea: Governance Approaches and Actors

Through UNCLOS, the international community successfully developed a set of rules and regulations about the use and management of the ocean, using both spatial (e.g., jurisdiction) and functional (e.g., pollution and conservation) approaches. Over time, these approaches expanded to explicitly link ocean governance with the 2030 Global Sustainable Development Agenda. The implementation of those rules and regulations is shared by governance actors (or institutions of governance), including international agencies, individual countries, environmental non-government organizations, and, to varying degrees, a broadly defined civil society.

Governance Approaches

Spatial Governance

The *spatial* governance approach refers to the establishment of the previously defined jurisdictional zones in ocean and coastal spaces. These zones essentially determine what can and cannot be done by whom and where. For instance, internal waters, territorial seas, international straits, and archipelagic waters are considered areas under national sovereignty, or in other words, areas over which countries have rights and responsibilities similar to those on land. The contiguous zone, EEZ, and continental shelf are considered to be areas under national jurisdiction. This means that country actions are guided primarily by the ability to exercise rights over and to gain access to resources (e.g., fisheries management), as well as by a shared responsibility to collaborate with other countries over resource management and conservation according to the parameters of international law (Spalding & deYcaza, 2020).

Functional Governance

Jurisdictional zones provide clarity on individual and shared rights and responsibilities in space. However, these do not necessarily match up with ecologically defined zones (ecosystems, species ranges, etc.). Thus, in parallel to establishing jurisdictional zones, international law also created governance mechanisms to address issues that defy borders, such as fisheries management (which is crucial for species like tuna that move across jurisdictional lines), pollution, climate change, and conservation. This is what we call *functional* governance. Under the functional governance approach, countries or groups of countries align over shared interests. To do so, they must enter into negotiations to secure international treaties and agreements, including but not limited to UNCLOS (see Table 10.1).

Governance Related to the 2030 Sustainable Development Agenda

Toward the end of the first decade of the new millennium, international agencies increasingly identified ocean governance as a sustainable development concern.[4] Indeed, in 2015, in the context of the United Nations 2030 Agenda for Sustainable Development and its 17 Sustainable Development Goals (SDGs), the ocean and related issues became a cross-cutting issue in development discussions such as poverty reduction and food security. Specifically, ocean-related actions and coalitions formed around SDG 14, which is designed to conserve and sustainably use the oceans, seas, and marine resources. Additionally, recent initiatives focused on a sustainable ocean future include the annual UN Ocean Conference and the Our Ocean Conference, both of which increasingly call for voluntary commitments toward sustainable development in the ocean from governments, industry, and civil society.

The Blue Economy

Last, and certainly not least, we include the Blue Economy as an emerging, market-based governance approach. While not a traditional mechanism for governance per se, the Blue Economy, defined here as an approach that promotes economic growth through investment in marine-based industries, and fosters sustainable development, supports healthy marine ecosystems (Silver et al., 2015), has permeated government, industry, and some civil society attitudes and perceptions of the ocean. While driven primarily by private capital interests in the ocean (Campbell et al., 2016), the Blue Economy draws on the spatial approach to governance by relying on clear property rights to marine space, expands on the functional approach by incorporating new uses that cross jurisdictional boundaries, and – at least as originally defined by Silver et al. (2015) – is focused on sustainability. In practice, this translates into the development of national-level policies that promote Blue Economy industries, such as aquaculture, marine renewable (or conventional) energy, transportation and logistics, and tourism, among others. Some of these policies rely on existing ocean governance approaches described previously, whereas others guide or influence market-based management of spaces and resources. Studies on the impacts of Blue Economy–inspired governance are relatively new and beg the question of whether and how equitable distribution of benefits for all ocean stakeholders can be maintained in the context of historically dominant economic sectors such as oil and gas.

Actors of Environmental Governance

A diversity of actors plays a crucial role in shaping and implementing ocean governance. Here we describe the key international, national, intergovernmental, industry, and civil society actors engaged in ocean issues. Multi-party conventions, like UNCLOS, and other mechanisms used for functional ocean governance, such as bi- or multilateral agreements, directly involve the most traditional actors of ocean governance: international institutions and sovereign nations. However, other actors, such as environmental NGOs, philanthropies, Indigenous and other local coastal communities, academia, and industry, play key roles in generating information, building community around ocean issues, and representing a diversity of stakeholders (Spalding & deYcaza, 2020).

The most recognized international institution for ocean governance is the United Nations System. Notably, the UN hosts UNCLOS and includes several other agencies, commissions, programs, and committees that oversee maritime concerns, oceanographic research, environment and development issues, and regional organizations. Some well-known UN agencies include the United Nations Educational, Scientific and Cultural Organization (UNESCO) and its Intergovernmental Oceanographic Commission (IOC), the UN Food and Agriculture Organization (FAO), and the International Maritime Organization (IMO), among others.

Nation states or countries, including sovereign Indigenous nations, also play an important role in ocean governance. First, they determine governance and management rules within their jurisdiction, putting into practice rules established by UNCLOS and other international agreements to which they subscribe. Nations also implement these international commitments, both within their own borders and through engagement with other members of the international community. And finally, they support mediation, communication, and engagement about ocean issues internally between national and local governments, as well as across different sectors of society.

Additional actors involved in ocean governance include organizations whose role in governance may not be specific to the design of rules and regulations, but are instead more focused on influencing policy, generating the scientific information necessary to design those rules and regulations, or building the awareness and community around issues related to the ocean. These include intergovernmental organizations that are comprised of government and civil society organizations (e.g., International Union for Conservation of Nature – IUCN), science for decision-making organizations (e.g., the International Council for the Exploration of the Sea – ICES), and other thematic (e.g., Interamerican Tropical Tuna Commission – IATTC) or regional (e.g., European Commission of the European Union – EC of EU) organizations.

Finally, civil society and industry actors play a critical role in representing public and private interests. Ocean-related civil society actors typically include nongovernmental organizations (NGOs) such as The Nature Conservancy and Conservation International; philanthropies such as the Pew Charitable Trust and the Ocean Foundation; and Indigenous communities and other local communities. Civil society actors tend to focus on concerns related to site-specific sustainable development and livelihoods at both local and international scales and can sometimes also support coalition-building among diverse actors. In contrast, industry represents private business interests in activities such as energy production and transportation. These actors most often are concerned with access to resources and the generation of capital gains from the use and production of ocean resources.

In sum, these actors influence ocean governance by designing, implementing, and overseeing policies; developing science to influence policies; and representing public and private interests. Although each of the actors effectively has their own role, they also interact with each other in various national, regional, and international fora, together comprising the ocean governance landscape of today.

Case Study: As the Ice Melts, What Becomes of the Arctic Ocean?

Climate change, increasing ocean temperatures, and resulting loss of sea ice are leading to a changing Arctic. Most visibly, the loss of sea ice means that key shipping routes through the Arctic, including the Northern Sea Route and Northwest Passage, are becoming available for more days during the year, and new routes are emerging. Recent studies suggest that the Arctic may be entirely ice-free in September by 2050 (SIMIP Community, 2020), suggesting that a new trans-Arctic passage cutting directly across the North Pole will likely become available in the near term (Bennett, 2019). Arctic routes will grow increasingly less expensive than other options, particularly as alternative routes such as the Suez or Panama Canal reach capacity (Khon et al., 2010). But, unless there is a large-scale commitment to decarbonizing shipping, they will also create more emissions and pollutants

(Gallucci, 2018), including black carbon from the heavy fuel oil used to power ocean tankers, threatening delicate Arctic ecosystems (Cho, 2016).

Other changes are coming for the Arctic itself: As climate change raises the temperature of the ocean, fish species migrate northward, raising the possibility of new fisheries in the Arctic Ocean (Rayfuse, 2019). The decrease in ice cover will also make sections of the Arctic seabed available for oil, gas, and mineral exploitation (Willaert, 2020). But who controls these potential fisheries, should they develop? Who controls the Arctic seabed? And how do we limit pollution in this ecologically sensitive area?

Governance of the Arctic is enormously complex. Five nations (Canada, Denmark [via Greenland], Russia, Norway, and the United States) claim part of the Arctic Ocean as their territorial waters – some of which conflict. Canada, for example, claims that portions of the Northwest Passage fall within its territorial waters, while the United States claims that these waters are "high seas" and not subject to Canadian control (Schlanger, 2019). Under UNCLOS, four of the Arctic coastal nations (not including the United States, which has not ratified UNCLOS) have formally claimed rights to develop and exploit the seabed to the limits of an extended continental shelf – again overlapping in some areas – also leading to conflict over who will control mining, and oil and gas extraction in those areas (Skaridov, 2020). Multiple nongovernmental actors are also at work in the Arctic, including private companies, such as multinational fossil fuel corporations, mining, and fishing companies (Nordea, 2017), and nonprofit conservation groups such as Oceana (Arctic NGO Forum, 2020). A group of Arctic nations, Arctic Indigenous people, and residents of the Arctic also participate in the Arctic Council, an intergovernmental forum that works to promote "cooperation, coordination, and interaction" among participants (Arctic Council, 2021).

Beyond jurisdictional claims, the non-jurisdictional areas of the Arctic Ocean also present challenges. As noted above, under UNCLOS, areas of the Arctic that are beyond the jurisdiction of any single state are part of the "common heritage of mankind" (UNCLOS, Art. 136–137) and states have a responsibility to carry out activities in those areas "for the benefit of mankind" (UNCLOS, Art. 140(1)). Moreover, Article 192 of UNCLOS provides that "States have the obligation to protect and preserve the marine environment". Following a functional approach to governance, other international agreements, including the International Convention for the Prevention of Pollution from Ships (MARPOL), and more general agreements regarding environmental protection, such as the Convention on International Trade in Endangered Species of Wild Fauna and Flora (CITES), also apply to provide some protection to Arctic waters (Molenaar, 2007). Yet international law does not forbid fishing, deep sea mining, or oil and gas extraction in the Arctic, and does not provide a clear path to limiting developing fisheries, the use of heavy fuel oil, or resource extraction in sensitive ecological areas.

The world still has limited knowledge and understanding of the Arctic ecosystem, so application of the precautionary principle[5] and an ecosystems management[6] approach is essential to protecting the pristine waters of the Arctic (De Lucia, 2020). In a hopeful sign of early commitment to this approach, the five Arctic coastal nations plus four additional non-Arctic nations (China, Iceland, Japan, and Korea) and the European Union negotiated the Central Arctic Ocean Fisheries Agreement (CAOFA), which prevents unregulated fishing in the Central Arctic Ocean. The CAOFA, which entered into force in March 2019, also creates responsibility for parties to work together to facilitate research and monitoring of potential new fisheries (Department of State, 2019). Another promising development which would affect the Arctic and other ocean and coastal issues is the draft *Agreement under the United Nations Convention on the Law of the Sea on the Conservation and Sustainable Use of Marine Biological Diversity of Areas Beyond National Jurisdiction*, better known as biodiversity beyond national jurisdiction (BBNJ). While the precise language of the BBNJ is still being negotiated, the goal of the agreement is for the parties to take on affirmative duties to build the resilience of the ecosystem and help protect it against the adverse effects of climate change (Craig, 2020).

Conclusion

As has been documented from the earliest human civilizations, societies organize around a set of rules, regulations, processes, and institutions, as well as agreed-upon moral and ethical values. In turn, how these societies interact with the natural world is also based on a set of agreements about the use, distribution of, and access to nature and its resources. In the case of the ocean, it is our relationship to these resources and spaces, as mediated by technology (e.g., boats, fishing gear, oil platforms, etc.) and nature itself, that has influenced those rules and regulations. We can think of the ocean as a socially constructed space that people use to transport and trade, as a resource base, or as a place of enjoyment and reflection. This way of thinking guided the approach to governing the ocean outlined in this chapter.

Here we have shown how humans simultaneously rely on ocean resources, contribute to their demise through unsustainable behaviors, and are tasked with developing ways to secure their long-term viability and health. UNCLOS represents a critical international legal tool that has informed and guided subsequent treaties and domestic regulations. Whether and how UNCLOS, and associated sources of and approaches to law in the sea, are able to keep up with the complex issues facing our global ocean remains to be seen. It is clear, however, that elevating ocean concerns to the UN Sustainable Development platform is an important step in considering the ocean and its resources as a partner in addressing some of the grand environmental challenges of our time.

Professional Pathways

Students interested in a career in ocean governance can follow a variety of paths. Those who are interested in the scientific application of resource conservation and management may focus on studying marine biology, oceanography, or engineering. Experts in these fields conduct research at universities and in private laboratories, studying topics ranging from the way pollution impacts marine organisms to the impacts of changing ocean currents on fisheries. They might work for governments developing management plans for protected ocean waters or study the best ways to protect endangered species, like whales and sharks, or fragile coral reefs. They may also work in industry, developing technology for shipping or fossil fuel extraction, or for NGOs or conservation groups.

For those interested in the law and policy side of governance, careers may include a variety of ways in which the human element of ocean governance is studied and managed. Policy analysts can come from a variety of disciplines, including economics, finance, psychology, political science, and law. They analyze, make recommendations, and develop ways to implement policies, and work for governments, NGOs, or businesses. Those interested in the law could pursue work in private firms advising clients on compliance with maritime law and regulations related to the ocean, or could advise governments on compliance with, or the creation and negotiation of new domestic laws and international conventions. Lawyers and policy experts may also work internationally for organizations like the United Nations, or in diplomatic positions as representatives of their home nation.

Questions for Reflection

1. What makes ocean and coastal spaces unique? And why do they require this particular governance structure?
2. What are the major sources of law in the sea?
3. How do interactions between actors of environmental governance influence governance approaches?
4. Given what you know about the current state of the environment, is the current ocean governance regime sufficient? What might ocean governance actors consider for a more sustainable and future of the ocean?
5. If you had the power to design a sustainable ocean governance regime, who would you include? And why?
6. How accountable do you think international organizations are to ensuring the sustainable and equitable management of the ocean and its resources?

Notes

1 A wicked problem, as it relates to the environment (and in the context of this book, to the ocean), is a long-standing and complex policy dilemma that is often driven by deep-seated conflicts over resource management and contradicting values around economic, social, and ecological priorities (Balint et al., 2011).

2 For more information on this subject, see William Gallo, "Why Hasn't the US Signed the Law of the Sea Treaty?" (Jun 26, 2016), https://www.voanews.com/usa/why-hasnt-us-signed-law-sea-treaty#:~:text=The%20U.S.%20has%20not%20accepted,linked%20the%20issue%20to%20China. See also De Tolve, Robert. 2012. "At What Cost? America's UNCLOS Allergy in the Time of Lawfare." *Naval Law Review, 61*:1-16.
3 Innocent passage generally refers to the right of a vessel to pass through the territorial waters of a nation, subject to certain restrictions as established in UNCLOS. *See* UNCLOS, Art. 17–32.
4 Sustainable development is understood as that which upholds human needs, as well as ensuring the sustainability of natural resources and ensuring services that support societies and economies. For more information, see: https://www.un.org/sustainabledevelopment/ (accessed September 24, 2020).
5 The precautionary principle, which has been applied in a wide variety of environmental contexts, including in international climate law, generally holds that a lack of scientific certainty around the precise nature of an environmental threat should not be used as a reason to avoid taking action. International Union for the Conservation of Nature. 2007. "Guidelines for Applying the Precautionary Principle to Biodiversity Conservation and Natural Resource Management." https://www.iucn.org/sites/dev/files/import/downloads/ln250507_ppguidelines.pdf (accessed September 20, 2020).
6 An ecosystems management approach "is a strategy for the integrated management of land, water and living resources that promotes conservation and sustainable use in an equitable way". Secretariat of the Convention on Biological Diversity. 2004. "The Ecosystems Approach." https://www.cbd.int/doc/publications/ea-text-en.pdf (accessed September 20, 2020).

References

Arctic Council. (2021). *About the Arctic Council.* Retrieved April 4, 2021 from https://arctic-council.org/en/about/

Arctic NGO Forum. (2020). *Partners.* Retrieved September 20, 2020 from http://www.arcticngoforum.org/partners.aspx

Balint, P. J., Stewart, R. E., Desai, A., & Walters, L. C. (2011). *Wicked environmental problems: Managing uncertainty and conflict.* Island Press.

Baty, T. (1928). The three mile limit. *American Journal of International Law, 22*(3), 503–537. https://doi.org/10.2307/2188741

Bennett, M. (2019, May 8). The arctic shipping route no one's talking about. *The Maritime Executive.* https://www.maritime-executive.com/editorials/the-arctic-shipping-route-no-one-s-talking-about

Campbell, L. M., Gray, N. J., Fairbanks, L., Silver, J. J., Gruby, R. L., Dubik, B. A., & Basurto, X. (2016). Global oceans governance: New and emerging issues. *Annual Review of Environment and Resources, 41*(1), 517–543. https://doi.org/10.1146/annurev-environ-102014-021121.

Cho, R. (2016, March 22). *The damaging effects of black carbon.* State of the Planet. Retrieved September 20, 2020 from https://blogs.ei.columbia.edu/2016/03/22/the-damaging-effects-of-black-carbon/

Craig, R. K. (2020). The new United Nations high seas treaty: A primer. *Natural Resources & Environment, 34*(4), 48–50.

De Lucia, V. (2020). The BBNJ negotiations and ecosystem governance in the arctic. *Marine Policy, Forthcoming.* https://doi.org/10.1016/j.marpol.2019.103756

De Tolve, R. (2012). At what cost? America's UNCLOS allergy in the time of lawfare. *Naval Law Review*, 61, 1–16.

Department of State. (2019). The United States ratifies central Arctic Ocean fisheries agreement. Retrieved September 20, 2020 from https://translations.state.gov/2019/08/27/the-united-states-ratifies-central-arctic-ocean-fisheries-agreement/

Gallo, William. (2016, June 26). Why hasn't the US signed the law of the sea treaty? *Voice of America*. https://www.voanews.com/usa/why-hasnt-us-signed-law-sea-treaty#:~:-text=The%20U.S.%20has%20not%20accepted, linked%20the%20issue%20to%20China

Gallucci, M. (2018, September 26). As the arctic melts, the northern sea route opens for business. *Wired*. https://www.wired.com/story/as-the-arctic-melts-the-fabled-northwest-passage-opens-for-cargo-ships/

Grant, J. (2010). *International law essentials*. Dundee University Press.

Head, B., & Alford, J. (2013). Wicked problems: Implications for public policy and management. *Administration and Society*, *47*(6), 711–739. https://doi.org/10.1177/0095399713481601

International Court of Justice. (2020). https://www.icj-cij.org/en

Intergovernmental Panel on Climate Change. (2019). Summary for Policymakers. In H. -O. Pörtner, D. C. Roberts, V. Masson-Delmotte, P. Zhai, M. Tignor, E. Poloczanska, K. Mintenbeck, A. Alegría, M. Nicolai, A. Okem, J. Petzold, B. Rama, N. M. Weyer (Eds.), *IPCC Special report on the ocean and cryosphere in a changing climate*. https://doi.org/10.1017/9781009157964

Kramer, A. (2020, June 9). Major fuel spill in Russia's north spreads toward Arctic ocean. *New York Times*. https://www.nytimes.com/2020/06/09/world/europe/russia-arctic-oil-spill.html

Khon, V. C., Mohkov, I. I., Latif, M., & Semeov, V. A. (2010). Perspectives of northern sea route and northwest passage in the twenty-first century. *Climate Change*, *100*(3), 757–768. https://doi.org/10.1007/s10584-009-9683-2

Khurshudyan, I., & Freedman, A. (2020, July 28). An oil spill in Russia's Arctic exposes risks for Moscow's Far North plans. *Washington Post*. https://www.washingtonpost.com/climate-environment/2020/07/28/an-oil-spill-russias-arctic-exposes-problems-moscows-big-plans-far-north/?arc404=true

Kjær, A. M. (2004). *Governance*. Polity Press.

Kobayashi, L. (2006). Lifting the international whaling commission's moratorium on commercial whaling as the most effective global regulation of whaling. *Environs*, *29*(2), 177–219. [Juris Doctoral Dissertation, UC Davis]. https://environs.law.ucdavis.edu/volumes/29/2/kobayashi.pdf.

Kulp, S., & Strauss, B. (2019). New elevation data triple estimates of global vulnerability to sea-level rise and coastal flooding. *Nature Communication*, *10*, 4844. https://doi.org/10.1038/s41467-019-12808-z

Levine, A. S., Richmond, L., & Lopez-Carr, D. (2015). Marine resource management: Culture, livelihoods, and governance. *Applied Geography*, *59*, 56–59. https://doi.org/10.1016/j.apgeog.2015.01.016

Miles, E. L. (1999). The concept of ocean governance: Evolution toward the 21st century and the principle of sustainable ocean use. *Coastal Management*, *27*(1), 1–30. https://doi.org/10.1080/089207599263875

Molenaar, E. (2007). Managing biodiversity in areas beyond national jurisdiction. *International Journal of Marine and Coastal Law*, *22*(1), 89–124. https://doi.org/10.1163/157180807781475263

Nordea. (2017). *Analyses of key companies having business operations in the Arctic*. Retrieved September 20, 2020 from https://insights.nordea.com/wp-content/uploads/2019/02/Analyses-of-Key-Companies-having-Business-Operating-In-the-Arctic_0.pdf

Nunez, C. (2014, April 24). What happens when oil spills in the Arctic? *National Geographic*. https://www.nationalgeographic.com/news/energy/2014/04/140423-national-research-council-on-oil-spills-in-arctic/#close

Osborn, A., & Balmforth, T. (2020, July 10). Russia's Nornickel fights cover-up accusations over Arctic oil spill. *Reuters*. https://www.reuters.com/article/us-russia-pollution-insight/russias-nornickel-fights-cover-up-accusations-over-arctic-oil-spill-idUSKBN24B0QH

Rayfuse, R. (2019). The role of law in the regulation of fishing activities in the Central Arctic Ocean. *Marine Policy*, *110*, 103562. https://doi.org/10.1016/j.marpol.2019.103562

Sand, P. (2008). Japan's 'research whaling' in the Antarctic Southern Ocean and the North Pacific Ocean in the face of the endangered species convention (CITES). *Review of European Community & International Environmental Law*, *17*(1), 56–71. https://doi.org/10.1111/j.1467-9388.2008.00587.x

Schlanger, Z. (2019, June 27). The US is picking a fight with Canada over a thawing Arctic shipping route. *Quartz*. https://qz.com/1653831/the-us-is-picking-a-fight-with-canada-over-an-arctic-shipping-route/

Silver, J. J., Gray, N. J., Campbell, L. M., Fairbanks, L. W., & Gruby, R. L. (2015). Blue economy and competing discourses in international oceans governance. *The Journal of Environment & Development*, *24*(2), 135–160. https://doi.org/10.1177/1070496515580797

SIMIP Community. (2020). Arctic sea ice in CMIP6. *Geophysical Research Letters*, *47*(10), e2019GL086749. https://doi.org/10.1029/2019GL086749.

Skaridov, A. (2020). The sea bed in the high north—How to address conflict? In C. Banet, (Ed.), *The law of the sea bed* (pp.104–124). Brill.

Smith, J. (2014). Evolving to conservation? The international court's decision in the Australia/Japan whaling case. *Ocean Development & International Law*, *45*(4), 301–327. https://doi.org/10.1080/00908320.2014.957965

Sohn, L. B., Juras, K. G., Noyes, J. E., & Franckx, E. (2010). *Law of the sea in a nutshell* (2nd ed.). West Publishing.

Solano, G., & Torchia, C. (2020, July 30). 260 Chinese boats fish near Galapagos; Ecuador on alert. *Washington Post*. https://www.washingtonpost.com/world/the_americas/260-chinese-boats-fish-near-galapagos-ecuador-on-alert/2020/07/30/01b0d98e-d29f-11ea-826b-cc394d824e35_story.html

Song, Y. (2005). Declarations and statements with respect to the 1982 UNCLOS: Potential legal disputes between the United States and China after U.S. accession to the convention. *Ocean Development & International Law*, *36*(3), 261–289. https://doi.org/10.1080/00908320591004405

Spalding, A. K., & DeYcaza, R. (2020). Navigating shifting regimes of ocean governance: From UNCLOS to Sustainable Development Goal (SDG) 14. *Environment and Society: Advances in Research*. https://doi.org/10.3167/ares.2020.110102

Telesetsky, A. (2019). Managing marine resources: Can the law of the sea treaty adapt to climate change? In P. Harris, (Ed.), *Climate change and ocean governance: Politics and policy for threatened seas* (pp.325–339). Cambridge University Press. https://doi.org/10.1017/9781108502238.020

Treves, T. (n. d.) *Introductory Note: 1958 Geneva Convention on the Law of the Sea*. Audiovisual Library of International Law. Retrieved September 20, 2020 from https://legal.un.org/avl/ha/gclos/gclos.html#:~:text=The%20Conventions%20and%20Protocol%20are,

February%20to%2027%20April%201958.&text=It%20had%20its%20precedents%20in, of%20the%20League%20of%20Nations

Tuerk, H. (2012). *Reflections on the contemporary Law of the Sea*. Martinus Nijhoff Publishers.

United Nations. (2020). *Oceans and the Law of the Sea*. Retrieved September 20, 2020 from https://www.un.org/en/sections/issues-depth/oceans-and-law-sea/#:~:text=The%20 oceans%20had%20long%20been, all%20and%20belonged%20to%20none

Willaert, K. (2020). Crafting the perfect deep sea mining legislation: A patchwork of national laws. *Marine Policy*, 119, 104055. https://doi.org/10.1016/j.marpol.2020.104055

Wold, C. (2020). Japan's resumption of commercial whaling and its duty to cooperate with the international whaling commission. *Journal of Environmental Law & Litigation*, 35, 87–142.

Zacharias, M. (2014). *Marine policy: An introduction to governance and international law of the oceans*. Routledge. https://doi.org/10.4324/9780203095256

Zhang, L. (2014). A *guide to Chinese legal research: Who makes what?* Library of Congress, Law Library Blog. Retrieved September 20, 2020 from https://blogs.loc.gov/law/2014/01/a-guide-to-chinese-legal-research-who-makes-what/

11 Stewardship and Conservation of the Marine Environment

Marine Protected Areas

M. Shivlani and D. O. Suman

The global ocean faces serious problems. Over a third of marine resources are harvested unsustainably, and another 60% are at maximum sustainable levels (Food and Agriculture Organization of the United Nations, 2020). Many fishing fleets are overcapitalized, resulting in excessive fishing effort. Improved technologies permit more efficient capture and at ever greater depths. Marine biodiversity is under threat, and many large marine animal populations have been driven to low levels (Sala & Knowlton, 2006). Existing and new sources of marine pollution continue to degrade water quality and adversely impact marine living resources. In short, the human footprint on the global ocean is ubiquitous and growing (Halpern et al., 2019). Conserving areas of the ocean as Marine Protected Areas (MPAs) has gained international recognition as one solution to the ocean's problems.

A central aspect of understanding MPAs is through their formal definition and how they can be distinguished from the surrounding seascape. The International Union for the Conservation of Nature (IUCN) defines a protected area as "a clearly defined geographical space, recognized, dedicated and managed, through legal or other effective means, to achieve the long-term conservation of nature with associated ecosystem services and cultural values" (Dudley, 2008). The IUCN recognizes six MPA categories, ranging from strict nature reserves that are established primarily for scientific research/monitoring or wilderness protection, to managed resource protection areas established mainly for sustainable use (Day et al., 2019).

Recently, The MPA Guide (n.d.), developed by a consortium of international organizations, scientists, and environmental groups, redefined existing language, clearly recognizing that the overarching purpose of an MPA is conservation of nature. MPAs' level of protection and the resources covered vary based on the objective of the designation. The MPA Guide classifies Level of Protection into four categories: Fully Protected, Highly Protected, Lightly Protected, and Minimally Protected. These categories define the level of extraction or destructive activities that are allowed. For example, a Fully Protected MPA would prohibit destructive activities and extraction of marine resources; these MPAs have also been referred to as "no-take MPAs" or "marine reserves". At the other extreme, a Minimally Protected MPA would allow extensive extraction of marine resources.

DOI: 10.4324/9781003058151-14

In addition, The MPA Guide categorizes areas by their Stage of Establishment: Proposed; Designated by legal or regulatory means; Implemented with defined boundaries, objectives, and management strategies; and Actively Managed with enforceable rules, monitoring, evaluation, and adaptive management. A matrix of Level of Protection vs. Stage of Establishment should theoretically lead to conservation outcomes that we will discuss further in this chapter. The strongest conservation outcomes should result from Fully Protected MPAs that are Actively Managed. Other marine-managed areas, referred to as OECM (Other Effective Area-Based Conservation Measures), may offer conservation benefits although conservation is not their primary objective. Examples of OECM include military restriction areas, buffer zones around aquaculture or ocean energy facilities, or marine transportation routes.

Although MPAs are an important conservation tool, alone they cannot achieve optimal conservation objectives and must be implemented as part of an integrated approach comprised of coastal and marine zone management and marine spatial planning (marine zoning) that addresses ecosystem-based management (Fraschetti et al., 2011; Halpern et al., 2010). In addition, MPAs may lead to numerous wellbeing outcomes in economic, social, and governance dimensions, as we will discuss (Ban et al., 2019).

History of MPAs

Humans have a long history of managing the marine environment for conservation objectives, exemplified by customary marine tenure systems established by seafaring communities in the western Pacific Ocean (Cinner, 2005; Cinner & Aswani, 2007; Wells et al., 2016). The modern MPA movement initially mirrored park enclosure systems – also known as fortress conservation – established by European states (including in their colonies), mainly for the purposes of marine resource conservation and habitat protection (Blaustein, 2007; Brockington, 2002; Wells et al., 2016).

Recreational demand for MPAs grew out of the invention and proliferation of SCUBA and snorkeling in the post–World War II era (Wells et al., 2016). This was coupled with an increasing concern for the stress on coastal and marine resources resulting from damaging fishing activities, pollution, and development, which researchers like Jacques Cousteau identified as long-term threats to the ocean (Cicin-Sain & Knecht, 2000). In 1960, the State of Florida (USA), for example, established one of the first underwater parks, John Pennekamp Underwater Coral Reef State Park, due to the threat of uncontrolled hard-coral harvest in the coral reefs of the Upper Florida Keys (Florida Department of Environmental Protection, 2019).

The watershed designation occurred in 1975 when Australia established the Great Barrier Reef Marine Park (GBRMP), which became the largest MPA in the world (344,000 km^2) (Wells et al., 2016). While it would take over a decade for similarly large MPAs to become established elsewhere, the GBRMP established various principles that are still applicable today, including ecosystem protection (later, ecosystem-based management) and zoning based on compatible uses by habitat type and vulnerability. In 1990, the US Congress designated the Florida Keys National Marine Sanctuary, which supported large-scale zoning and no-take marine reserves as part of an ecosystem-based approach to coral reef management. Other large MPAs were established via the National Marine Sanctuary Act in the US throughout the 1990s, designated to protect a variety of resources and with a focus on ecosystem protection (National Academy of Public Administration, 2000).

By the early 21st century, international organizations and agreements recognized the lack of representation and overall failure in protecting a meaningful percentage of the world's oceans (Humphreys & Clark, 2020). Member States of the Convention on Biological Diversity agreed to the 2010 Aichi Targets, whose Target 11 called for the establishment of MPAs in 10% of the world's coastal and marine areas. Regional initiatives such as the Caribbean Challenge set up targets for individual states to achieve a 20% target of MPA designation within their jurisdictional boundaries. Also, in 2015 the United Nations General Assembly (UNGA) finalized 17 Sustainable Development Goals (SDGs) to be achieved by the year 2030 to ensure a sustainable future for humanity. SDG 14 (Life Below Water) consists of ten targets to conserve the world ocean, and Target 14.5 calls for the protection of 10% of global coastal and marine areas by the year 2020 (United Nations, 2015). The push to protect ever-wider areas of the ocean continues to grow. A campaign to protect 30% of the world's oceans by 2030 is now supported by over 70 nations and is a goal of the Post-2020 Global Biodiversity Framework being negotiated under the Convention on Biological Diversity. In January 2021, US President Joe Biden issued Executive Order 14008, which set a goal to preserve 30% of US lands and waters by 2030. This campaign is often referred to as "30 by 30".

To address these and other national and international initiatives, nations worked unilaterally and, in some cases, via international agreements to increase the total percentage of MPAs. While the drive to increase MPA coverage did not reach the proposed SDG target of 10% by the year 2020, the call did lead to a protection of 6.4% of the world ocean, including 2.6% that is fully protected and 3.8% that is partially protected (Marine Conservation Institute, 2020). Part of the reason for the growth in MPA coverage in the 21st century was the rise of Very Large MPAs (VLMPAs), which are MPAs greater than 100,000 km^2. Overall, VLMPAs comprise two-thirds of total MPA area in the ocean, and while these offer ecological benefits, such as ecosystem integrity and function, biodiversity protection, and ecological spillover potential, critics have argued that VLMPAs provide a false sense of progress while protecting fewer representative areas, that the large remote areas are often established at the neglect of protecting smaller nearshore ones, and that these designations do little to address global stressors (O'Leary et al., 2018).

Another related, significant initiative that commenced in the 2010s was the 2017 UNGA resolution and subsequent progress toward the development of a global treaty under the umbrella of the UN Convention on the Law of the Sea (UNCLOS) to protect and manage biodiversity in areas beyond national jurisdiction (ABNJ) (De Santo et al., 2020). The *high seas* comprise about two-thirds of the world's ocean, and this area is afforded some protection via several fisheries agreements, pollution and mining regimes, as well as limited efforts in establishing MPAs in the Northwest Atlantic and the Southern Ocean.

Ecological Dimensions of MPAs

The conservation outcomes and effectiveness of MPAs relate to their design, which in turn should be dependent on the ecological characteristics of a given location. The key ecological dimensions of importance to MPAs are: siting/location and size, connectivity, spillover effects, and larval export. To appreciate the potential that MPAs offer as a management tool, it is important to review their ecological dimensions, especially as these relate to the ecological and fishery benefits that well-designed and managed MPAs produce. Ecological theory establishes the tenets for MPA function, contending that biodiversity protection and ecological integrity will have positive effects on ecosystem function and resilience (and hence, sustainability in the face of ecological disturbances) (Claudet et al., 2011).

Location

By setting aside areas, especially Fully Protected MPAs, where no extractive activities are allowed, biodiversity and ecological relationships can be protected where they remain intact or be allowed to recover where they have been disrupted. Recovery over time can lead to a functional and more resilient ecosystem that is better able to withstand disturbances than an altered ecosystem (Worm et al., 2006). Generally, a positive relationship exists between biodiversity and ecosystem function; ecosystem resiliency is tied to functional redundancy and intact trophic levels. These are often the very components that are rendered vulnerable due to anthropogenic impacts, such as overfishing. Well-designed and implemented MPAs can thus restore and protect biodiversity (Worm et al., 2006).

Considerable heterogeneity exists between ocean areas, and certain ecosystems are more biodiverse or productive than others (Roff & Zacharias, 2011). Ecosystems, like coral reefs, present unique characteristics that result in higher levels of biodiversity or production, as do other discrete sites, such as ecological hotspots, endemism centers, and aggregation sites (Claudet et al., 2011; Edgar et al., 2008). Spawning aggregation sites, for example, serve as sources for larval production and "reseed" sinks across a region, comprising a metapopulation; if such sites are exhausted (via serial overfishing), the associated populations dependent on the sites may eventually be eliminated (Grüss et al., 2014). These are examples of ocean ecosystems that are high-priority MPA sites.

Connectivity

Considerable connectivity occurs across ocean areas, such that ecosystems are invariably connected to each other across horizontal and vertical, permeable boundaries. An important component of ecological theory concerns connectivity and relates to the size and/or number of protected sites required to fully protect biodiversity and ensure ecosystem function (Claudet et al., 2011). Species-area relationships dictate that protected areas should be large enough to contain all the species that characterize that area (habitat), but within marine environments, the mobility of species varies considerably even when they might occur together within the same habitat (Grüss et al., 2011). For example, Caribbean spiny lobster (*Panulirus argus*) has a pelagic larval duration (PLD) of up to nine months, such that a phyllosoma (larva) hatched in the water column in the western Caribbean may settle in shallow water habitats off southern Florida (Kough et al., 2013). Also, some species, such as the spiny lobster, change habitat preferences over their lifespans, switching from pelagic to demersal phases based on life history stage. Yet there are other species, such as salmonids or certain reef fish, that migrate on a seasonal basis for spawning or other purposes, greatly expanding their ranges for part of a year/season. Finally, there are wholly pelagic species, such as tuna, that occupy large ranges across entire ocean basins. These examples suggest that MPA networks may present better solutions than a single MPA. Establishment of MPA networks requires identification and protection of habitats that serve as stations for these species at their most vulnerable life history stage (Food and Agriculture Organization of the United Nations, 2011; Grüss et al. 2011). MPA networks can cover large expanses of the ocean if well-designed, i.e., if based on the species' ranges, movement patterns, and other ecological relationships that affect mobility. Networks may also protect certain areas that serve as sources for species export and others that are sinks via species import (Cowen et al., 2006).

Spillover

Another ecological aspect of MPA theory is species' net movement outside the MPA through a process known as ecological spillover (Di Lorenzo et al., 2016). Factors driving spillover include density-dependent movements resulting from density-related conditions (Grüss et al., 2011). Movements into open areas resulting from effective MPA establishment are related to greater intra-specific competition, increased prey shortages, and lack of suitable habitat (for refuge, courtship, etc.), due to increases in both total population and population density. The net effect expected is a gradient of higher species abundance within the MPA to decreasing totals farther away from the MPA (Goñi et al., 2011).

Larval Export

MPAs may confer what is referred to as larval subsidy, resulting from greater target species and larger average sizes of individuals within MPAs that can result in higher larval production to repopulate open areas outside the MPA. However, successful larval subsidy, like fisheries spillover, is subject to several factors, including variability in recruitment and settlement and questions of whether recruits end up in open areas (Goñi et al., 2011). Larval export may benefit regional fisheries and biodiversity.

The potential conservation outcomes for MPAs are many and include: greater abundance and size of species previously exploited, increased biodiversity and habitat recovery, and enhanced reproductive output due to larger body size of many species. As a result, MPAs may improve or restore ecological indicators and provide ecosystem "insurance", or resilience, in the event of disruptions (ecological, physical, or anthropogenic).

Economic Dimensions of MPAs

Whether they are established for ecosystem protection, fisheries management, or ecotourism, research suggests that MPAs provide economic value through the sustained delivery of ecosystem services. Ecosystem protection results in ecosystem service benefits, including sustainable fisheries (provisioning services), habitat protection (supporting services), primary production (regulating services), and tourism (cultural services) (Leenhardt et al., 2015). Fisheries management refers to the spillover and larval subsidy and insurance policies that MPAs provide in the form of fishery catches, livelihoods, and coastal community health. Ecotourism in MPAs, as a cultural service, provides income to local communities and industries (including fisheries) and helps to divert pressure from extractive uses when marine reserves are established to accommodate divers' and snorkelers' demands for larger-sized species, higher biodiversity, and unfished habitat.

In economic terms, MPAs provide market and non-market benefits, the former of which can be measured via market transactions, while the latter cannot (and thus require non-market valuation methods) (Davis et al., 2019). By contrast, MPAs also incur costs related to establishment, compliance, and maintenance. The total economic value is the difference between the total benefits and total costs associated with MPA establishment. Different types of use and non-use values incorporate total economic value. Use values are comprised of direct use, indirect use, and option value. *Direct use* refers to those outputs or services used directly, such as fisheries or tourism-related activities. An MPA may produce *indirect use* benefits via ecosystem function, such as ecosystem resilience. *Option value* is the combined, future direct and indirect use. Non-use values consist of those values that conserve resources for future generations (*bequest value*) and those that are related to the knowledge that a resource may continue to exist (*existence value*), such as the protection of endangered species.

The beneficiaries of the economic value provided by MPAs are businesses that rely on higher or less variable catches that fisheries spillover can offer; tourism operators who benefit from a higher demand for services due to MPA establishment and related biodiversity changes; consumers who benefit from a higher supply and quality of fisheries and tourism-related resources due to MPA establishment; and local communities that enjoy increases in public revenues via fees or taxes due to greater economic activity following MPA establishment (Davis et al., 2019).

While MPAs provide benefits, the restrictions on resource access and use may negatively affect some stakeholders and thus be socially problematic. Operational costs associated with MPAs relate to establishment, maintenance, and compliance costs (see Box 11.1). In fact, the main factor associated with MPA success is adequate staff capacity (Gill et al., 2017), indicating the importance of building, training, and maintaining a strong staff presence for interpretation, monitoring, and compliance purposes. Also important from an economic cost perspective is the relationship between MPA size, distance from the coast, and the MPA country's development status and cost per unit area (Balmford et al., 2004). Generally, costs for MPAs tend to decrease in a nonlinear manner per unit area protected, suggesting that larger, more remote MPAs may provide the best return on investment in terms of establishment and maintenance costs (Edgar et al., 2014; McCrea-Strub et al., 2011).

BOX 11.1 Funding for MPAs

Established in 1979, Bonaire National Marine Park (BNMP) includes all the waters surrounding the Caribbean Island of Bonaire from high water to 60 m water depth. The park is a well-known dive destination with high-quality coral reefs, many of which are accessible from shore. BNMP is also considered to be a model in the Caribbean for effective MPA management. Anchoring anywhere in the park is prohibited, and mooring buoys are abundant. In large part, the effective management in BNMP stems from its funding mechanism. Implementation of a fee system for SCUBA divers began in 1992 and, as a result, the park became largely self-financing that year. The fees support park maintenance, research and monitoring, educational activities, and law enforcement. The original diver fee was $10 per year. The Bonaire government approved a new fee structure in 2019. Today the "Nature Fee" costs $45 per year for SCUBA divers and $25 per year for other water activities. It is possible to pay the Nature Fee online and purchase the required "Nature Tag". In recent years, the number of cruise ship visitors has grown, and controversy surrounds visitors' payment of the Nature Fee. Cruise ship visitors who SCUBA dive must pay the fee. However, passengers are exempt from the non-SCUBA activity fee.

Social Dimensions of MPAs

Over the past three decades, studies in the social dimensions of MPAs have found that MPAs can provide ecological and economic benefits and yet be rejected by coastal communities and stakeholders (Christie, 2004; Suman et al., 1999). This is because numerous social issues often remain unresolved prior to MPA implementation, including the use of local ecological knowledge in MPA design, imbalanced distributional benefits, opacity in the planning process, lack of meaningful participation (especially engagement with those stakeholders who stand to lose access rights or areas), and the lack of establishment of effective conflict resolution mechanisms. The sense of distrust and the perceived illegitimacy of the planning process surrounding the designated MPA may hinder maintenance, monitoring, and compliance; increase distrust (erode social capital); and endanger the long-term sustainability of the MPA. Effectively, MPAs are spatial management tools developed to address the status of marine resources by managing uses and behaviors to address societal goals such as food security, sustainable livelihoods, and distributional benefits; the long-term sustainability of MPAs can only be ensured if these societal goals are effectively addressed to modify uses and behaviors (Pomeroy et al., 2007).

A common concern among communities, particularly fishery stakeholders, is that MPAs are often implemented without meaningful consultation with local ecological knowledge related to high-value areas, use areas, and areas representing conflict (Christie, 2004; Pita et al., 2011). Other MPA skeptics perceive that the process is closed to them and is a means to decrease or eliminate their presence (Suman et al., 1999). Studies have demonstrated that when channels of dialogue and consultation are opened to stakeholder input throughout the MPA process, while certain groups may still not consider the outcomes to be beneficial, perceptions concerning the validity of the process and its legitimacy increase (Dalton, 2005; Shivlani et al., 2008). (See Box 11.2.)

BOX 11.2 Stakeholder Involvement in MPA Formation

Designated by the US Congress in 1990, the Florida Keys National Marine Sanctuary (FKNMS) is one of the largest MPAs administered by the National Oceanic and Atmospheric Administration (NOAA); the actual area is 7,800 km^2. Several years of contentious debate surrounded the release of the Final Sanctuary Management Plan in 1997. NOAA had proposed a large no-take marine reserve in the extreme western region of the FKNMS, called the Dry Tortugas Replenishment Reserve. However, NOAA opted to postpone creation of this no-take reserve due to the extreme opposition from organized commercial and recreational fishers who argued the closure would impact their economies and, moreover, that the boundaries were incorrect.

NOAA returned to this issue several years later, adopting a well-orchestrated participatory management strategy called "Tortugas 2000" (Delaney, 2003). First, the agency changed the name of the no-take area to "Dry Tortugas Ecological Reserve" (DTER) to lessen debate about the effectiveness of spillover results. NOAA formed a working group of 25 representatives of various target fisheries (shrimp, finfish, crabs, lobster), members of the Sanctuary Advisory Council, as well as federal and state officials exercising authority in the FKNMS. The group was charged with defining the boundaries of the DTER. Representatives noted specific areas in the Dry Tortugas region that were extremely important to their fishing operations and, subsequently, in May 1999 reached consensus on boundaries of the DTER that would not have extreme adverse impacts on any of the fishing interests. The outcome was the designation of the no-take Tortugas Ecological Reserve (comprising two sectors: Tortugas North and Tortugas South) that together limit extractive activities over 635 km^2 of high-quality marine habitat (coral reefs, seagrass beds, hardbottom).

Another aspect of the social dimensions of MPAs relates to the post-implementation stage. Unfortunately, due to a combination of inadequate capacity and funding, imbalances in distributional benefits, and lack of effective conflict resolution mechanisms, MPAs can and do sometimes fail after enjoying initial success. These so-called "paper parks" remain on the books and may even count toward percentage targets, but they do not contribute toward any meaningful conservation outcomes because they are not actively managed (Watson et al., 2014). While capacity remains a key factor in determining MPA success (Gill et al., 2017), continued stakeholder engagement, socioeconomic monitoring, and outreach and education also affect MPA outcomes (Pomeroy et al., 2007).

Governance Dimensions of MPAs

MPA governance dimensions that comprise these mechanisms are best exemplified as top-down approaches, bottom-up approaches, hybrid approaches that utilize both top-down and bottom-up approaches, and community-based approaches (Jones et al., 2019). Governance by itself refers to how societies reach decisions, and this involves a continuous negotiation between different stakeholders concerning legal and customary norms and economic influences. All approaches utilized to establish and manage MPAs require good governance characterized by transparency, participation, and respect for the stakeholders and communities affected by the decisions (World Bank, 2006). The United Nations Environment Programme (UNEP) states that good governance must be inclusive and promote stewardship with a focus on economic and social benefits and environmental improvements, and the governance structure must adapt to changing circumstances (Jones et al., 2019).

Top-down approaches are centralized, and decisions are government-driven (Vann, 2010). Bottom-up approaches are decentralized, and decisions are devolved to the local level (Bown et al., 2013). Community-based approaches involve governance at the local level under either formal or informal collective-management agreements. In these arrangements, communities may wholly manage MPAs and their resources with minimal central government oversight. Hybrid approaches incorporate elements of top-down and community-based approaches as discussed below. They also allow for a certain level of decentralization, and decision-making may be hierarchically organized (Gaymer et al., 2014; see Box 11.3.).

BOX 11.3 Community Participation in MPA Management

Brazil created a category of protected area in 2000 called Extractive Reserves (RESEX) that today are administered by the Chico Mendes Institute for Biodiversity Conservation (ICMBio). The objective of these reserves is promotion of sustainable use of natural resources and protection of traditional livelihoods. Brazil has designated over 90 Extractive Reserves, about one-quarter of which are marine and coastal. Canavieiras is a 1,006 km^2 marine Extractive Reserve located in Bahia State (East-central Brazil) that was created in 2006. About 85% of the reserve is marine, while a coastal mangrove forest forms the remainder. Canavieiras is managed by a Deliberative Council presided over by a representative of ICMBio. The Council is a good example of co-management, with council members representing the national environmental agencies, municipal governments, the state university, tourism businesses, environmental groups, and numerous subsistence farmers and artisanal fishing organizations. Only recognized community organizations receive traditional use rights to extract natural resources (mangrove wood, marine resources) and engage in tourist-support activities inside the reserve under strictly defined rules that guarantee sustainable use. The reserve supports livelihoods of an estimated 1,300 families.

Coiba National Park in the Pacific Ocean near southwest Panama protects a total of 2,701 km^2: marine waters (2,165 km^2) and insular lands (536 km^2) that include the island of Coiba, a former penal colony, and some 38 smaller islands. The park's marine and terrestrial habitats feature high biodiversity and marine and terrestrial endemism, as well as numerous threatened species. Created by Panamanian law in 2004, UNESCO inscribed Coiba National Park as a natural World Heritage Site the following year. Coiba is an important component of the Eastern Tropical Pacific Marine Corridor, which is formed of the oceanic areas of Panama, Costa Rica, Colombia, and Ecuador. Although it composes part of the Ministry of the Environment's National System of Protected Areas, the park is managed by means of an innovative governance model with coordination and participation of both authorities and stakeholders. The Coiba Directive Council is the ultimate authority over the national park and has members representing national and local authorities, environmental NGOs, universities and scientific research institutions, and the productive sector. Some areas of the park are no-take marine reserves, while others allow for highly regulated fishing that provides an important source of income for several nearby coastal communities.

Top-down approaches concentrate decision-making power within the central government, and these tend to work best in the cases where there are national MPAs (i.e., those located in an exclusive economic zone, or EEZ) or MPAs in ABNJ (i.e., those located in the high seas and overseen by international regimes that have multilateral representation) (Jones et al., 2019; Petra, 2012). For example, the US has established a series of VLMPAs that encompass the entire EEZs surrounding its Pacific Ocean territories, and these are managed by the federal government under the Department of Commerce (with input from territorial governments). Parties to the OSPAR Convention, a UN Regional Seas Programme that oversees maritime and coastal activities in the Northeast Atlantic Ocean, have established a series of MPAs in ABNJ over the past decade, and they jointly coordinate MPA management activities (mainly research and monitoring).

Bottom-up approaches rely on local governments and institutions to manage MPAs, and these succeed in areas where centralized approaches may be too weak, far away, or otherwise non-responsive to local needs (Bown et al., 2013). Bottom-up approaches also are advantageous where there is an underlying tradition of decentralization. In the US, local and state agencies manage local MPAs that are located totally within their respective jurisdictional boundaries. In several cases, however, a scalar approach may be required to achieve full protection (Dietz et al., 2003). This requires coordination between local and centralized authorities, such as is exercised between the California Coastal Commission (a state agency) and NOAA (a federal agency) to jointly manage the Channel Islands National Marine Sanctuary off Southern California.

Community-based approaches, which range from consultative to fully community-based management, are those in which communities self-organize to manage MPAs with minimal governmental oversight (Bown et al., 2013). These approaches work well in situations similar to those of bottom-up approaches, but community-based approaches elevate community-level initiatives and regulations over those implemented by a governmental entity. These approaches have been utilized successfully in areas that have a long history of self-organization at the local level, especially in coastal developing states, as well as in areas that lack effective government capacity. The Philippines, for example, has numerous local MPAs that are co-managed, with assistance from local governments or non-governmental organizations. The Locally Managed Marine Areas (LMMA) Network is an NGO that promotes and informs community-based MPA management in Southeast Asia, the Pacific, and the Americas. These and other, related national and international efforts seek to reestablish past practices, such as customary marine tenure or strengthening community capacity, empowering local entities to conserve and manage their own resources (Bown et al., 2013).

MPA Design and Planning

MPA design is invariably influenced by numerous factors, including ecological requirements, economic impacts, social equity considerations, governance capacity, and political will, but design should be grounded by the MPA's primary objectives (Day et al., 2019; Salm et al., 2000). Once an MPA has been designed to meet its objectives, the next step in MPA implementation is MPA planning. Effective planning should employ an adaptive, long-term strategy, devised to set a direction for MPA growth and maturation toward objective completion, as well as operational tactics, consisting of ongoing operations and actions in support of effective MPA implementation and maintenance.

The first step in MPA planning is site characterization, including a full assessment of the MPA, its biophysical conditions, ecological attributes, social setting, and economic sectors. Stakeholder and community engagement and participation should be prioritized throughout the planning process, commencing with site characterization, and the emphasis should be on the inclusion of all stakeholders to obtain local ecological knowledge, identify use areas, and assess MPA-related impacts. Engagement permits planners to identify the most effective modes of dialogue, learn of any conflict issues, build trust between stakeholders, and determine means to implement conflict resolution mechanisms.

The primary strategic document for an MPA is its management plan, consisting of the MPA characterization, goals and objectives, and proposed actions to be taken (zoning, outreach and education, public use, enforcement) to achieve the MPA goals and objectives. The management plan is not a tactical document identifying daily operations, nor should the management plan be a static document, as both ecological and socioeconomic environments are dynamic and may require strategic changes. For example, the 2019 Florida Keys National Marine Sanctuary Habitat Restoration Blueprint revised the Sanctuary's management plan, emphasizing a greater focus on resource rehabilitation and recovery, especially concerning the long-term decline of the region's coral reef ecosystem.

MPA planning must include an outreach and education effort that continues past the planning stage to provide the necessary information to stakeholders to build and reinforce support; provide updates on management goals, objectives, and activities; and dispel any inaccuracies or counter-narratives that may otherwise engender opposition. MPA advocates, such as non-governmental organizations, government-funded advisory groups, and citizens' groups, can all play important, supporting roles in broadcasting critical MPA information to others and, especially, to specific stakeholders who may otherwise be unreceptive to management agency sources.

Another critical aspect of MPA planning is the creation of monitoring and compliance strategies. Monitoring MPA trends and outcomes is the best way to determine the degree to which the MPA is in fact effective, and to identify deficiencies. Compliance refers to the extent to which stakeholders obey MPA rules. Enforcement is used when rules are broken with prior knowledge and cause major impacts. Thus, within an MPA, compliance can be used to estimate the level and type of rule breaking, and enforcement measures can be applied appropriately to address these infractions.

MPA Effectiveness

Ongoing discussions about global conservation targets raise questions about how to assess and measure the effectiveness of existing areas. MPA effectiveness, itself a vast field of study, is subject to how effectiveness is defined, and it can refer to whether an MPA (1) meets its defined objectives, (2) provides expected or ancillary benefits to other MPAs and the overall region as a node in an MPA network, (3) is not a "paper park" (i.e., it has the necessary governance, funding, compliance, and enforcement required to prevent violations), or (4) has gained stakeholder acceptance. We suggest three areas to measure MPA effectiveness: (1) the overall effectiveness of MPAs as a global management tool, (2) the effectiveness of individual MPAs relative to their goals and objectives, and (3) the perceived effectiveness of individual MPAs relative to stakeholder knowledge, attitudes, and perceptions.

As of 2020, just over 7.5% of the world's coasts and ocean was designated as MPAs, and less than half of that total was set aside as no-take marine reserves (Marine Conservation Institute, 2020). Perhaps an even greater indictment than the global shortfall is that the vast majority of the total MPA coverage is contained in a few VLMPAs. These nation-sized MPAs are not representative, as most are located in the Pacific Ocean (with the exception of the Chagos MPA in the Indian Ocean, the OSPAR MPA network in the Northeast Atlantic Ocean, and the South Orkney Islands MPA and Ross Sea MPA in the Southern Ocean). Several have also come under criticism for having employed a fortress conservation approach, where native (often underrepresented) communities have been marginalized in the implementation process. Thus, while global initiatives, such as the Aichi Targets and SDG 14 goals call for greater coverage, the areas protected must be effectively managed, established through democratic principles (such as open and full participation), and accepted by stakeholders.

MPAs must also meet their stated goals and objectives to be considered effective, and this requires a clear and targeted design, an open and comprehensive planning process, and consistent monitoring to identify trends and outcomes that may require changes in management strategies. MPAs cannot be expected to achieve their objectives without an adaptive framework, especially considering the rapidly changing climate. For example, coral reef MPAs may require enhanced restoration efforts in areas where more frequent, warm water episodes have worsened coral bleaching, disease, and mortality. Similarly, fishery spillover may be affected by changes in the mean temperature at which target species are concentrated such that target species may migrate completely out of a warming MPA. There are no preestablished protocols on how to address these types of unprecedented, climate-induced changes, but MPAs accounting for the likelihood of such change can integrate adaptations to remain effective into the foreseeable future.

Finally, the social aspects of MPA effectiveness are the most important. MPAs, after all, represent a social agreement decided upon by stakeholders to alter human behavior and attitudes. If stakeholders feel alienated by the MPA, whether due to misinformation, lack of expected revenue, or changes in distributional benefits, an MPA may be subject to long-term failure resulting from a lack of support and compliance. An MPA must measure effectiveness not solely in terms of ecological success but also by stakeholder engagement, acceptance, and support. This requires the MPA management agency proponents to provide vehicles to receive stakeholder input and criticism (via advisory bodies), act on stakeholder concerns where these warrant changes in management strategy, and partner with stakeholders in monitoring MPA trends and arriving at mutually acceptable solutions.

Conclusion

MPAs are excellent, flexible, place-based conservation tools that are increasingly adopted throughout the world for fishery management (spillover of adults, larval export, larger and more abundant fish, increased biomass) and protection of biodiversity. Recognition has grown of the importance in siting MPAs in areas of enhanced biological importance (high biodiversity, spawning aggregation sites) and configuring them in ecologically meaningful networks. MPAs may lead to many positive wellbeing outcomes, as well. The socioeconomic and governance aspects of MPAs are of utmost importance, but research in these areas is lacking. Local communities and marine resource users must be involved in MPA creation and management. Emerging areas in the science of MPAs concern their relevance in times of climate change that is altering environmental conditions and causing geographical shifts in species of interest. The creation and management of MPAs on the high seas (areas beyond national jurisdiction) present additional challenges.

Professional Pathways

Numerous professional opportunities exist in marine conservation and MPA arenas. Many marine policy graduates find employment at local, state/provincial, or national MPAs with responsibilities related to education and outreach, scientific research and monitoring, and enforcement. Environmental and conservation organizations offer other possibilities in support of marine conservation projects and campaigns, lobbying decision-makers for increasing support of conservation policies. Many business ventures (tourism, diving, fishing) leverage MPAs as part of their attractions, employing professionals trained in marine science and policy. Academic research groups often address a variety of questions related to MPA ecological benefits, size and location, socioeconomic impacts, governance models, and public perceptions.

Questions for Reflection

1. Even if the international community agrees to a goal of protecting 30% of the ocean by 2030, implementation of this goal will be difficult. What obstacles might stand in the way to effective implementation of the "30 by 30" goal?
2. What implications might global climate change (sea level rise, changing ocean currents, increasing ocean temperatures, ocean acidification) have for MPAs?
3. Do you think that it is generally easier to manage a MPA or a terrestrial protected area? Why or why not?
4. How can marine resource users, such as fishers, become part of MPA planning and management? Why might this be beneficial?
5. What factors would you consider in designing an MPA?
6. Make arguments in favor of and against allowing tourism (fishing, diving) in MPAs.
7. What are "paper parks" and how can these be avoided when establishing MPAs?
8. Is fortress conservation a suitable approach to establish MPAs? Why or why not?

9. Restricted-entry zones, such as areas around petroleum platforms, ocean wind turbines, ocean aquaculture pens, and military use areas, may provide some conservation benefits. Do you think that they could be considered MPAs?
10. If an MPA is established on the high seas, who would manage it?

References

Balmford, A., Gravestock, P., Hockley, N., McClean, C. J., & Roberts, C. M. (2004). The worldwide costs of marine protected areas. *Proceedings of the National Academy of Sciences of the United States of America, 101*(26), 9694–9697. https://doi.org/10.1073/pnas.0403239101

Ban, N. C., Gurney, G. G., Marshall, N. A., Whitney, C. K., Mills, M., Gelcich, S., Bennett, N. J., Meehan, M. C., Butler, C., Ban, S., Tran, T. C., Cox, M. E., & Breslow, S. J. (2019). Well-being outcomes of marine protected areas. *Nature Sustainability*, 2(6), 524–532. https://doi.org/10.1038/s41893-019-0306-2

Blaustein, R. J. (2007). Protected areas and equity concerns. *BioScience, 57*(3), 216–221. https://doi.org/10.1641/B570303

Bown, N., Gray, T. S., & Stead, S. M. (2013). *Contested forms of governance in marine protected areas: A study of co-management and adaptive co-management.* Routledge.

Brockington, D. (2002). *Fortress conservation: The preservation of the Mkomazi game reserve, Tanzania.* Indiana University Press.

Christie, P. (2004). Marine protected areas as biological successes and social failures in Southeast Asia. *American Fisheries Society Symposium*, 42, 155–164.

Cicin-Sain, B., & Knecht, R. (2000). *The future of US ocean policy: Choices for the new century*. Island Press.

Cinner, J. (2005). Socioeconomic factors influencing customary marine tenure in the Indo-Pacific. *Ecology and Society, 10*(1), 36. https://www.jstor.org/stable/26267728

Cinner, J. E., & Aswani, S. (2007). Integrating customary management into marine conservation. *Biological Conservation, 140*(3–4), 201–216. https://doi.org/10.1016/j.biocon.2007.08.008

Claudet, J., Guidetti, P., Mouillot, D., Shears, N. T., & Micheli, F. (2011). Ecological effects of marine protected areas: Conservation, restoration, and functioning. In J. Claudet (Ed.), *Marine protected areas: A multidisciplinary approach* (pp. 37–71). Cambridge University Press.

Cowen, R. K., Paris, C. B., & Srinivasan, A. (2006). Scaling of connectivity in marine populations. *Science, 311*(5760), 522–527. https://doi.org/10.1126/science.1122039

Dalton, T. M. (2005). Beyond biogeography: A framework for involving the public in planning of US marine protected areas. *Conservation Biology, 19*(5), 1392–1401. https://doi.org/10.1111/j.1523-1739.2005.00116.x

Davis, K. J., Vianna, G. M., Meeuwig, J. J., Meekan, M. G., & Pannell, D. J. (2019). Estimating the economic benefits and costs of highly protected marine protected areas. *Ecosphere, 10*(10), e02879. https://doi.org/10.1002/ecs2.2879

Day, J., Dudley, N., Hockings, M., Holmes, G., Laffoley, D., Stolton, S., Wells, S., & Wenzel, L. (Eds.). (2019). *Guidelines for applying the IUCN protected area management categories to marine protected areas* (2nd ed.). IUCN.

Delaney, J. M. (2003). Community capacity building in the designation of the Tortugas Ecological Reserve. *Gulf and Caribbean Research, 14*(2), 163–169. https://doi.org/10.18785/gcr.1402.13

De Santo, E. M., Mendenhall, E., Nyman, E., & Tiller, R. (2020). Stuck in the middle with you (and not much time left): The third intergovernmental conference on biodiversity beyond national jurisdiction. *Marine Policy, 117*, 103957. https://doi.org/10.1016/j.marpol.2020.103957

Dietz, T., Ostrom, E., & Stern, P. C. (2003). The struggle to govern the commons. *Science, 302*(5652), 1907–1912. https://doi.org/10.1126/science.1091015

Di Lorenzo, M., Claudet, J., & Guidetti, P. (2016). Spillover from marine protected areas to adjacent fisheries has an ecological and a fishery component. *Journal for Nature Conservation, 32*, 62–66. https://doi.org/10.1016/j.jnc.2016.04.004

Dudley, N. (Ed.). (2008). *Guidelines for applying protected area management categories*. IUCN.

Edgar, G. J., Langhammer, P. F., Allen, G., Brooks, T. M., Brodie, J., Crosse, W., De Silva, N., Fishpool, L. D., Foster, M. N., Knox, D. H., McCosker, J. E., McManus, R., Millar, A. J., & Mugo, R. (2008). Key biodiversity areas as globally significant target sites for the conservation of marine biological diversity. *Aquatic Conservation: Marine and Freshwater Ecosystems, 18*(6), 969–983. https://doi.org/10.1002/aqc.902

Edgar, G. J., Stuart-Smith, R. D., Willis, T. J., Kininmonth, S., Baker, S. C., Banks, S., Barrett, N. S., Becerro, M. A., Bernard, A. T., Berkhout, J., Buxton, C. D., Campbell, S. J., Cooper, A. T., Davey, M., Edgar, S. C., Försterra, G., Galván, D. E., Irigoyen, A. J., Kushner, D. J. … Thomson, R. J. (2014). Global conservation outcomes depend on marine protected areas with five key features. *Nature, 506*(7487), 216–220. https://doi.org/10.1038/nature13022

Food and Agriculture Organization of the United Nations. (2011). Fisheries management. 4. Marine protected areas and fisheries. *FAO Technical Guidelines for Responsible Fisheries*. Retrieved July 6, 2022 from https://www.fao.org/3/i2090e/i2090e.pdf

Food and Agriculture Organization of the United Nations. (2020). *The State of World Fisheries and Aquaculture 2020: Sustainability in Action*. https://doi.org/10.4060/ca9229en

Florida Department of Environmental Protection. (2019, February 15). *John Pennekamp Coral Reef State Park: Approved Unit Management Plan*. DRP/FDEP. Retrieved July 6, 2022 from https://floridadep.gov/sites/default/files/JPCRSP_02.15.2019%20Approved%20UMP.pdf

Fraschetti, S., Claudet, J., & Grorud-Colvert, K. (2011). Transitioning from single-sector management to ecosystem-based management: What can marine protected areas offer? In J. Claudet (Ed.), *Marine protected areas: A multidisciplinary approach* (pp. 11–36). Cambridge University Press.

Gaymer, C. F., Stadel, A. V., Ban, N. C., Cárcamo, P., Ierna, J., & Lieberknecht, L. M. (2014). Merging top-down and bottom-up approaches in marine protected areas planning: Experiences from around the globe. *Aquatic Conservation: Marine and Freshwater Ecosystems, 24*(S2), 128–144. https://doi.org/10.1002/aqc.2508

Gill, D. A., Mascia, M. B., Ahmadia, G. N., Glew, L., Lester, S. E., Barnes, M., Craigie, I., Darling, E. S., Free, C. M., Geldmann, J., Holst, S., Jensen, O. P., White, A. T., Basurto, X., Coad, L., Gates, R. D., Guannel, G., Mumby, P. J., Thomas, H. … Fox, H. E. (2017). Capacity shortfalls hinder the performance of marine protected areas globally. *Nature, 543*(7647), 665–669. https://doi.org/10.1038/nature21708

Goñi, R., Badalamenti, F., & Tupper, M. H. (2011). Fisheries – Effects of marine protected areas on local fisheries: Evidence from empirical studies. In J. Claudet (Ed.), *Marine protected areas: A multidisciplinary approach* (pp. 72–98). Cambridge University Press.

Grüss, A., Kaplan, D. M., Guénette, S., Roberts, C. M., & Botsford, L. W. (2011). Consequences of adult and juvenile movement for marine protected areas. *Biological Conservation, 144*(2), 692–702. https://doi.org/10.1016/j.biocon.2010.12.015

Grüss, A., Robinson, J., Heppell, S. S., Heppell, S. A., & Semmens, B. X. (2014). Conservation and fisheries effects of spawning aggregation marine protected areas: What we know, where we should go, and what we need to get there. *ICES Journal of Marine Science*, *71*(7), 1515–1534. https://doi.org/10.1093/icesjms/fsu038

Halpern, B. S., Frazier, M., Afflerbach, J., Lowndes, J. S., Micheli, F., O'Hara, C., Scarborough, C., & Selkoe, K. A. (2019). Recent pace of change in human impact on the world's ocean. *Scientific Reports*, 9(1), 1–8. https://doi.org/10.1038/s41598-019-47201-9

Halpern, B. S., Lester, S. E., & McLeod, K. L. (2010). Placing marine protected areas onto the ecosystem-based management seascape. *Proceedings of the National Academy of Sciences*, *107*(43), 18312–18317. https://doi.org/10.1073/pnas.0908503107

Humphreys, J., & Clark, R. W. (2020). A critical history of marine protected areas, In J. Humphreys & R. W. Clark (Eds.), *Marine protected areas: Science, policy, and management* (pp. 1–12). Elsevier.

Jones, P. J. S., Murray, R. H., & Vestergaard, O. (2019). *Enabling effective and equitable marine protected areas: Guidance on combining governance approaches* (Regional Seas Reports and Studies No. 203). Ecosystems Division/UNEP. Retrieved July 6, 2022 from https://apo.org.au/sites/default/files/resource-files/2019-04/apo-nid228726.pdf

Kough, A. S., Paris, C. B., & Butler IV, M. J. (2013). Larval connectivity and the international management of fisheries. *PLoS One*, 8(6), e64970. https://doi.org/10.1371/journal.pone.0064970

Leenhardt, P., Low, N., Pascal, N., Micheli, F., & Claudet, J. (2015). The role of marine protected areas in providing ecosystem services. In A. Belgrano, G. Woodward, & U. Jacob (Eds.), *Aquatic Functional Biodiversity* (pp. 211–239). Academic Press.

Marine Conservation Institute. (2020). *The marine protected atlas*. https://mpatlas.org/

McCrea-Strub, A., Zeller, D., Sumaila, U. R., Nelson, J., Balmford, A., & Pauly, D. (2011). Understanding the cost of establishing marine protected areas. *Marine Policy*, *35*(1), 1–9. https://doi/10.1016/j.marpol.2010.07.001

National Academy of Public Administration. (2000). *Protecting Our National Marine Sanctuaries: A Report by the Center for the Economy and Environment*. Retrieved July 6, 2022 from https://montereybay.noaa.gov/research/techreports/trnapa2000.html

O'Leary, B. C., Ban, N. C., Fernandez, M., Friedlander, A. M., García-Borboroglu, P., Golbuu, Y., Guidetti, P., Harris, J. M., Hawkins, J. P., Langlois, T., McCauley, D. J., Pikitch, E. K., Richmond, R. H., & Roberts, C. M. (2018). Addressing criticisms of large-scale marine protected areas. *Bioscience*, 68(5), 359–370. https://doi.org/10.1093/biosci/biy021

Petra, D. (2012). Marine protected areas in areas beyond national jurisdiction. *The International Journal of Marine and Coastal Law*, *27*(2), 291–350. https://doi.org/10.1163/157180812X637975

Pita, C., Pierce, G. J., Theodossiou, I., & Macpherson, K. (2011). An overview of commercial fishers' attitudes towards marine protected areas. *Hydrobiologia*, *670*(1), 289–306. https://doi.org/10.1007/s10750-011-0665-9

Pomeroy, R. S., Mascia, M. B., & Pollnac, R. B. (2007). Marine protected areas: The social dimension. In *FAO expert workshop on marine protected areas and fisheries management: Review of issues and considerations* (pp. 149–275). FAO. Retrieved July 6, 2022 from http://citeseerx.ist.psu.edu/viewdoc/download?doi=10.1.1.120.4804&rep=rep1&type=pdf#page=157

Roff, J. C., & Zacharias, M. (2011). *Marine conservation ecology*. Earthscan.

Sala, E., & Knowlton, N., (2006). Global marine biodiversity trends. *Annual Review of Environmental Resources*, *31*(1), 93–122. https://doi.org/10.1146/annurev.energy.31.020105.100235

Salm, R. V., Clark, J. R., & Siirila, E. (2000). *Marine and coastal protected areas: A guide for planners and managers*. IUCN.

Shivlani, M., Leeworthy, V. R., Murray, T. J., Suman, D. O., & Tonioli, F. (2008). *Knowledge, attitudes and perceptions of management strategies and regulations of the Florida Keys National Marine Sanctuary by commercial fishers, dive operators, and environmental group members: A baseline characterization and 10-year comparison*. ONMS/NOS/NOAA/DOC. http://hdl.handle.net/1834/20077

Suman, D. O., Shivlani, M., & Milon, J. W. (1999). Perceptions and attitudes regarding marine reserves: A comparison of stakeholder groups in the Florida Keys National Marine Sanctuary. *Ocean & Coastal Management, 42*(12), 1019–1040. https://doi.org/10.1016/S0964-5691(99)00062-9

The MPA Guide. (n.d.). Retrieved July 20, 2022, from https://mpa-guide.protectedplanet.net/

United Nations. (2015). *Transforming Our World: The 2030 Agenda for Sustainable Development* (A/Res/70/1). Retrieved July 6, 2022 from https://sdgs.un.org/sites/default/files/publications/21252030%20Agenda%20for%20Sustainable%20Development%20web.pdf

Vann, A. (2010). Marine protected areas: Federal legal authority. In F. B. Mayr (Ed.), *Marine protected areas* (pp. 25–50). Nova Science Publishers, Inc.

Watson, J. E., Dudley, N., Segan, D. B., & Hockings, M. (2014). The performance and potential of protected areas. *Nature, 515*(7525), 67–73. https://doi.org/10.1038/nature13947

Wells, S., Ray, G. C., Gjerde, K. M., White, A. T., Muthiga, N., Bezaury Creel, J. E., Causey, B. D., McCormick-Ray, J., Salm, R., Gubbay, S., Kelleher, G., & Reti, J. (2016). Building the future of MPAs–lessons from history. *Aquatic Conservation: Marine and Freshwater Ecosystems, 26*, 101–125. https://doi.org/10.1002/aqc.2680

World Bank. (2006). *Scaling up marine management: The role of marine protected areas*. http://hdl.handle.net/10986/8152

Worm, B., Barbier, E. B., Beaumont, N., Duffy, J. E., Folke, C., Halpern, B. S., Jackson, J. B., Lotze, H. K., Micheli, F., Palumbi, S. R., Sala, E., Selkoe, K. A., Stachowicz, J. J., & Watson, R. (2006). Impacts of biodiversity loss on ocean ecosystem services. *Science, 314*(5800), 787–790. https://doi.org/10.1126/science.1132294

12 Social Justice in Coastal Spaces

M. M. Maldonado and B. Boovy

Social science fields like environmental sociology have centered the inextricable linkages and interdependencies between humans and the biophysical environment since the 1970s (Catton & Dunlap, 1978, 1980). Since then, increasing numbers of natural science and resource management fields have come to recognize the connections between the natural and social worlds. Marine scientists, as well, increasingly acknowledge that the world's oceans and coasts are "peopled seascapes", that is, people (and social concerns, values, and practices) are integral to marine environments (Bennett, 2019; Shackeroff et al., 2009).

Humans use, interact with, manage, attribute, and add different kinds of value to marine environments. Humans also shape and become part of marine environments, for example, when they recreate at the beach or build oceanfront houses whose economic value is determined precisely by their location and views. Coastal locations and ocean vistas not only provide attractive settings for tourists and residents; they are also abiding sources of creative and artistic inspiration. Additionally, humans rely on marine environments for income and work, such as those involved in fishing and seafood processing industries. We do not just live, work, and play in marine environments. We consume marine resources both literally and figuratively. We eat the fish and develop coastal cuisines and a taste for seafood. We celebrate special occasions at seafood restaurants. We go for walks at the beach with our significant others and dogs. We tell stories about the sea, and in that process, the marine environment becomes part of our cultural imaginary, linked to our individual and collective memories, histories, and identities. Those cultural imaginaries, in turn, mediate our relationship to coastal spaces.

Human dimensions research has attended to a wide range of human experiences, including culture, policy and governance, health, and economics (Barreto et al., 2020; Christie et al., 2017; Levine et al., 2015). Here, we present an approach to marine studies that engages conceptual tools from the humanities and social sciences to call for a nuanced understanding of the terms "human" and "social" that recognizes people's embeddedness in historically specific and geographically contingent *social relations of power*. We explore key concepts and questions that pertain to the critical study of social relations of power in coastal spaces. Our discussion offers a foundation and opportunities for reflection on how to develop an orientation toward marine studies that is informed by a commitment to social

DOI: 10.4324/9781003058151-15

justice. We do not regard such an orientation as specific to a particular area or application of marine science or policymaking. Rather, it is pertinent across fields. We conclude by discussing how collaborations between marine sciences and critical liberal arts, social sciences, and humanities can lead to research and applied projects that derive the broadest social impacts, and that join marine and coastal health and vitality concerns with social justice concerns.

Foundational Concepts

Throughout this chapter, we use the terms *power* and *oppression* as foundational analytic categories that are central to any discussion of social justice in marine studies. We define these terms to also allow readers to begin making the invisible visible for themselves. Power and oppression point to social processes that tend to be elusive. A shared vocabulary is also important for the kinds of translational work across disciplines that marine studies students are increasingly called to perform. This brief overview of key terms helps readers to build a conceptual toolkit that will enable them to engage marine studies through a critical lens attentive to social justice. We then provide an example of how people can incorporate analyses of social relations of power into their work, referencing examples from our own research into demographic changes in the seafood processing industry. Our research illustrates the many ways in which social categories, such as race, class, gender, and ability, shape not only marine and coastal communities but also our own perspectives on what we study. We conclude with a call for increased and sustained multidisciplinary collaborations, as well as the hope that readers will continue learning about movements for social and environmental justice that intersect with the study of oceans and society.

When power and oppression are named, there is a tendency to first consider *individuals* as actors who exercise power over others. We follow van Dijk (1995) in focusing our attention on "social power" as opposed to "individual power". Young (2009) suggests that power and oppression do not merely, or even primarily, refer to the dominance of a single individual or ruling group, as popular understandings of these terms tend to suggest. Rather, when we refer to structural oppression, we are speaking of

> the vast and deep injustices some groups suffer as a consequence of often unconscious assumptions and reactions of well-meaning people in ordinary interactions, media and cultural stereotypes, and structural features of bureaucratic hierarchies and market mechanisms – in short, the normal processes of everyday life.
>
> (p. 56)

These dynamics can be elusive to the casual observer – sometimes even to the well-trained social scientist who might not focus on questions of power. In this way, power and oppression are reduced to individual acts of discrimination on the interpersonal level: racist or sexist epithets, acts of violence, segregation, or exclusion.

Related terms include "diversity", "inclusivity", "tolerance", and "acceptance". Although these terms have become very common in social justice–oriented research and engagement, in some cases they have come to suggest a watered-down analysis of social relations that imagines all people and groups to be on an equal playing field (Ahmed, 2012). These terms form part of some well-worn analogies of US culture, including the "melting pot", "different but equal", and the "self-made man". By failing to consider how power and oppression infuse our socialization and everyday interactions – from the gender roles we learn in our families to whether we have access to education, stable housing, health care, a secure food supply, and other necessities – we imagine that all social groups have the same opportunities in life. In fact, social relations are relations of power, shaped by histories and ongoing processes of racialization, gendering, class stratification, ideas about ability, and other social and cultural processes.

An important first step to becoming critically aware of power, and learning how to name and locate it, is to interrogate everyday social arrangements. Power often works to obscure the realities of oppression and systemic inequity, taking the shape of familiar narratives. In the United States, those narratives include the myths of meritocracy, equal opportunity, and colorblindness, among others. Power often also gets embedded in "business as usual", rendered invisible as it becomes part of the everyday practices of institutions and communities. We do not need to look far to begin to discern the day-to-day workings of power and oppression – the political nature of everyday life.[1] Critical approaches to inquiry and knowledge production often begin with understanding our own positionality. By *positionality*, we mean how the social categories that we inhabit as individuals position us in relation to power, within the structure and culture of our society. What is our social location? How does the history of the place and people we come from condition our interaction in society and with others who do not share the same origins or experiences? How can a deep understanding of our stories and social locations help us become critical of systemic power and oppression? What social relations and processes remain invisible to us based on our social location, and how we have learned to interact with others? How do our social locations prevent us from perceiving systemic power and oppression?

Social Relations of Power

The key argument is that the study of the human dimensions of marine environments must center an analysis of social relations of power. Humans are members of many *social categories*. Sociologists define a social category as a collection of people who have certain characteristics or traits in common, but generally have no connection beyond that. This differentiates social categories from *social groups*, which are collections of people with shared characteristics who interact and acknowledge their connections to each other. Power is embedded in the social categories we inhabit because social positions are both relational and hierarchical. The meaning and value associated with different positions within a social category are determined in relation to one another. Additionally, different social positions are valued more or less, or are granted more or less access to resources and advantages relative to one another.

Social relations of power have several key characteristics. First, they are historically specific and dynamic. Second, they are geographically contingent, meaning that, while shaped by large-scale (global and national) forces, they also tend to take on different inflections in specific spaces and places. *Space* thus constitutes one tangible canvas upon which social relations of power become traceable. We draw from critical geographer Massey's (2005) conceptualization of space as socially constituted, relational, and always under construction. Space is both *constitutive of* and *constituted by* social relations of power. In other words, space is political. Anchored in this framework, we approach coastal spaces not simply as a natural backdrop, or a collection of geographic and marine features and resources. Instead, we attend empirically to *how* coasts – those spaces where land meets the sea or the ocean – are produced and shaped by humans and by the power relations among them. We also seek to understand how coasts, in turn, produce and reproduce (or allow interruption of) social relations of power.

Another key characteristic of social relations of power is that they are mutually imbricated and mutually shaping. Humans embody and give life to multiple social relations of power simultaneously, and power and oppression operate through multiple categories and relations at once. You might have heard of this as "intersectionality", which we discuss in more detail below.

Finally, history has shown us that where there is power, there is resistance. Power is not absolute; there is always the potential to interrogate, interrupt, and intervene, individually and collectively, for social change.

Analyzing Social Relations of Power

We briefly consider some specific social categories and relations of power that ought to be centered in the study of coastal spaces and offer additional resources for further engagement and reflection.

Race/Ethnicity

Simply put, *race* refers to physical differences that groups and cultures consider or treat as socially significant. *Ethnicity*, on the other hand, refers to shared cultural elements, such as language, ancestry, practices, and beliefs. Race and ethnicity are often connected in people's minds and are sometimes even used interchangeably. For this reason, we refer to "ethno-racial groups". The United States is a *racialized society*[2] in that its institutions, culture, and the distribution of wealth and resources have long been and continue to be shaped by race. In the post–Civil Rights era of the United States, an ideology of colorblindness has become predominant. There is a tendency for people to not see race, and/or to explain racial phenomena in non-racial terms (Bonilla-Silva, 2017). The casual observer might imagine that race is absent or irrelevant in coastal communities. However, coastal communities, like all geographically bound communities, are always already racialized. Communities that are racially homogeneous (for example, all-white communities) are imbued with racial meaning, content, and politics because spatial segregation and displacement are racialized phenomena. All-white communities are socially produced as such, with sociopolitical implications.

Given broad demographic trends in the United States, many coastal communities are becoming increasingly diverse ethno-racially. For example, the state of Oregon (in the Pacific Northwest United States) and other coastal regions of the United States have seen remarkable growth of Latino populations (both immigrant and US-born) in the past few decades. Latino individuals and their families embody particular cultural identities, values, and practices – and engage, relate to, and use marine resources in particular ways. To the extent that spaces are produced by the relationships among those who inhabit them, coastal community identities are in flux, and demographic change entails the potential for changes in identity and politics. Latinos and other non-white groups who live in coastal communities also fill jobs in specific coastal industries, like seafood processing and restaurants. Research that attends to social justice in coastal spaces must consider the racialized histories of place, as well as the racialized dynamics unfolding in the present, even if they might be invisibilized by normative presumptions of whiteness or by perceptions of racial homogeneity or de-raciality.

In our initial research in the Coos Bay area of the Southern Oregon Coast, we suspected that Latinos might be part of the local seafood processing workforce.[3] This was based on preliminary conversations with several non-Latino white residents of Coos Bay, whose anecdotal perception was that Latinos were beginning to move into Coos Bay and were possibly taking jobs in seafood processing. Our research revealed that, in fact, Latinos had lived in the area and had been working in seafood processing for several generations. Why had their presence for such extended time been mostly unnoticed by white individuals from the community? Research has documented how white and non-white segments of rural communities tend to exist in parallel worlds (Chávez, 2005; Nelson & Hiemstra, 2008). The timing of circulation in public spaces of different ethno-racial segments of community might be different, and there are typically few, if any, instances of meaningful interaction across ethno-racial lines. Segments of community are thus invisible to one another in their particularities, needs, and contributions. This is just one example of the ways in which coastal spaces are routinely racialized.

Gender

Just as the United States is a racialized society, so, too, is US society stratified in relation to gender.[4] The binary division of gender into *female/male* and *feminine/masculine* represents a dominant Western understanding of gender that has historically assigned different social roles to people socialized into the categories of *woman* and *man*. Of course, these conventional notions of binary gender roles do not hold, in many cases, when we observe actual social relations of power. For one, binary gender fails to capture the complex workings of systemic power and oppression, of which binary gender is but one example – what Butler (2006) refers to as "one tactic among many ... in the service of expanding and rationalizing the masculinist domain" (p. 19). The binary logic that organizes gender roles in dominant society serves not only to maintain and naturalize social phenomena, such as a gendered division of labor and the gender–wage gap, but also excludes the multiplicity and complexity of gender expression.

To say that gender is a social construct, however, does not diminish its impact in everyday social practices and interactions, from the way spaces are designed (e.g., bathrooms and dressing rooms) to what jobs and roles we assume are appropriate for women in contrast to jobs that we tend to think are best reserved for men. Indeed, it is difficult to think of any aspect of US society that is not structured in part by binary gender. Binary gender shapes the kinds of tasks we associate with the conventional roles of *woman* and *man*. Dominant understandings of the role of women in US society have been tied to *reproductive* labor, that is, the kinds of work and everyday tasks that make it possible for societies to reproduce themselves – such as food preparation, health care, childcare, and other kinds of work that have come to be associated with the domestic sphere. By contrast, the gender binary also dictates masculine gender norms, where men's roles are associated with *productive* labor or wage work. In concrete terms, we see this in what social scientists refer to as "occupational segregation by gender", a phenomenon that persists despite the increasing numbers of women in the US workforce over the past several decades (Cohen, 2013; Hegeswich & Hartmann, 2014; Mastracci, 2005). Moreover, as other studies have shown (Cohen & Huffman, 2003), women's labor is consistently valued less than men's in the labor market, as evidenced by lower wages. Additionally, race and ethnicity in conjunction with gender norms further exacerbate occupational segregation and devaluing of jobs that are typically occupied by women of color (Mintz & Krymkowski, 2010; Reskin & Cassirer, 2012).

The US labor force has changed considerably over the past several decades, and coastal spaces are no exception. We can observe how gender norms structure society, and at the same time notice changes in how people interact with those norms. Historically, men have worked in fishing, crabbing, and other activities related to catching seafood, while women have found more work in seafood processing, as well as the associated activities of food preparation in the home. Our research in Coos Bay revealed that the gender makeup of the seafood processing force has changed over time, with more men entering the industry. Nevertheless, we also saw trends that suggest that seafood processing is considered a source of supplemental income due to the seasonality of the work and the presumption that it is primarily women's work. Here, we can consider how binary gender stereotypes continue to influence people's perceptions of who performs what kinds of jobs in coastal and maritime economies.

Socioeconomic Class

Class is a much-debated concept in the social sciences and humanities. Scholars have defined it in a multiplicity of ways in connection to different theoretical frameworks. While there is little consensus around a single definition, by returning to space as the canvas upon which power becomes tangible and observable, we can identify some class-related dynamics that routinely shape coasts. Some of the class content of coasts is relatively easy to ascertain. For example, we know that coastal real estate is highly coveted and valued in economic terms. Homes in certain coastal areas come with a hefty price tag, as do lodging accommodations that give visitors ocean views and/or beach access. Further, many coastal communities that were at some point dependent on fishing have experienced economic transformation and demographic shifts, becoming whiter and more affluent through gentrification, often under the auspices of "waterfront revitalization". Sharp contrasts are perhaps the easiest spatial/physical manifestations of class dynamics in coastal spaces. In Coos Bay, Oregon, for example, an interesting class contrast in housing is apparent. On the one hand, there is a dearth of affordable housing in the area, as well as a high rate of homelessness. Meanwhile, existing condos and apartment complexes, and a handful of luxurious homes, are devoted to recreational tourism through the Airbnb market. These class contrasts, though all too often presumed normal, represent the tension between possession and dispossession, wealth and poverty, as connected relational class processes on the coast.

At the macro level, coasts are shaped and animated by class dynamics of investment and disinvestment, as well as development and underdevelopment. For instance, some critical geographers refer to "ocean grabbing" (Bennett et al., 2015), a term used to describe actions, policies, or initiatives that deprive small-scale fishers of resources, dispossess vulnerable populations of coastal lands, and/or undermine historical access to areas of the sea. This phenomenon constitutes an example of power-imbued coastal transformation – a simultaneously classed and racialized spatial dynamic. Ocean grabbing is not always readily legible as such, because it often occurs through legal institutional practices, and even through policies and practices developed in response to ecological concerns.

However, coasts are not all the same; different coasts have different characteristics and historical and present-day uses, so we can see different class articulations and dynamics in different places. For example, examining the dynamics of class might require more nuanced exploration of the "working" aspects of "working waterfronts", that is, detailed exploration of what different kinds of work happen in a setting, who the workers are, and how wages and conditions vary across the board (see Doyle et al., 2018). Whether we work, the work we do, the general level of prestige associated with employment, the wages we earn, and the conditions we face in jobs are partly what places us within classed hierarchies and relationships. What types of occupations and jobs are associated with different coastal and marine spaces? Also, as noted earlier, class intersects with gender and race/ethnicity. Occupations and jobs within the coastal economy, as elsewhere, are typically *segmented* by race and gender. Men and women and people from different ethno-racial groups tend to work in different jobs, associated with different levels of wages and job conditions. People of color – women, especially – tend to be concentrated in the lowest-paid and most precarious jobs. Articulations of race and immigrant illegality in the United States create additional vulnerabilities for unauthorized immigrant workers, resulting in even more precarious conditions and the baked-in potential for exploitation of a racialized and gendered workforce.

The need to pay attention to the intersections of class with other relations of power becomes clear if one examines research on natural disasters in coastal areas. One consistent finding of this literature, in the United States and across the globe, is that the impacts of natural disasters are not equally distributed across the population. The poor tend to be most heavily impacted. Precarity and poverty in US coastal areas cut across racial lines; there are poor people of all races in coastal communities. Yet, given the intersectional nature of power, poor people of color – and, in particular, poor women of color, along with gender-nonconforming and nonbinary people of color – are especially vulnerable, and it is more difficult for them to recover from the devastation caused by natural disasters. Sociopolitical relations and arrangements, in this way, exacerbate impacts for some. This has been abundantly documented in the US Gulf Coast region following Hurricane Katrina (Gault et al., 2005; Rodríguez & Dynes, 2007).

Intersectionality

The three categories outlined above do not represent the totality of social relations of power. Just as race, gender, and class overlap and interlock, so do other social categories and relations structure access to resources and opportunity, and are integral to the social production of advantage and disadvantage. For example, when we speak of the social and cultural biases that limit access to the natural and built environments, we refer to *ableism* as a system of oppression that creates and sustains inequities in relation to physical, psychological, or developmental impairments. Here, too, differences in bodies are interpellated into a larger cultural and social system that disadvantages people defined as divergent from a presumed set of socially and culturally constructed norms about bodies and minds. Disability studies scholars (Swain et al., 2013) refer to disabling environments and the ways in which society "enables" certain bodies to move in society without difficulty. For example, are places of work in fisheries and the seafood processing industry accessible? What norms about ability and the body inform how equipment is designed and organized in processing plants? How do the tasks and work processes of fishers and seafood processors impact workers' bodies, from sore backs and feet after long days of standing, to injuries on their hands from knives and crab shells?

We can interrogate the ways in which many other social categories shape our relationship to power. For example, sexual orientation, religious affiliation, language and racialized notions of linguistic ability, national origin, and citizenship, to name a few, might shape human interaction with and within coastal and marine environments in particular places. The concept of "intersectionality", first developed by Black feminist scholars (Crenshaw, 1993), helps us name the dynamic of interlocking systems of power and oppression that constitute social relations of power (Carastathis, 2016; Cho et al., 2013; May, 2015). In reflecting on how to develop and apply a social justice approach to marine studies, students should not regard the social relations of power we define and center here as discrete categories, but rather should focus on the *relationships among* these categories that shape and are shaped by systems of power and oppression. Box 12.1 includes several case studies that are helpful in exploring and understanding the importance of an intersectional lens when considering coastal communities.

BOX 12.1 Coastal Change across the Globe

The social dimensions of power discussed in this chapter combine with historical, geographic, and ecological change to impact communities differently. Coastal communities are also resourceful in different ways in their responses to phenomena, such as hurricanes, sea level rise, and, most recently, the COVID-19 pandemic. The resources below are from a range of sources, from research reports to media coverage. As you read, pay close attention to where authors bridge perspectives from humanities, social sciences, and natural sciences to address human dimensions.

- "Small-Scale Fisheries in Southeast Asia See Harsh Impacts of COVID-19" illustrates the vulnerability of artisanal fisheries to impacts from the pandemic (reduced demand, collapse of prices, lockdown measures). https://seagrant.uconn.edu/wp-content/uploads/sites/1985/2021/06/wracklines-S-2021-Small-Scale-article.pdf
- "America's Sordid Legacy on Race and Disaster Recovery" describes the vulnerability of people of color to environmental disasters and hazards via the case of Hurricane Maria in Puerto Rico. https://www.americanprogress.org/issues/race/news/2018/04/05/448999/americas-sordid-legacy-race-disaster-recovery/
- "Gender Matters in Coastal Livelihood Programs in Indonesia" critiques investments in coastal community development projects that fail to empower women. https://theconversation.com/gender-matters-in-coastal-livelihood-programs-in-indonesia-127747
- "On the Louisiana Coast, an Indigenous Community Loses Homes to Climate Change" illustrates the vulnerability of a poor indigenous coastal community to coastal erosion. https://www.scientificamerican.com/article/on-the-louisiana-coast-an-indigenous-community-loses-homes-to-erosion1/
- "Decades After Clashing with the Klan, a Thriving Vietnamese Community in Texas" chronicles the Vietnamese immigrant fishing community's struggle against racism and anti-immigrant sentiment. https://www.npr.org/2018/11/25/669857481/decades-after-clashing-with-the-klan-a-thriving-vietnamese-community-in-texas

In our research on the Oregon coast, for example, intersectionality helps us understand the different experiences, challenges, and opportunities facing people from various ethno-racial groups who are part of the seafood processing workforce. Although all workers in the seafood processing industry face vulnerabilities associated with seasonal and precarious work, Latino workers face the added vulnerability that results from widespread racialized narratives that dehumanize them and disable them from developing a sense of belonging in community. Some segments of the Latino population face further disadvantages from racialized immigration policies that restrict their job options and their ability to circulate freely and without fear in community contexts. While there is much economic need in the Coos Bay region, race and ethnicity function to deepen vulnerability and also to shape the coping mechanisms and resilience of different groups. For instance, we found that livelihood strategies and networks vary across ethno-racial groups in Coos Bay. Latino households routinely rely on extended family networks to deal with many of the challenges associated with the seasonality and precarity of work, and, additionally, have developed strategies that enable them to cope when work in seafood processing is scarce or not available. White workers in our study seemed to either lack access or not rely on similar strategies.

The Need for Multidisciplinary Collaborations

The terms and concepts we highlight might be novel, but this is not surprising. The *siloing* of disciplines within academic contexts often means that students of the natural and physical sciences have limited opportunities to engage in the social sciences and the humanities, let alone study power as a routine feature of social and cultural life. Furthermore, questions of sociocultural values and social justice are often presumed to be extraneous to the natural or physical sciences, fields that are supposedly about facts. Those other fields – in particular, critical interdisciplinary fields such as ethnic studies, women and gender studies, queer studies, and disability studies – are the ones that most often deal with social justice issues. In the sciences, studies outside the core curriculum are often minimal and intended to fulfill general education requirements. This structural characteristic of academic contexts has consequences for all involved in knowledge production, and for the sciences' ability to ensure engagement of a broad public with scientific concerns, securing the broadest impacts of scientific research and innovation.

Increasingly, grant agencies and entities that promote and support scientific research expect scientists to attend to the social dimensions of the issues they study. Along with intellectual merit, solid "broader impacts" rationales are routinely expected in research and application projects sponsored in the United States by the National Science Foundation (NSF) and entities like the National Oceanic and Atmospheric Administration (NOAA). Further, NSF has recently also called for a "convergence research approach" that considers technical and social issues simultaneously through multidisciplinary team development. Therein lies part of the promise and potential of multidisciplinary collaborations linking marine sciences with the critical social sciences and the humanities.

Scholars who work in critical fields, such as ethnic studies and women and gender studies, are particularly well-equipped to collaborate with marine scientists in the study of social relations of power in coastal spaces, for several reasons. First, their approach to research starts from the premise that *all* knowledge systems are rooted in social relations of power. This, in turn, entails a questioning of the possibility of objectivity/neutrality that is central to the natural sciences and mainstream social science. While it might appear that these contradictory assumptions make collaboration across disciplines impossible, acknowledging the foundational assumptions of our respective fields enables open and honest discussion of the implicit values in how research questions are framed, how data is collected, and how findings are interpreted. It requires intentionality on the part of all those involved in multidisciplinary research projects to engage in that conversation.

Beyond producing sound projects that are capable of securing funding, there are other important reasons for the concerned scientist to pursue research and scholarship that is critical and attentive to power dynamics. Just as marine studies researchers are always embedded in the social and cultural networks that comprise marine environments, so, too, do they approach their research questions from a multiplicity of backgrounds and experiences. In marine studies, this means adopting and developing a critical perspective on the systems and institutions that mediate our engagement with marine environments, ecologies, and species. To begin to interrogate these systems, we can use some of the same observational tools that we apply in research, but we must also reflect critically on our social location. This means acknowledging how we, ourselves, are implicated in systems of power and oppression, and questioning conventional claims of objectivity in our research and study of marine environments. We all bring our own set of biases, preconceptions, stereotypes, and modes of interacting with others to our work as marine studies practitioners. Recognizing how we are implicated in social relations of power is a critical first step in transforming marine studies and developing a social justice orientation in our work.

Often, when we begin to take stock of how we are implicated in systems of power and oppression, we might feel as though social injustice is simply too widespread to do anything about it. In our work, it is not uncommon for students to confront feelings of anger, betrayal, sadness, guilt, detachment, disbelief, and frustration, and these feelings will likely vary depending on lived experience and social location. Although these feelings are an important part of the process of developing a social justice orientation in marine studies, they are not the process itself. We encourage readers to acknowledge these feelings and build a community of peers and friends with whom to communicate and discuss feelings. Nevertheless, remember that the focus is on the social relations of power. Although our individual feelings are not insignificant, it is important not to get weighed down by them or allow them to dissuade readers from engaging in this critical work.

Coastal communities around the world have mobilized to confront the challenges of social inequality and injustice. The work of residents in these communities illustrates how everyday people can resist and act effectively against pervasive and systemic injustice. In many cases, their efforts have been successful. Box 12.2 presents references to several recent movements for social justice in coastal communities.

BOX 12.2 Movements for Social Justice in Coastal Spaces

Las playas son del pueblo (Beaches Are for the People) – As privatization of coastal properties makes public access increasingly difficult, Puerto Ricans organize to defend beach access and turtle nesting habitats.

- #PlayasPalPueblo – hashtag used by the movement on social media platforms
- "Tension Rises in Rincón Due to Police Force and Construction on Beach" –https://www.latinorebels.com/2021/07/27/tensionrisesinrincon/
- "Protestors Resist Militarized Gentrification of Public Beaches in Puerto Rico" –https://www.commondreams.org/views/2021/08/13/protestors-resist-militarized-gentrification-public-beaches-puerto-rico
- "'The beaches belong to the people': Inside Puerto Rico's anti-gentrification protests" – https://www.theguardian.com/us-news/2022/jul/23/puerto-rico-beach-anti-gentrification-protests?CMP=Share_AndroidApp_Other

Indigenous/First Nations fishing rights – These cases illustrate Indigenous Peoples' struggles to protect their fishing rights guaranteed by tradition, treaties, constitutions, or the courts.

- Northwest Indian Fisheries Commission – https://nwifc.org
- "The Long, Expensive Fight for First Nations' Fishing Rights" – https://www.hakaimagazine.com/news/the-long-expensive-fight-for-first-nations-fishing-rights/
- "Historic Court Ruling Upholds Sami Rights in Sweden" – https://www.culturalsurvival.org/news/historic-court-ruling-upholds-sami-rights-sweden

Movement against slavery and human trafficking in the seafood processing and fishing industries – These websites illustrate human rights violations that occur on industrial fishing vessels and in seafood processing plants throughout the world.

- Slave Free Sees – http://www.slavefreeseas.org
- "Forced Labor and Human Trafficking in Fisheries" – https://www.ilo.org/global/topics/forced-labour/policy-areas/fisheries/lang--en/index.htm
- "Tun Lin Was Trapped for 11 Years on a Slave Ship. Now He's Fighting to End Slavery" – https://www.globalcitizen.org/en/content/slavery-in-thai-fishing-industry

Conclusion

Developing a social justice orientation toward marine studies does bring with it certain challenges. First, the need for an approach oriented toward social justice is not self-evident. Often, critical approaches are considered add-ons to degree programs in the natural sciences, rather than being foundational to the ways students are taught to think about their discipline. Developing an approach informed by social justice also requires reaching out to people in different fields and to individuals and communities outside of universities. Moreover, it can be challenging to think relationally and consider that focusing on the human dimensions of marine studies is not an opportunity to map humans as data points, or an end in itself. Instead, our discussion of social relations of power is intended to generate discussion and reflection on how marine studies as a field and an approach to learning about marine systems is always embedded in historical, social, and cultural contexts. Moreover, the questions that we raise here are intended to encourage readers to think of their engagement with marine studies as an opportunity to transform the field and advance social justice through one's work as a student, researcher, policymaker, scientist, artist, or in whatever path pursued.

Many scientists now acknowledge the complexities of pursuing their work in sociocultural contexts imbued with power. National and international organizations have emerged that call for acknowledgment of the power implications of scientific work. For example, Science for the People organizes scientists, activists, students, and scholars to address the connections between science and power in research institutions, universities, and communities. Similarly, the stated mission of Free Radicals is to "create a more socially just, equitable, and accountable science" (Free Radicals, n.d.). These organizations engage their members and the public in dialogues about connections and disconnections between science and social justice, both past and present. Recently, *Scientific American* published a cover story by sociologist Aldon Morris titled "The Power of Social Justice Movements", which focuses on the history of racial justice movements. Morris (2021) ends the piece with a call to scholars to "face the challenge of keeping pace with social justice movements as they develop" and to "do more … to illuminate the paths that movements should traverse in their journeys to liberate humanity". Will the sciences concerned with marine and coastal spaces keep up with this historic call?

Professional Pathways

Students interested in working at the juncture of social justice concerns and coastal spaces can pursue studies combining ocean sciences or marine studies with co-majors, minors, or certificates in critical interdisciplinary fields such as Ethnic Studies, Indigenous Studies, and Women's Gender and Sexuality Studies. These fields may have other names across academic institutions. However, the key is to identify interdisciplinary units that take a critical approach to space, natural resources, environment, and to race/ethnicity, gender, and other social relations of power, as we outline in this chapter. This sort of interdisciplinary training equips students to consider issues of inequity and power along with specific coastal contexts and/or natural resource issues; and provides a critical lens, an analytical framework, and methodological approaches that are relevant and can be applied in any work setting. Many workplaces with a focus on oceans and coasts, and/or on natural resources more broadly, have increasingly come to recognize the importance of bringing on board personnel who are trained and equipped to engage questions of social and cultural difference and power. Students of ours (Maldonado and Boovy) who have combined academic work in ocean sciences and/or marine studies fields with work in Ethnic Studies and/or Women's, Gender, and Sexuality Studies have gone on to take jobs with entities such as NOAA, non-governmental organizations focused on marine conservation, community resource centers and non-profits, and as academics doing research and teaching at various universities.

Questions for Reflection

1. What communities and commitments motivate your engagement with marine studies and marine systems?
2. How does the work you do align with, call into question, or challenge power and systemic oppression?
3. What does an approach to marine studies that prioritizes social justice look like?
4. What learned assumptions within your field or discipline might be in tension with critical acknowledgment of power?
5. How does a social justice–oriented approach shift our understanding of some of the central focus areas in marine studies, many of which are represented by other chapters in this book?
6. Who is active and visible in the fields represented by marine studies? Who are your classmates? Who are your teachers? Who is not in your class?

Notes

1 Throughout this chapter, we use the term *political* to mean *having to do with power.*

2 Race and ethnicity are features of the modern world system, not just the United States. They are integral to power dynamics within and across countries. Different countries and regions have their own racialized histories and narratives, and there are also commonalities in the racial dimensions of life in different parts of the world. For a discussion of race and ethnicity in different parts of the world, see Bacchetta et al., 2019; Winant, 2002. For further examples of the role of race and ethnicity in coastal spaces globally, see Boxes 12.1 and 12.2.

3 For a helpful map along with images of the Coos Bay/North Bend area, see: https://seagrant.oregonstate.edu/sites/seagrant.oregonstate.edu/files/m15001_1.pdf

4 Gender, like race/ethnicity and other social relations of power, is integral to how societies are structured and how hierarchies are created and maintained across the globe (see Runyan & Peterson, 2018). For further examples of the role of gender in coastal spaces, see Boxes 12.1 and 12.2.

References

Ahmed, S. (2012). *On being included: Racism and diversity in institutional life*. Duke University Press.

Bacchetta, P., Maira S., & Winant, H. (Eds.). (2019). *Global raciality: Empire, postcoloniality, decoloniality*. Routledge.

Barreto, G. C., Di Domenico, M., & Medeiros, R. P. (2020). Human dimensions of marine protected areas and small-scale fisheries management: A review of the interpretations. *Marine Policy*, *119*, 104040. https://doi.org/10.1016/j.marpol.2020.104040

Bennett, N. J. (2019). Marine social science for the peopled seas. *Coastal Management*, *47*(2), 244–252. https://doi.org/10.1080/08920753.2019.1564958

Bennett, N. J., Govan, H., & Satterfield, T. (2015). Ocean grabbing. *Marine Policy*, *57*, 61–68. https://doi.org/10.1016/j.marpol.2015.03.026

Bonilla-Silva, E. (2017). *Racism without racists: Color-blind racism and the persistence of racial inequality in America* (5th ed.). Rowman and Littlefield.

Butler, J. (2006). *Gender trouble: Feminism and the subversion of identity*. Routledge. (Original work published 1990) https://doi.org/10.4324/9780203824979

Carastathis, A. (2016). *Intersectionality: Origins, contestations, horizons*. University of Nebraska Press.

Catton, W. R., Jr., & Dunlap, R. E. (1978). Environmental sociology: A new paradigm. *The American Sociologist*, *13*(1), 41–49. https://doi.org/10.1177/000276428002400103

Catton, W. R., Jr., & Dunlap, R. E. (1980). A new ecological paradigm for post-exuberant sociology. *American Behavioral Scientist*, *24*(1), 15–47. https://doi.org/10.1177/000276428002400103

Chávez, S. (2005). Community, ethnicity, and class in a changing rural California town. *Rural Sociology*, *70*(3), 314–335. https://doi.org/10.1526/0036011054831224

Christie, P., Bennett, N. J., Gray, N. J., Wilhelm, T. A., Lewis, N. A., Parks, J., Ban, N. C., Gruby, R. L., Gordon, L., Day, J., Taei, S., & Friedlander, A. M. (2017). Why people matter in ocean governance: Incorporating human dimensions into large-scale marine protected areas. *Marine Policy*, 84, 273–284. https://doi.org/10.1016/j.marpol.2017.08.002

Cho, S., Crenshaw, K. W., & McCall, L. (2013). Toward a field of intersectionality studies: Theory, applications, and praxis. *Signs*, 38(4), 785–810. https://doi.org/10.1086/669608

Cohen, P. N. (2013). The persistence of workplace gender segregation in the US. *Sociology Compass*, *7*(11), 889–899. https://doi.org/10.1111/soc4.12083

Cohen, P. N., & Huffman, M. L. (2003). Occupational segregation and the devaluation of women's work across U.S. labor markets. *Social Forces*, *81*(3), 881–908. https://www.jstor.org/stable/3598179

Crenshaw, K. (1993). Mapping the margins: Intersectionality, identity politics, and violence against women of color. *Stanford Law Review*, *43*(6), 1241–1299.

Doyle, J., Boovy, B., Maldonado, M. M., & Conway, F. D. (2018). Understanding the working in working waterfronts: The hidden faces of the industries that make up the working waterfront. In L. Price & N. Narchi (Eds.), *Coastal heritage and cultural resilience* (pp. 223–242). Springer.

Free Radicals. (n.d.). *Mission*. https://freerads.org/mission/

Gault, B., Hartmann, H., Jones-DeWeever, A., Werschkul, M., & Williams, E. (2005, October). *The women of New Orleans and the Gulf Coast: Multiple disadvantages and key assets for recovery part I: Poverty, race, gender and class*. (IWPR # D464). Institute for Women's Policy Research. Retrieved July 7, 2022 from https://katrinareader.cwsworkshop.org/sites/katrinareader.org/files/iwpr.pdf

Hegeswich, A., & Hartmann, H. (2014). *Occupational segregation and the gender wage gap: A job half done*. Institute for Women's Policy Research. https://hdl.handle.net/1813/79410

Levine, A. S., Richmond, L., & Lopez-Carr, D. (2015). Marine resource management: Culture, livelihoods, and governance. *Applied Geography*, 59, 56–59. https://doi.org/10.1016/j.apgeog.2015.01.016

Massey, D. (2005). *For space*. Sage.

Mastracci, S. H. (2005). Persistent problems demand consistent solutions: Evaluating policies to mitigate occupational segregation by gender. *Review of Radical Political Economics*, *37*(1), 23–38. https://doi.org/10.1177/0486613404272326

May, V. (2015). *Pursuing Intersectionality, unsettling dominant imaginaries*. Routledge.

Mintz, B., & Krymkowski, D. H. (2010). The intersections of race/ethnicity and gender in occupational segregation: Changes over time in the contemporary United States. *International Journal of Sociology*, *40*(4), 31–58. https://doi.org/10.2753/IJS0020-7659400402

Morris, A. (2021, February 3). *From civil rights to Black Lives Matter*. Scientific American. https://www.scientificamerican.com/article/from-civil-rights-to-black-lives-matter1/

Nelson, L., & Hiemstra, N. (2008). Latino immigrants and the renegotiation of place and belonging in small town America. *Social & Cultural Geography*, 9(3), 319–342. https://doi.org/10.1080/14649360801990538

Reskin, B., & Cassirer, N. (2012). Occupational segregation by gender, race, and ethnicity. *Sociological Focus*, *29*(3), 231–243. https://doi.org/10.1080/00380237.1996.10570642

Rodríguez, H., & Dynes, R. R. (2007). Finding and framing Katrina: The social construction of disaster. In D. L. Brunsma, D. Overfelt, & J. S. Picou (Eds.), *The sociology of Katrina: Perspectives on a modern catastrophe* (pp. 23–33). Rowman & Littlefield.

Runyan, A. S., & Peterson, V. S. (2018). *Global gender issues in the new millennium* (4th ed.). Routledge.

Shackeroff, J. M., Hazen, E. L., & Crowder, L. B. (2009). The oceans as peopled seascapes. In K. McLeod & H. Leslie (Eds.), *Ecosystem-Based management for the oceans* (pp. 33–54). Island Press.

Swain, J., French, S., Barnes, C., & Thomas, C. (2013). Introduction. In J. Swain, S. French, C. Barnes, & C. Thomas (Eds.), *Disabling barriers – enabling environments* (pp. xvii–xxi). Sage.

Van Dijk, T. A. (1995). Discourse, power, and access. In C. R. Caldas-Coulthard & M. Coulthard (Eds.), *Texts and practices: Readings in critical discourse analysis* (pp. 84–104). Routledge.

Winant, H. (2002). *The world is a ghetto: Race and democracy since World War II*. Basic Books.

Young, I. M. (2009). Five faces of oppression. In G. L. Henderson & M. Waterstone (Eds.), *Geographic thought: A praxis perspective* (pp. 55–71). Taylor & Francis. (Original work published 1990)

13 Truth-Telling

Understanding Historical and Ongoing Impacts to Traditional Ecological Knowledge (TEK)

C. B. LaPorte

Introduction

We start this chapter, which is focused on traditional ecological knowledge (TEK), with a comparison of worldviews. It is difficult to address shared issues if we do not have a foundational understanding of where we come from and how we got here. *Here* is a place that is wrought with oppression, with tension, with uncertainty, and with urgent concern for our shared space going forward. *Here* is a place that bears the wounds of settler-violence and requires healing. Healing does not occur without truth-telling. To that end, this chapter will endeavor to address what TEK is, what it is not, and, most importantly, the historical context in which it managed to survive in all of its forms. This is less a discussion on the scientific impacts of TEK or on the ways in which TEK is beneficial to occupied spaces facing climate change impacts (specifically around coastal areas), and more a story of resistance. In particular, this is a story of resistance against and resourcefulness around the ways in which federal Indian law and environmental policy has impacted, supported, or abrogated TEK altogether.

A discussion of this resistance is best chronicled by remembering creation stories and oral tradition, as Indigenous worldviews are shared and shaped by stories and language. The role of creation stories is to act as a pathway for various cultures to transmit original values and fundamental teachings. While the stories are as varied as are the Indigenous cultures globally and within the United States, these stories are valid expressions of instructional living. They do more than give us a starting place. They provide us with a lens and a reminder of who we are, where we come from, and, most importantly, the ways in which we set ourselves about the critical work of belonging and of remembering.

Many Indigenous cultures do not treat land as a resource, do not treat water only for its use to humans or corporations, and do not view animals or plants as being less-than. Land, water, and non-human plants and animals are kin. This worldview is rooted in numerous creation stories. We can contrast this directly with Anglo-Christian-centric worldviews, in which we are told to have dominion and to subdue, and where morality is conflated with hierarchy, rule, and oppression. In the story of Skywoman, non-human agency not only predates the existence of land but also precipitates it. We start with these two stories because they

DOI: 10.4324/9781003058151-16

provide a valid contrast to consider, one that is extensively chronicled in Robin Wall Kimmerer's 2013 work, *Braiding Sweetgrass: Indigenous Wisdom, Scientific Knowledge, and the Teachings of Plants.*

The point to be made is that an Indigenous worldview is antithetical to that of the settler. While you read this chapter, you should seek out the creation stories and oral traditions of the people who originally inhabited the land you work or live on and think critically about your own worldview and its impact on Indigenous people.

What Is TEK?

Traditional ecological knowledge, sometimes called "Indigenous knowledge", is the "knowledge, practice, and beliefs concerning the relationship of living beings to one another and to the physical environment, which is held by [Indigenous peoples] with a direct dependence on local resources" (Berkes, 1993 as cited in Kimmerer, 2002, pp. 432–433). It arises when humans and their non-human kinship (including land, water, plants, and animals) are interconnected (Kimmerer, 2002). This interconnectedness is a belief system, a worldview rooted in spirituality, in culture, and in language. It is tethered to individual Indigenous worldviews. What is distinctive about it, as compared to other observational sciences whose principles can be extrapolated to fit the needs of the global community, is that it is unique to place and also intergenerational. It is deeply personal to the caretakers of a specific place. It does not reduce land, water, or non-humans to objects or specimens to be studied, but it is still a systematic cyclical observation of nature over a period of time (Kimmerer, 2002). TEK is fluid and not fixed (United States Caucus of the Traditional Ecological Knowledge Task Team, 2021). In its simplest form, it is a story: The story of the relationship between the Earth and her Indigenous inhabitants.

BOX 13.1 Case Study: Anishinaabe and Manoomin, a Treaty between People and Creator

Manoomin (good berry, or wild rice) is essential to Anishinaabe and Ojibwe culture, and Tribes have been respectfully harvesting it for thousands of years. Its use for subsistence and ceremony cannot be understated and is even central to migration prophecies of Ojibwe people.

> As the prophecies foretold, however, the bands were reunited several generations later at Manitoulin Island, forming a union known as the Three Fires Confederacy that remains to this day. In the time of the Third Fire, they found the place foretold in prophecy, 'where the food grows on the water', and established their new homelands in the

> country of wild rice. The people lived well for a long time under the care of maples and birches, sturgeon and beaver, eagle and loon. The spiritual teachings that had guided them kept the people strong and together they flourished in the bosom of their nonhuman relatives.
>
> (Kimmerer, 2013)

The process of harvesting Manoomin is TEK, but not only because of the observational benefits derived from years of this practice. It is TEK because the practice itself is ceremony. Harvesting Manoomin is rooted in gratitude. The practice centers on an understanding that you will not take more than you need and that you will harvest in a way that does not disturb the life process of the plant.

> You bring in the wind; you bring in the motions of everything, the heat, sun, all that, you sing about that. When you dance it, you're asking the plant to give up its fruit, so you dance on it gentle. Good dancers, traditional Indian dancers, they don't stomp their feet on the ground. They're real light when they dance. Just like we dance rice, because we don't want to break the kernels. You got to get the husk off without breaking the kernels, so you got to dance real light. You got to be related, thinking in your mind and your body and that's when we sing that song.
>
> (Fred Ackley, Jr., The Ways, 2019)

This practice of only taking what was needed, rather than harvesting all that there was, seemed foreign and lazy to settlers who first encountered Ojibwe people (Kimmerer, 2013). Land itself under a Western worldview is merely a commodity. This is one of the biggest threats to Manoomin. A recent report from the Great Lakes Indian Fish and Wildlife Commission found that Manoomin was extremely vulnerable to climate change. Specifically, many Tribal members noted the decrease of Manoomin in the region due to the installation of dams, brown spot infestation, changes in water levels, stronger and more frequent weather events, pollution, and competition from invasive species (Great Lakes Indian Fish & Wildlife Commission, 2018).

Oil pipelines and copper-nickel mines constitute a particular threat to Manoomin in the region. In response, the White Earth Band of Ojibwe passed laws recognizing the "Rights of Manoomin" and establishing an enforcement mechanism. It was the first Tribal law to pass and grant legal status for a plant or animal species (Douglas, 2021).

> Manoomin, or wild rice, within all the Chippewa ceded territories possesses inherent rights to exist, flourish, regenerate, and evolve, as well as inherent rights to restoration, recovery, and preservation.
>
> (1855 Treaty Authority, Resolution 2018–2005, White Earth Band of Ojibwe)

These rights include the right to clean water, the right to a natural environment free from industrial pollutants, the right to a healthy, stable climate free from human-caused climate change impacts, the right to be free from patenting, and the right to be free from contamination by genetically engineered organisms (Stopline3, 2019).

In August 2021, Manoomin brought suit as a lead plaintiff in a Tribal court case (Douglas, 2021) seeking to stop Enbridge from pumping up to 5 billion gallons of ground and surface water for construction and alleges the Minnesota Department of Natural Resources has violated multiple treaty rights as well as the rights of Manoomin by allowing the extractive industries to persist despite the clear impacts (Douglas, 2021).

BOX 13.2 Case Study: The Guna Peoples and Sea Turtles

The Guna-Yala *Comarca* in Panama was established as a semi-autonomous Indigenous territory in 1954 and today has a population of over 35,000 persons residing on some 400 coralline islands and the Caribbean mainland in northeastern Panama. *Bab Igar* (Way of the Great Father) is the Guna cultural and spiritual framework that guides collective and individual decision-making (Apgar et al., 2015). All material things (plants, animals, rocks, people) have an essential spirit called *burba* that is interconnected. The Guna people call May "*Yauk Nii*", the Month of the Hawksbill Sea Turtle, because during that time there is significant nesting of sea turtles on the beaches of the Guna-Yala Comarca. Traditionally, the Guna did not kill the turtles, and local rules prohibited harvesting more than half of the turtle eggs. Guna believe that turtles were humans who were punished by *Bab Dummad*, the Great Father or Creator. Persons who killed sea turtles would risk contracting tuberculosis because the turtles' spirits would invade the person's burba (Ventocilla et al., 1995). Others believe that, in revenge, a turtle's spirit will drag the fishermen's boat out to sea.

In 2012 the Guna General Congress declared that leatherback sea turtles could not be hunted, exploited, or sold at any point during their life cycle, to protect and conserve this endangered species that is symbolic in Guna culture. Some Guna communities have been especially active in the promotion of sea turtle traditions and have created linkages to conservation efforts. The community of Armila has declared itself to be a Sacred Sea Turtle Sanctuary (*Yauk Galu*) and monitors sea turtle nesting. In May, Armila celebrates a sea turtle festival to promote sea turtle conservation and strengthen cultural traditions of music, dance, storytelling, and creation of *molas* (an article of women's clothing). The Armila community has created the Yauk Galu Foundation to support their efforts (United Nations Educational, Scientific and Cultural Organization, 2018).

What Is TEK Not?

Pan-Indigenous

Traditional ecological knowledge is not pan-Indigenous. In the United States alone there are 574 federally recognized Tribes and over 60 state-recognized Tribes – each with their own languages, customs, traditions, ways of being and knowing, and "significance of place". To this point, it bears noting that not all Tribes or other Indigenous peoples occupy their ancestral homelands. One need only look to the Tribes forcibly removed West to reservations as a result of violent colonial forces and an insatiable greed for Native land. What traditional ecological knowledge can be present in communities who no longer occupy their homelands? If language is tied to land and land tied to TEK, this is a question for settlers and beneficiaries of settler-violence to address. It is a question whose answer requires redress. Regardless, it is necessary to understand that TEK varies from community to community, that it can change with time, and that it is relative to place, as all stories are.

Limited to Addressing a Specific Problem

TEK is a way of life. It is not merely a scientific solution to a problem. TEK is a way of interacting with the land, water, and non-human kinship (including plants and animals) in a way that is not hierarchical (Kimmerer, 2002). It is a collection of observations that one often makes over the course of a relationship. This means that TEK cannot be separated from the individual and distinct worldviews of which it is a manifestation. To simply uplift TEK in the environmental justice or climate spaces but ignore Indigeneity, to still promote and uphold colonial world-views and privileges in all other contexts, is another form of settler-violence. It is necessary to constantly approach and reassess this work from an anti-oppression lens that centers decolonization.

Whole or Safe

The problem with talking about TEK as merely a solution is the assumption that it remains wholly intact despite ongoing efforts to eradicate Indigenous people from their lands and to systematically, intentionally, and violently strip Indigenous people of what it means to be Indigenous. Thinking of TEK as whole or safe ignores the fact that it is under constant threat. TEK is often rooted in language, story, and ceremony (Kimmerer, 2002). But in the United States, Indigeneity has been deeply disrupted by ongoing colonization and genocide. The next section of this chapter provides a brief overview of the relationship between the United States and Tribes, a relationship that is predicated on extreme oppression.

Foundational Knowledge

Three of the four case studies provided in this chapter are geographically based in what is now considered the United States – and the United States' relationship with Tribal Nations is rooted in genocide. Thus, it is necessary to understand the historical context within which TEK exists and resists. It is necessary to learn about this history because it is our shared responsibility to provide redress. But we also cannot expect Tribal Nations or Indigenous peoples to provide non-Native communities with the solutions to climate change when settler-violence intentionally sought out and systematically destroyed much of this knowledge. Indigenous ways of being and knowing were viewed as inferior, savage, primitive, or uncivilized, and in many spaces that thinking remains prevalent. They were certainly not favored by social or natural sciences. Much of our languages, ceremonies, stories, kinship, and clanship were stolen. And while there are certainly ongoing efforts, generationally, to restore Indigenous worldviews within Tribal communities or for individual Indigenous people living outside of Indian country, it would be naïve to imagine these worldviews as uninterrupted or fully intact. Simply put, settler-violence has produced consequences. These consequences were felt first and most acutely in Indigenous communities, but because we all live in shared spaces, the destruction of stewardship has impacted communities beyond those of the original inhabitants. Climate change, after all, is a threat to everyone.

Historical Relationship between the United States and Indian Nations

Most academic and legal scholars agree that several distinct eras of federal Indian law and policy provide context for the relationship between Tribes and the United States, at least in a general sense. These eras coincide with marked shifts in policy, and those policies influence the treatment of and ramifications for Native people. The following is a brief discussion of those eras, with an expanded focus on assimilation.

Pre-contact

This era is generally known as the one that ended in 1492 (Pevar, 2012). We document the end of this era with the arrival of Christopher Columbus. Plenty of scholars rightfully critique this demarcation, as Columbus more accurately arrived in what is now Puerto Rico, and his contact with North American Tribes as they have come to be defined was nil. Indigenous ways of life and being flourished in this early era, uninterrupted by the genocidal impact of settler arrival.

Tribal Independence

This era (1492–1787) is best defined by the arrival of other sovereigns (in particular, the British, the French, and the Spanish) and the interactions between these colonizing nations and the Indigenous people of North America. It is also marked by increasing tension among Native people and their oppressors (Pevar, 2012). Particularly oppressive was the iron fist of Christianity, whose believers increasingly connected their views of Native people as non-human with their right to Indian land. Settler diseases had a major impact on Native populations as settlers began to occupy North America at alarmingly high rates. The prevailing mindset of this era was that Indians were an impediment to America's future.

The Necessity of Sovereignty

Between the years of 1787 and 1828, the founders of the United States realized that their mere existence on the land did not equate to sovereignty – something that is very much needed to govern oneself. Other sovereigns on US soil engaged in the practice of treaty-making with Tribes who were seen as independent nations, so the colonists followed suit. The first treaty signed by the United States after the adoption of its new constitution was with an Indian Tribe: The Delaware Indians (Pevar, 2012). Laws were passed to protect Indigenous sovereignty rights, particularly because that sovereignty established the basis for the United States' own sovereignty. Few of those laws, however, were enforced (Pevar, 2012).

Removal and "Relocation"

From 1828 to 1887, the United States engaged in the explicit policy of removal. Andrew Jackson, a one-man genocidal think-tank, and his Congress passed the Indian Removal Act in 1830. The result was the violent, forced, and deadly removal of many Indian tribes westward. Many treaties were signed and broken during this period, but in 1871, Congress passed a law that prohibited federal agents from making any new treaties with Indian tribes (Pevar, 2012).

Assimilation and Allotment

Between 1887 and 1934, the federal policy toward Indian people was shaped by two efforts: (1) take Indian lands for settlement by whites, and (2) take Indians, specifically children, and assimilate them into white society. As a result, the General Allotment Act (commonly referred to as the Dawes Act) was passed. The overall goal of the Dawes Act was "to extinguish tribal sovereignty, erase reservation boundaries, and force the assimilation of Indians into society at large" (US Supreme Court, 1992). Indians had no desire to become white because they were already Indians. Additionally, the allotments they received, upon which they were to farm and ranch, were not suitable for agriculture. Due to impoverishment, many Indians were forced to sell their allotments to settlers, and others lost their land to foreclosures. Of the nearly 150 million acres of land that Tribes owned in 1887, by the time the Dawes Act was repealed in 1934, less than 50 million acres remained (Pevar, 2012).

Devastatingly, on March 3, 1819, Congress passed the Civilization Fund Act. This Act authorized the President of the United States to "improve the habits and conditions of such Indians practicable" so as to "employ capable persons of good moral character". The fund paid missionaries and other church organizations and leaders to conspire with the federal government for the sole purpose of creating schools in Indian country to assimilate Native children into white Anglo-Christian society. The explicit federal policy was to "kill the Indian, save the man", as if what was Native was not human. The United States has a particularly sordid history of defining humanity by its proximity to whiteness.

By 1887, the federal government had established over 200 schools in an intentional and genocidal effort to strip Native people of what made them Native (Pevar, 2012). To "civilize" Native children, those in the employ of these institutions punished Native children for speaking their Native languages, refused them access to their siblings, gave them Christian names, and would not allow them to practice their traditions, customs, ceremonies, or religions (Pevar, 2012). Children were forced into unpaid labor, required to partake in Anglo-Christian religious practices, and were often punished by means of severe isolationist tactics (Pevar, 2012). Native children in these government-run schools were physically beaten, raped, and killed (Pevar, 2012). All of this violence was sanctioned by the state.

Indian Reorganization

Federal policy toward Tribes shifted in 1930 and trended toward basic humanity. The Meriam Report, published in 1928, documented reservation life (Pevar, 2012; Brookings Institution & Institute for Government Research, 1928). Hearing about the oppressive conditions under which Natives were living prompted public criticism of the federal government's treatment of Natives. In June 1934, Congress passed the Indian Reorganization Act (IRA) to "rehabilitate the Indian's economic life and to give him a chance to develop the initiative destroyed by a century of oppression and paternalism" (US House of Representatives, 1934). The IRA was enacted to restore Tribal governments through various provisions to protect Tribal land and encourage Tribes to adopt their own constitutions. The IRA, though well-meaning, was paternalistic. Tribes were not consulted in its development, and it promoted a Western version of Tribal government that did not equate to traditional Tribal governance (Pevar, 2012).

Termination

In 1953, the treatment of Tribes and Native people in the United States began to center on complete integration of Natives into white society. Also paternalistic in nature, the idea was that integration was in the best interests of the Natives. A primary goal of this legislation was to save money. By terminating the federal government's trust relationship with certain Tribes, all federal benefits and services could cease. Between 1953 and1966, Congress terminated its trust relationship with 109 Tribes. Another policy of this era was to remove Native people from reservations to urban areas by providing incentives, like job training and housing assistance (Pevar, 2012).

Era of Self-Determination

As of 1968, the official federal policy of the US government toward Indian tribes has been one of self-determination (Pevar, 2012). As enacted, it is not perfect, but the general goal of supporting Tribal sovereignty (or Tribal inherent authority to self-govern) has been a welcome change from the policy of termination.

How Has This Historical and Ongoing Context Impacted TEK?

The impact of the historical and ongoing relationship between the United States as a colonizing government and Tribal Nations as the colonized has been detrimental to TEK. While it is true that TEK exists and resists the impacts of genocide, it is necessary to discuss the effects of federal policies and acts to understand what needs to be healed.

Simply put, the process of making Indians not Indian was detrimental to the land. If Native people are the original caretakers of their ancestral homelands, removal of Native people from the land and from themselves constitutes settler-state violence against land. All of these policies have resulted in distrust; historical trauma; loss of language, culture, and tradition; loss of generations; loss of land; and loss of traditional food. The loss of Indigeneity has been the impact of these policies, which directly impacts TEK.

We can clearly see this impact when we consider the removal of Native people from their ancestral lands. Without connection to a place of cultural significance, what can be said for an impacted Tribe's TEK? Another example is boarding schools: The loss of language and cultural teachings, in addition to the physical removal of entire generations of children from their Tribes, clearly impacts TEK. To provide some context to boarding schools, recall that the Civilization Fund Act, passed in 1819, encouraged "benevolent societies" (i.e., religious institutions), via state funding, to provide education for American Indians and Alaska Natives to assimilate and civilize them into white society (Native American Rights Fund [NARF], 2019). The process of civilizing Native children was an act of genocide that persisted into the 1980s (NARF, 2019). Native children were forcibly removed from their homes and sent to schools, oftentimes very far from their Tribal communities (NARF, 2019). These children suffered unspeakable violence in the forms of sexual and physical abuse, murder, and intentional removal of their identities as Indigenous people. Their names were changed to Christian names, they were punished for speaking their language or practicing their ceremonies, they were separated from sibling groups, forced to practice Christianity, and their hair was cut. Entire generations of children were removed over a sustained period of time. Canada had a similar process. As of November 2021, efforts (led by Indigenous activism, and often in direct opposition to the messaging of power sources such as the Catholic Church and two world sovereigns) are underway in both countries to address the murders that occurred at these schools and the repatriation of remains. Of course, these efforts only received mainstream media attention when, in the summer of 2021, the first mass grave was discovered at Kamloops Indian Residential School in British Columbia. The bodies of 215 children, lost to their parents, their families, and their communities, were found callously discarded beneath the school (Austen, 2021). It is worth noting that Kamloops continued to operate until the late 1970s (Austen, 2021). The link between TEK and these institutions should be relatively clear. Just as with other cultural practices and traditions, boarding schools cut off transmission of TEK from generation to generation by prohibiting and punishing children for their Indigeneity, which is a necessary component of TEK.

What Legal Frameworks Seek to Protect Indigenous TEK and Traditional Relationships with Land and Water within the United States, and Why Do They Inevitably Fail?

Types of Laws Specific to the United States

Treaties

Many people are unaware that there are three distinct sovereigns in the United States: the federal government, the individual states, and Indian tribes (Article I, Section 8 of the US Constitution). Most are additionally unaware that the United States' own sovereignty is predicated upon Tribal sovereignty, as discussed in a previous section. What is clear, however, is that while treaty law is the law of the land in the United States, it is rarely upheld.

> Treaties were originally viewed as contracts. Many treaties contain the phrase 'contracting parties' and specify that each party must agree to the terms of the treaty for it to be valid.
>
> (Deloria, 1988)

Regarding TEK, we see the impact of the continuous and blatant abrogation of treaties in relation to upholding Tribal rights to whaling (as in the case of the Makah Tribe of the US Pacific Northwest), Manoomin (wild rice), and fishing rights. We also see how the practice of ignoring treaties results in a deprioritization of consultations with Tribal communities over proposed permitting; the TEK within these groups continues to go ignored in favor of private or governmental interests (which are detrimental to land, water, and non-human kin).

BOX 13.3 Case Study: Makah Whaling – Subsistence Hunting and Treaty Rights vs. the Marine Mammal Protection Act

> The thunderbird brought us the whale and fed us.
>
> (Patrick DePoe in Rust, 2019)

The Makah Tribe of the Pacific Northwest traditionally occupied the northwest border of the Olympic Peninsula, until 300,000 acres of their land were ceded to the United States under the 1855 Treaty of Neah Bay (Rust, 2019). Part of the uniqueness of this treaty is that it expressly reserved the Tribe's right to practice whaling in their traditional waters (Rust, 2019).

> [T]he right of taking fish and of whaling and sealing or sealing at usual and accustomed grounds and stations is further secured to said Indians in common with all citizens of the United States.
>
> (The Treaty of Neah Bay, 1855)

The Makah Tribe has been practicing whaling for over 2,000 years, as is central to their culture, but they stopped in the 1930s after observing through their TEK that the population of the gray whale had been significantly reduced by the American capitalist whaling industry (Rust, 2019).

In the 1970s, the US Congress passed the Endangered Species Act (ESA) and the Marine Mammal Protection Act (MMPA), which drastically impacted the ways in which the Makah people could exercise their treaty rights (despite the absence of clear congressional intent for either act to supersede these rights). The gray whale was listed as endangered, and then as threatened, under the ESA. By the mid-1990s the gray whale population had significantly recovered, and in 1995, the Tribe notified the United States of its intent to resume its cultural practice (Eligon, 2019). In 1999, the Tribe was able to successfully hunt a gray whale for the first time since the 1930s (Eligon, 2019). Following the hunt, however, anti-treaty and animal rights activists launched a series of bigoted attacks against the Makah. In one such example, the Sea Shepherd Conservation Society posted their ship in Neah Bay and sounded a horn for the entirety of a Makah community celebration (Hopper, 2019). Through the combined efforts of anti-treaty activists, animal rights activists, and far-right congressional members, the Makah were ordered to suspend whaling by the Ninth Circuit Court of Appeals in *Metcalf v. Daley* (The Last Real Indians, 2020). The Court held that the National Oceanic and Atmospheric Administration's (NOAA) National Marine Fisheries Service (NMFS) would have to complete a new environmental assessment before the Tribe could resume its 2,000-year-old practice (The Last Real Indians, 2020). A subsequent suit was filed by a similar cohort in 2002 in *Anderson v. Evans*, where the Court held that the environmental assessment needed in *Metcalf v. Daley* was no longer sufficient and that an Environmental Impact Statement would need to be filed (National Oceanic and Atmospheric Administration, 2021). Perhaps most damaging was the Court's holding that the MMPA applied to the Makah's treaty-reserved rights. This was unusual, considering Congress's determination that nothing in the 1994 amendments to the MMPA "alters or is intended to alter any treaty between the United States and one or more Indian tribes" (Last Real Indians, 2020; Public Law, 1994).

As a result of this ruling, the Tribe had to apply for a waiver under the MMPA to resume treaty-based hunting of gray whales for ceremonial and subsistence purposes in the Tribe's traditional fishing area. In 2005, the Tribe did just that – and to this day remains in the process of acquiring the waiver it filed 16 years ago (National Oceanic and Atmospheric Administration, 2021).

As of September of 2021, an administrative law judge issued a recommendation to the US Department of Commerce stating that the Makah Tribe should be granted a waiver under the MMPA (Mark, 2021). The final decision rests with NMFS (Mark, 2021).

Lesser Laws

One of the first acts of Congress was the Northwest Ordinance of 1787 (reaffirmed in 1789), which states, "The utmost good faith shall always be observed towards Indians; their land and property shall never be taken without their consent." Of course, the Northwest Ordinance of 1787 encompasses far more than the several lines dedicated to American Indians, but the entirety of the phrase quoted here is important for the legal obligations established by the terms of art used. For example, the phrase "utmost good faith" has a specific legal meaning, as does "consent". Both are contract terms, meaning that a breach by the US government, or a failure to assure consent prior to taking land and property, should be problematic. Yet, the language of the treaty has been ignored with no ramifications for the United States. Some scholars have argued that the genocidal violence against American Indians that occurred within the United States was an intended consequence of this language, and that "dispossession with gratitude" of Indian land was the preferred policy because it allowed "US Americans to enjoy the benefits of territorial expansion at minimal expense, their consciences soothed and their sense of themselves as exceptional, nourished by the fantasy that Indians endorsed their claim to have acted with 'utmost good faith'" (Ostler, 2016). We discuss this here to understand that where there are laws specific to Indian country, they are often not enforced, are overly paternalistic, or were enacted or executed in bad faith.

Readers should also understand that the relationship between federally recognized Tribes and the United States is legally characterized as that of trustee and ward. This characterization has led to rampant paternalism under the guise of federal policy for the benefit of Native people. It is also necessary to know that because Congress has plenary power over Indian country, it has the ability to establish what constitutes Indian country (as it has done since the founding of the United States). Of course, exercising this power can be exclusionary and can impact communities in disparate ways. For example, though there are 229 federally recognized Tribes in Alaska, the federal definition of Indian country is only applicable to the Metlakatla. Thus, the protections that apply to Indian country leave out nearly half of all federally recognized Tribes.

Several of the case studies presented in this chapter discuss the impacts of relevant US environmental laws on Tribal rights, culture, and sovereignty (e.g., the Marine Mammal Protection Act and the case of the Makah Tribe whaling, and the case of the Miccosukee Tribe in South Florida and the Clean Water Act).

BOX 13.4 Case Study: The Federal Clean Water Act and the 404 Assumption

The Miccosukee and Seminole Tribes' cultural, spiritual, and traditional practices are dependent on a healthy Everglades ecosystem. Using traditional teachings like those found in Betty Mae Jumper's 1994 text, *Legend of the Seminoles*, and through language and the passing of oral traditions, both

Tribes have significant TEK and use it in conjunction with other advancements. For example, the Miccosukee Community and the Miccosukee Environmental Protection Agency (MEPA) have been documenting traditional ecological knowledge as well as peer-reviewed data that centers on the overall environmental health of the Everglades (Love the Everglades, 2017). As a sovereign, the Tribe is able to – and does – create policies that positively impact the Everglades' ecosystem.

This practice is important not just for its Tribal TEK, which is culturally rooted and based in kinship with the Everglades, but also for the Nation-to-Nation relationship that the Miccosukee Tribe rightfully expects as a federally recognized Tribe. Tribal sovereignty, however, is perpetually under attack, and even the federal Clean Water Act of 1972 creates colonial and paternalistic problems for that sovereignty.

> The best way it could be for the environment was how the creator originally intended it to be. But the white man came and thought he knew better.
>
> (Betty Osceola in *The Swamp*, MacLowry, 2019)

The Clean Water Act establishes the basic structure for regulating discharge of pollutants into US waterways, as well as surface water quality standards. This legislation expressly authorizes the Environmental Protection Agency (EPA) to treat Tribes as states for the purposes of grants, water quality standards, clean lakes, nonpoint source management, and more. Importantly, the EPA has agency policy that directly speaks to the necessity of consultation with the Tribes. Given the Nation-to-Nation relationship of Tribes and the US government, consultation on permitting is obligatory.

The US government previously ceded 90,000 acres of wetlands to the Seminole and Miccosukee Tribes around the same time that both Tribes acquired federal recognition (Schulman, 2020). Under this previous schematic, if an individual wished to dredge and fill in the Everglades, that individual would be required to apply for a permit from the US Army Corps of Engineers, which maintained oversight of wetland development (Schulman, 2020).

However, the Clean Water Act also has a provision for a process known as the 404 Assumption, which can be found in section 404(g). A 404 assumption permits US states to petition for control of federally protected land for leasing and permitting. In the Everglades, an already threatened wetland, opening these protected spaces to drilling, fracking, or development will inevitably lead to further degradation of the water quality and the habitat. For the Tribes, the impact of these potential activities cannot be understated.

> The tribe is deeply appalled about the loss of culturally sensitive sites and the potential destruction of the Miccosukee way of life. This way of life is integrally entwined within the Florida Everglades
>
> (Miccosukee Tribe of Florida, Schulman, 2020)

The lack of consultation has been appalling. The Tribe stated,

> The Miccosukee, along with several other affected tribes, have engaged in limited government-to-government consultation with the EPA and Army Corps of Engineers regarding the State of Florida's application for overall assumption. However, but when the Miccosukee requested specific consultation on this critical interpretation, we were denied.
> (Williams, 2020)

Alongside advocacy groups arguing that the state assumption over these lands will result in a significant decrease in oversight to favor development, the Tribe argues that the impact to the wetlands will directly impact the quality of water, biodiversity, and existing wildlife habitats – all three of which have already been detrimentally impacted by increased human migration to the area.

The Plenary Authority of Congress

Congress has plenary authority over Indian tribes under the US Constitution. This means that Congress has the ability to terminate a Tribe's sovereign status, ending the trust relationship and legal obligations with that individual Tribe. In this way, laws that are specific or relevant to Tribal land interests and protections become expendable.

International Human Rights Law

In theory, international human rights law is a legal framework that anticipates the whole of a people: their cultural, social, economic, and political rights, and, in some instances, their Indigeneity. The United Nations Declaration on the Rights of Indigenous People (UNDRIP) contains 46 articles, at least six of which explicitly state that Indigenous peoples have specific rights to their traditional homelands:

- Article 10 states that Indigenous people shall not be forcibly removed from their lands.
- Article 20 provides that Indigenous people have the right to engage freely in their traditional and economic activities, and that if deprived, they are entitled to just and fair redress.
- Article 24 declares Indigenous people the right to their traditional medicines.
- Article 25 states they have the right to maintain their distinctive spiritual relationship with their traditionally owned or otherwise occupied and used lands, territories, and coastal seas.
- Article 26 provides that Indigenous people have the right to the lands, territories, and resources which they originally owned, occupied, or otherwise used; and

- Article 29 states that Indigenous people have the right to the conservation and protection of the environment and the productive capacity of their lands (United Nations, 2007).

Where international human rights law falls short is in implementation. To be blunt, UNDRIP has no teeth because it lacks an enforcement mechanism. It is not legally binding on states and does not impose any legal obligations on governments. In the United States, this reality is particularly relevant. While UN-based advocacy often results in additional awareness around a particular issue and memorializes Tribes' engagement with the UN as sovereign bodies, best efforts result in shaming. It is not a coincidence that the only four original non-signatories of UNDRIP were Australia, New Zealand, Canada, and the United States. All four have taken a similar approach to their Indigenous populations: genocide, massive land theft, assimilation/cultural genocide, and ongoing present-day occupation.

Intellectual Property Law

Arguments exist that TEK can (and should) be protected by intellectual property law. But intellectual property law is rooted in profit, or, more specifically, in quantifiable losses that the owner of the intellectual property can prove. In that way, intellectual property law falls short because it is a Western construct, rooted in a Western worldview that prioritizes individual ownership and capital. Additionally, intellectual property law fails to consider generations past and future. It is difficult to have legal standing (a necessary component to bringing a lawsuit) if the person or persons intended to benefit from the protection are not in physical existence as pertains to Western concepts of *being* and *owning*. Furthermore, TEK is not individually oriented. It is collective and cannot belong to one person or one group – especially where regionally, Tribes share similar ways of being and knowing. To be clear, TEK is culture.

A Discussion on the Challenges Facing Social Justice Movements

It can be tempting to adopt TEK into mainstream spaces that address coastal areas and the impact of climate change without addressing the impact of settler-violence on the TEK and on the Indigenous communities who hold it. This amounts to a taking, and it frequently occurs in social justice movements. Sometimes we may think we can overcome the historical relevance of settler-violence through things like land acknowledgments, or through adding an Indigenous component to our work. Both approaches are inadequate. Land acknowledgments are ceremonial expressions of acknowledging the ancestral land of the Tribal nation we occupy, occasionally acknowledging the historical atrocities that occur, and almost always abrogating responsibility or complicity. They are rarely accompanied by an action. More specifically, land acknowledgments fail to give land back.

There can be, additionally, a rush to respond to urgent issues, and to do so by merging TEK with other scientific advancements absent an understanding of the historical violence and occupation that continues to impact TEK as it exists today. We cannot attempt to reconcile our treatment of land, water, and other non-human kin if we do not understand the reasons TEK has been historically degraded. The main reason is simple: TEK is rooted in Indigeneity.

In failing anti-oppression work, which is certainly applicable to marine and coastal climate advocacy spaces, we are often more interested in how to maintain power under the guise of addressing root causes of harm in our communities than we are in actually doing that work. Instead of understanding how oppression manifests even in well-meaning spaces, we focus on how well-meaning our space is. But rarely does a shift in power result, nor is there an understanding about what a power shift looks like. Most social justice movement spaces are adept at acknowledging marginalized people, but most individuals who work in those movements fail to acknowledge how oppression is often mirrored in those spaces and built into hierarchies of organizations, into the ways social justice spaces assess legislative priorities, or into how they leverage and acquire funding.

Finally, social justice movements are centered on reform and incremental change. Given that an Indigenous worldview is antithetical to a settler one, it is impossible to reform settler-created systems and expect Indigeneity to thrive. Indigenous governance and worldviews are not at all compatible with current systems in the United States. Reforms fall short because they fail to dismantle. Ever-present in reform are colonization, occupation, oppression, patriarchy, paternalism, and hierarchy.

Conclusion

So, the question becomes, what is the role of the non-Indigenous person with regard to TEK? Naturally, the answer to this question is layered and depends on who, and when, you ask. There are, clearly, urgent needs that TEK can and does address. In those instances, TEK should be utilized. But it should not be co-opted, nor should it be elevated without meaningful moves toward decolonization. To do so would be an empty gesture reliant on the expenditure of Indigenous knowledge to address problems that an Indigenous worldview did not create. In that sense, part of the responsibility of non-Indigenous people is to engage in truth-telling with Indigenous communities. Whether that process manifests as formal truth and reconciliation, or through the process of ongoing education on the history of the treatment of Indigenous people on their homelands, is entirely up to each Indigenous community.

Additional roles can include supporting Tribal sovereignty, seeing the benefits of Indigenous governance for all, learning of, and helping to advocate for the priorities of Indigenous communities, and advocating for land-back movements. These are examples of places to begin, but they will not be sufficient to address historical and ongoing genocide, colonization, and occupation. Going forward, the hard work of decolonization is essential to TEK.

Professional Pathways

Students interested in a career related to traditional ecological knowledge have the benefit of being able to pursue multiple paths. Some of these paths are met through conventional modes of study as mentioned in this chapter. Because Tribal governments are varied in terms of location (even from a purely geographical sense), students could study marine biology, engineering, political science, environmental science and resource management, biochemistry, botany, environmental studies, aquatic and fishery sciences, law, and so on.

Students could work for Tribal governments in a variety of capacities. Because there are 574 federally recognized Tribes in the United States, there are plenty of opportunities for students who are interested in a career related to TEK. Tribal governments have their own natural resources departments, offices of environmental management, their own historic preservation offices, their own legislative/executive/judicial bodies, etc. Additional opportunities include working within academia (particularly within Native studies programs or at Tribal colleges/universities), working for an NGO, and working for other governments (such as federal agencies or even Congress).

Questions for Reflection

1. What makes TEK unique? How is it different from local knowledge? Who can have TEK? How can it be equitably utilized?
2. Please carefully reflect on and consider the ways in which your everyday actions continue to perpetuate colonial and exclusionary practices. What would it look like to address these worldviews and to reconcile with Indigenous people/Tribes?
3. Is the legal framework of settler-colonial nations antithetical to Indigenous ways of being and knowing? Can Western systems ever be utilized to uphold certain rights for Indigenous people?
4. Who were the original inhabitants on the land you currently reside on? What processes of recognition must they go through? How do those processes impact their relationship with other governments?
5. What is governance? Explain how Tribal sovereignty, particularly in the United States, is a protective factor for TEK.
6. What role does culture play in TEK (think about kinship systems, language, beliefs, etc.). What impact has settler-colonialism and genocide had on culture for Indian country in the United States and globally?

References

Apgar, M. J., Allen, W., Moore, K., & Ataria, J. (2015). Understanding adaptation and transformation through indigenous practice: The case of the Guna of Panama. *Ecology and Society*, *20*(1), 45. http://dx.doi.org/10.5751/ES-07314-200145

Austen, I. (2021, May 28). 'Horrible History': Mass Grave of Indigenous Children Reported in Canada. *New York Times*. https://www.nytimes.com/2021/05/28/world/canada/kamloops-mass-grave-residential-schools.html

Berkes, F. (1993). Traditional ecological knowledge in perspective. In T. J. Inglis (Ed.), *Traditional ecological knowledge: Concepts and cases* (pp. 1–9). Canadian Museum of Nature and International Development Research Centre.

Brookings Institution, & Institute for Government Research. (1928). *The problem of Indian administration.* Johns Hopkins University Press.

Deloria, V. (1988). *Custer died for your sins: An Indian manifesto.* University of Oklahoma Press.

Douglas, J. (2021, September 2). Wild rice sues to stop oil pipeline. *High Country News.* https://www.hcn.org/articles/latest-justice-wild-rice-sues-to-stop-oil-pipeline#:~:-text=It%20declared%20that%20within%20White, %2C%20recovery%2C%20and%20preservation.%E2%80%9D

Eligon, J. (2019, November 14). A native Tribe wants to resume whaling. Whale defenders are divided. *New York Times.* https://www.nytimes.com/2019/11/14/us/whale-hunting-native-americans.html

Great Lakes Indian Fish, & Wildlife Commission. (2018). Integrating Scientific and Traditional Ecological Knowledge. http://glifwc.org/ClimateChange/GLIFWC_Climate_Change_Vulnerability_Assessment_Version1_April2018.pdf

Hopper, F. (2019, May 7). Makah one step closer to hunting whales: Animal rights extremists continue to oppose it. *Indian Country Today.* https://indiancountrytoday.com/news/makah-one-step-closer-to-hunting-whales-animal-rights-extremists-continue-to-oppose-it

Jumper, B. M. (1994). *Legends of the seminoles.* Pineapple Press.

Kimmerer, R. W. (2002). Weaving traditional ecological knowledge into biological education: A call to action. *BioScience, 52*(5), 432–438. https://doi.org/10.1641/0006-3568(2002)052[0432:WTEKIB]2.0.CO;2

Kimmerer, R. W. (2013). *Braiding Sweetgrass: Indigenous wisdom, scientific knowledge, and the teachings of plants.* Milkweed Editions.

Love the Everglades. (2017). Miccosukee Tribe requests public comment on new non-point source pollution plan. https://www.lovetheeverglades.org/blog/miccosukee-tribe-requests-public-comment-on-new-nonpoint-source-pollution-plan

Maclowry, R. (2019). *The American experience: The swamp* [Film]. PBS Video.

Mark, J. (2021, September 29). A tribe has not hunted whales in decades. Now, it might have a chance — and animal rights groups aren't happy. *Washington Post.* https://www.washingtonpost.com/nation/2021/09/29/makah-whaling-judge-recommendation/

Native American Rights Fund. (2019, November). *Trigger points: Current state of research on history, impacts, and healing related to the United States' Indian industrial/boarding school policy.* https://secureservercdn.net/198.71.233.187/ee8.a33.myftpupload.com/wp-content/uploads/2019/12/trigger-points.pdf

National Oceanic and Atmospheric Administration. (2021). Makah tribal whale hunt frequently asked questions. https://www.fisheries.noaa.gov/west-coast/makah-tribal-whale-hunt-frequently-asked-questions.

Ostler, J. (2016). 'Just and lawful war' as genocidal war in the (United States) Northwest Ordinance and Northwest Territory, 1787–1832. *Journal of Genocide Research, 18*(1), 1–20. https://doi.org/https://doi.org/10.1080/14623528.2016.1120460

Pevar, S. L. (2012). *The rights of Indians and Tribes.* Oxford University Press.

Public Law No. 103–238, § 14, 108 Stat. 532, 558 (1994).

Resolution establishing rights of Manoomin, 1855 Treaty Authority. (2018, December 5). Retrieved July 21, 2022 from https://static1.squarespace.com/static/58a3c10abebafb5c4b3293ac/t/5c3cdbc940ec9ab9b9ffde9d/1547492298497/1855+Treaty+Authority+Resolution+for+2018-05+Rights+of+Manoomin+12-5-18.pdf

Rust, S. (2019, December 1). A U.S. Tribe wants to resume whale hunting. Should it revive this tradition? *Los Angeles Times*. https://www.latimes.com/environment/story/2019-12-01/whale-hunting-makah-tribe-tradition-washington-state

Schulman, S. (2020, December 24). Florida Tribes deeply appalled by wetlands deal. *Indian Country Today*. https://indiancountrytoday.com/news/florida-tribes-deeply-appalled-by-wetlands-deal

Stopline3. (2019). White Earth Nation and 1855 Treaty Authority pass laws to protect the "Rights of Manoomin." https://www.stopline3.org/news/rightsofmanoomin

The Ways. (2019). *Manoomin*. Retrieved June 29, 2022 from https://theways.org/story/manoomin.html

Treaty of Neah Bay. (1855, January 31). https://goia.wa.gov/tribal-government/treaty-neah-bay-1855

United Nations. (2007). United Nations Declaration on the Rights of Indigenous Peoples, General Assembly Resolution No. 61/295, adopted on Sept. 13, 2007. https://www.un.org/development/desa/indigenouspeoples/wp-content/uploads/sites/19/2018/11/UNDRIP_E_web.pdf

United Nations Educational, Scientific and Cultural Organization. (2018). *Community consent and participation*. https://ich.unesco.org/doc/src/40082.pdf

United States Caucus of the Traditional Ecological Knowledge Task Team. (2021). *Guidance document on traditional ecological knowledge pursuant to the Great Lakes water quality agreement*. https://www.bia.gov/sites/bia.gov/files/assets/bia/wstreg/Guidance_Document_on_TEK_Pursuant_to_the_Great_Lakes_Water_Quality_Agreement.pdf

US House of Representatives. (1934). H.R. Report No. 1804, at 6.

US Supreme Court. (1992). *County of Yakima v. Confederated Tribes and Bands of Yakima Indian Nation*, 502 U.S. 251, 254.

The Last Real Indians. (2020, February 18). *Makah whaling and the anti-treaty mobilization by chuck tanner*. Retrieved June 3, 2021 from https://lastrealindians.com/news/2020/2/11/makah-whaling-and-the-anti-treaty-mobilization.

Ventocilla, J., Herrera, H., & Núñez, V. (1995). *Plants and animals in the life of the Kuna*. University of Texas Press.

Williams, A. B. (2020). "Runaway train" policy change will hurt Florida's environment, advocates say. *News-Press*. https://www.news-press.com/story/tech/science/environment/2020/12/17/florida-native-americans-and-environmental-activists-decry-wetland-permitting-shift/3925599001/

14 Coastal Community Development

A. N. Doerr, C. Pomeroy, and F. Conway

Introduction

One might ask, "Why should we care – locally to globally – about coastal communities? Are coastal communities more important than other communities?" Answering these questions requires some consideration. For instance, while the coastal landscape where these communities are located is understood to be the physical interface between the land and the sea, the ocean itself is also now considered a peopled seascape (Shackeroff et al., 2009).

One must also consider that there are coastal communities of place (geographic, e.g., Newport, Oregon or Gold Coast, Queensland) and coastal communities of interest (occupation or activity, e.g., commercial fishing or tourism); and that coastal communities are interdependent ecologically, economically, and socially. Furthermore, coastal communities of place are often the fastest growing areas in the country. Situated at the land–sea interface, they play critically important roles in food systems, serving as a source of food and food security not only for local residents but also for larger regional or even nationwide populations. They also are at the forefront of climate threats, including threats to critical food systems. Similarly, coastal communities of interest are vital to coastal economies, and their role and requirements are important to consider when making ocean space use decisions. These kinds of connections capture our attention and play a role in marine policy and management – locally, nationally, and globally.

This chapter highlights community perspectives on the marine environment and is organized around the following three themes: considerations of coastal communities as a coupled natural–human system; understanding coastal community dynamics; and a discussion of the role of community engagement and partnerships in solving problems and harnessing opportunities along the coast.

A recurrent theme throughout this chapter is that places, people, and industries on the coast have always experienced change. However, climate change adds new challenges and requires new approaches and flexibility. The case studies presented here highlight the importance of community engagement and practical partnerships for developing and fortifying physical and social infrastructure, and the roles that people play in cooperatively discovering, learning, and innovating to work their way through emerging and chronic challenges and opportunities. In

DOI: 10.4324/9781003058151-17

so doing, they strengthen their resilience, their adaptive capacity, and, ultimately, their wellbeing.

Coastal Communities as Coupled Natural–Human Systems

Coastal communities, located at the land–sea interface, must be viewed as a coupled system composed of human and natural subsystems. The *human system* includes the people; the social, cultural, and economic relationships that connect them; and the formal and informal institutions that govern individual behavior and relationships in a given place. The *natural system* consists of the biophysical environment with its living (e.g., marine plants and animals) and non-living (e.g., substrate and water) components and the relationships among them. The human and natural subsystems are linked, or coupled, in many ways, so that events or changes in one have impacts throughout the system. Rather than being static or unchanging, coupled natural–human systems are dynamic, continually experiencing variability and change. This creates both opportunities and challenges, which in turn require adaptation, whether through incremental adjustment or wholesale change.

Coastal communities are typically engaged with and dependent on the marine environment in a variety of ways, such as for food production, livelihood, commerce, leisure, and spiritual pursuits. Together, these contribute to the social, cultural, psychological, economic, and physical wellbeing of individuals and communities. As such, they are affected by environmental and policy processes and events occurring at a larger scale. El Niño events, for example, are a broad-scale oceanographic phenomenon that can disrupt the availability, distribution, and quality of fishery resources. These events can also lead to increased storm activity in some areas, causing coastal flooding or posing hazards to safety at sea. Coastal and marine management policies, likewise, are designed and implemented through larger state, federal, and international governance systems that then play out in coastal communities.

Moreover, many coastal communities are interdependent with other coastal communities, often as a result of their connectedness to the sea. For example, fishermen based in a given coastal community may follow the fish up the coast, calling at other ports to land their catch and reprovision. They may count on access to other coastal communities for goods and services not available in their home port. Sometimes they are connected to other communities within a region through family networks. They might even be connected by social and cultural associations with coastal and marine environments, such as subsistence fishing and collecting, surfing, or recreational boating. As such, in addition to unique place-based characteristics, coastal communities can also be thought of as a network of places linked by common values, interests, and needs. The connectedness among these communities can take a variety of forms: a "hub and spoke" arrangement, where one community serves as the center of activity that extends to several other communities (and back again); "point-to-point", where multiple communities connect with one another; or some combination of the two. Whatever the form, this structure is important for understanding, anticipating, and adapting to system dynamics and change.

Even as such interconnectedness can be conducive to adaptation and resilience, there also can be tensions between communities of place and communities of interest. These tensions have become evident with increasingly variable and changing distributions of marine species due to climate change. In the state of California, along the west coast of the United States, shifting distributions of market squid have in turn led some fishery participants (a community of interest) to shift their effort from the fishery's historic centers of activity (communities of place) in Central and Southern California (e.g., Monterey, Ventura, San Pedro) to Northern California (e.g., Eureka) and even into Oregon (Chavez et al., 2017). These tensions arise not only on the water in terms of access to this limited-entry fishery, but shoreside as well. More southerly communities and their working waterfronts grapple with less (or less consistent) squid fishery activity, which translates into reduced demand for – and economic activity to support – infrastructure, goods, and services. At the same time, more northerly communities, in part, welcome the activity, as it bolsters the working waterfront and local economy; but it also contributes to tensions between incoming and established fisheries and working waterfront interests. Growing interest in expanded or new ocean and coastal uses, such as marine aquaculture and marine renewable energy, likewise provides examples of potential tensions among communities of interest and of place, with particular relevance to working waterfronts as well as ocean space.

Case Study: Navigating Working Waterfronts

There are urban waterfront cities, such as Stockholm, Sweden; Sydney, Australia; or Miami, United States, where citizens and visitors alike recreate, live, and work (see Figure 14.1). As with coastal communities themselves, not all waterfronts are the same. Working waterfronts are central to many coastal communities of place and of interest. They provide key infrastructure (e.g., wharves, berthing, hoists) and goods and services (e.g., fuel, ice, vessel maintenance and repair) that enable people to access and interact with the marine environment. They are also places where cargo and seafood make landfall for local processing, distribution, and transport to other communities along the coast and beyond. As coastal communities have grown and changed, many working waterfronts have become gentrified, often by replacing infrastructure necessary for historical ocean-dependent uses with housing, retail, and related facilities. In addition, amid the growing interest in the blue economy, some coastal communities are grappling with decisions about how to accommodate the infrastructure needs of new ocean uses – such as ocean renewable energy production – along with existing needs of fisheries, commerce, and recreation.

Learning how to balance industries on the waterfront takes a commitment to work together toward a common goal of community resilience. In the large and diverse US state of California, variability and change in the natural environment, ocean uses and their management, and coastal communities have led to changes and challenges to the state's working waterfronts. This is especially true for those waterfronts that have historically supported commercial fisheries. In particular, the downsizing of many commercial fisheries resulting from increased regulation, including restricted access, quotas, and time and area closures, has changed demand for and use of infrastructure, goods, and services at many ports. And although commercial fisheries have some specific needs (e.g., fish unloading sites and hoists, gear repair worksites, docks for loading and unloading gear and equipment), other amenities such as well-maintained (i.e., dredged) navigation channels, berthing, fuel, and ice are needed by a variety of port users. As coastal communities and port managers seek to adapt to changing conditions, they need information and understanding about the existing and potential infrastructure, goods, and services needs of both established and new ocean users. Collaborative efforts are helpful in building a complete understanding of such needs, through research to characterize working waterfronts, trends in fisheries and other uses, and associated waterfront requirements. Convening groups to discuss, validate, and develop information resources can be used to support decision-making. Such efforts in California's North Coast and Santa Barbara Channel port communities established a baseline (see Culver et al., 2007; Industrial Economics, Inc., 2012; Pomeroy et al., 2010) and continue to inform working waterfront decisions at ports throughout California. Similar conversations and collaborations are ongoing in coastal communities across the globe and are necessary for ensuring the effective support and continued resilience of working waterfronts.

Understanding Coastal Community Dynamics

When contemplating why and how one might seek to know and connect with coastal communities – and ultimately work to support their resilience and adaptation – it is important to acknowledge that the words used to describe the work of supporting communities have changed over time. The term "community development" has historically been used to describe various processes or programs where members of the community work together to identify and solve common problems. By contrast, the term "economic development" has been used to describe programs or processes where members of government or industry utilize the resources available to stimulate sustained growth through the production of goods and services. In the 1970s academics and practitioners alike initiated a movement to take a more holistic approach, often referred to as "community economic development". Community economic development blended efforts to address tenacious

Figure 14.1 Working waterfront: Unloading Dungeness crab on the San Francisco wharf. Photo credit: C. Pomeroy.

economic issues (such as poverty, good jobs, and affordable housing and health care) with quality-of-life matters (Clay & Jones, 2009). The goal of this work was to improve community *wellbeing*, *resilience*, and *adaptive capacity*, terms that are more commonly used today.

Community *wellbeing* has been defined in many ways but typically refers to the combination of "social, economic, environmental, cultural, and political conditions identified by individuals and their communities as essential for them to flourish and fulfill their potential" (Wiseman & Brasher, 2008, p. 358). In order for communities to flourish, particularly in a variable and changing system, they need to be resilient. A seminal paper by Holling (1973) focused on the ecological nature of *resilience* and defined it as a "measure of the persistence of systems and of their ability to absorb change and disturbance and still maintain the same relationships between populations or state variables" (p. 14). Walker and Salt (2006, p. 2) further defined resilience as "the capacity of a system to experience shocks while retaining essentially the same function, structure, feedbacks, and therefore identity". Sometimes coupled natural–human systems, including coastal communities, are substantially disrupted, or even destroyed, necessitating reorganization. It is important to include such disruptions or shocks along with growth and conservation because this provides a view of links among system organization, resilience, and other dynamics (Peterson, 2000). Social resilience, community resilience, and social–ecological resilience all can be considered in terms of: (1) the amount of disturbance a system can absorb and still retain the same structure and function, (2) the degree to which the system is capable of self-organization, and (3) the degree to which the system can build and increase the capacity for learning and adaptation (Carpenter et al., 2001; Folke, 2006; Holling, 1973).

Adaptation and *adaptive capacity* are also key components to consider in any discussion of community development, wellbeing, or resilience. Adaptation can be defined as the decisions and actions taken to maintain the capacity of the social–ecological system to deal with change, all the while maintaining basic structure and function (Berkes et al., 2003; Nelson et al., 2007; Walker & Salt, 2006). Adaptive capacity "refers to the preconditions that are necessary to enable adaptation" (Nelson et al., 2007, p. 397). Preconditions often include assets or types of capital – human (e.g., the knowledge and training associated with commercial fishing), social (e.g., close ties between individuals and businesses in a coastal community), political (e.g., the influence coastal industries, such as service or fishing, have on decision-makers), financial (e.g., cash or wealth possessed by coastal individuals and industry), physical (e.g., dive boats and fishing vessels), and natural (e.g., coral reefs and robust fish stocks) – that individuals and groups possess (Nelson et al., 2007; Whitney et al., 2017).

Community development, resilience, adaptation, and adaptive capacity can, therefore, be seen as interconnected, core concepts that are related to a community's ability to embrace change and move forward to the future in the best way they view possible. That said, development, resilience, and adaptation may be equally important, but they may not look the same across the globe. Culture and context matter. Coastal communities differ widely around the world, both within and across communities. They can be large or small, rural, or urban, and have a predominant population of historically marginalized or favored communities. Thus, resilience in a fishing community in Bangladesh will likely be manifested, described, and achieved differently than in a fishing community in Australia. It is important to consider this in understanding and supporting a coastal community's ability to embrace change.

Community Engagement and Partnerships

Many of the challenges facing coastal communities are best addressed through cooperative efforts. While much work is spearheaded and done *within* the community, many communities invite and depend on the support of and partnerships with others *outside* of it. Individuals working at the crossroads among these groups are well situated to help advance these efforts. Solutions or paths forward are often relevant to communities of interest, regardless of geography, although there may be regional differences in implementation. Entities such as governmental and nongovernmental organizations, academia, industry, religious and culturally based groups, along with Tribal or Indigenous leadership play important building and bridging roles. It takes dedicated effort from many actors to ensure that coastal communities, whether of place or of interest, become and remain viable and vital.

While a number of organizations work with communities to address challenges and opportunities, university-based Extension[1] agents are well-placed to advance community development and resilience efforts. University-based Extension programs use applied research, education, and outreach to inform, listen to, and engage with communities to identify challenges, opportunities, and information needs, as well as develop and implement strategies for addressing them. In the United States, Extension agents, sometimes called advisors, are often university-trained experts in specific areas such as, but not limited to, natural resources (e.g., agriculture, fisheries, forestry), youth development, family health, and economics or marketing. Whatever their area of expertise, Extension agents are responsible for connecting research-based knowledge with local practitioners in industry, government, and community. Extension agents are continually learning and updating their community outreach and engagement knowledge, skills, and abilities (see Table 14.1) and combining them with their expertise. They serve as nonpartisan, neutral brokers of information who share resources and science, rather than as advocates for specific policies or policy outcomes.

Sea Grant Extension programs in the United States use this approach, with a particular focus on coastal communities and coastal community development and are funded by the federal government. At the national level, Sea Grant's mission is to "enhance the practical use and conservation of coastal, marine and Great Lakes resources in order to create a sustainable economy and environment" (https://seagrant.noaa.gov/). There are Sea Grant educators, Extension agents and specialists, and researchers in every coastal state, as well as the Great Lakes states,

Table 14.1 Community outreach and engagement knowledge, skills, and abilities

Knowledge	*Skills*	*Abilities*
Organizational Process knowledge such as theories and strategies associated with educational program design, conflict management, team building, cooperation, science communication, and facilitation.	• Meeting or educational event or product design. • Translating technical and scientific information into public narrative (and vice versa). • Building consensus. • Scenario planning.	• Stakeholder identification and engagement. • Boundary-spanning and bridge-building. • Program planning, management, and evaluation.
Natural Science Content knowledge such as fisheries, coastal hazards, pollution, harmful algal blooms, or other natural science topics.	• Fisheries management. • Coastal hazards planning. • Pollution and invasive species management.	• Topical education and research. • Citizen science, cooperative and collaborative research.
Social Science Content knowledge such as economics, education, sociology, anthropology, geography, political science, or other social science disciplines.	• Cost–benefit analysis. • Evaluation/research methods. • Needs and impact assessment. • Policy and planning.	• Tradeoff analysis and communication. • Social/economic impact assessments. • Identification of unintended consequences.

Puerto Rico, and Guam. Sea Grant Extension personnel are a direct link among researchers, decision-makers, and community members, facilitating the transfer of knowledge from researchers to practitioners (e.g., the commercial fishing industry, policymakers) and from practitioners to researchers. This work creates an environment that inspires co-discovery and co-learning, with new insights and technologies provided to stakeholders, and traditional and local ("on-the-water") knowledge provided to researchers. As trusted neutral partners in learning, Extension agents are able to foster relationships among these different and highly diverse groups, encouraging iterative research and collaborative processes.

Extension work is not intended to be one-way, with the Extension agent providing information to the community or solving complex problems on its behalf. Rather, Extension agents learn from and work with community members and stakeholders to help the community build understanding and navigate complex issues, providing a neutral voice (and often, a location) in support of sometimes difficult conversations. Sea Grant Extension agents – whether they focus on aquaculture, coastal hazards, fisheries, renewable energy, or any of the other multitude of complex issues discussed in this text – are often called upon to facilitate conversations around scientific issues, management, or community economic development, particularly when the issues are contentious, or when disparate voices need to be brought together in the process. When dealing with a limited resource, such as a community's coastline, port space, or nearby fishing grounds, there may be multiple potential uses, users, values, and needs, and no way to accommodate all of them. However, in order to enable sustainable and equitable community resilience, it is important that community members and stakeholders have a voice in the decision-making process. Extension agents help to make this possible.

It is important to recognize that the role that Extension agents play in coastal community support is not exclusive to Extension. Many regions of the world do not have formal coastal and marine Extension programs, per se. They do, however, have people who serve as educational liaisons and relationship and/or knowledge bridge-builders. Think about who might play the role of a trusted neutral partner in learning within your region or country. How might one work (whether in a volunteer or paid position) to inspire co-discovery and co-learning? Who might one work with to provide new information, insights, and technologies? How might one bring together scientific, local, and traditional knowledge to support coastal community development and wellbeing? Think about some coastal communities in a region or country you know or would like to work in: What organizations in this place might be connected to supporting these communities? To help with this, below we have included some examples of practical partnerships – efforts of co-discovery that led to important innovations or advancements in coastal communities.

Improving Conservation Efforts and Local Economies

Bonefish (*Albula vulpes*) are popular in sports fishing because of the difficulty of catching and landing the wary, strong, fast fish (see Figure 14.2). The recreational bonefish industry is very important to the outlying Family Islands in the Bahamas, making up a large percentage of the service and tourism expenditures on some islands, and providing more than $100 million to the Bahamian economy (Sherman et al., 2018). Overfishing during the late 20th century led the Bahamian government to enact strict regulations on bonefish, prohibiting the commercial buying or selling of these species. Although consumptive catch was not prohibited, the economic importance of the recreational fishery to the Family Islands (responsible for up to 80% of tourism expenditures in Andros, Bahamas, alone) resulted in fishing guides instituting a catch-and-release fishery (Fedler, 2019). During the past 20 years, recreational fishing guides, anglers, and scientists have worked together to develop and implement best handling practices for caught fish, helping to ensure their survival after they have been released (Adams & Cooke, 2015). This proactive partnership continues to advance our understanding of bonefish species and has led to a sustained interest in protecting coastal habitats in order to maintain these species.

Partnership in Gear Development

The gear used to catch marine species commercially and recreationally is constantly evolving and improving. Over time, fishermen, Extension agents, and other members of the fishing industry have partnered to develop gear that is more efficient, easier to use, and more ecologically friendly (e.g., less damaging to the marine environment, or better at avoiding bycatch). Often, gear development is iterative and incremental, with smaller changes designed, tested, and implemented,

Figure 14.2 A bonefish swims through the flats in the Bahamas. Joint management by resource users and managers had led to healthier habitats and bonefish populations. Photo credit: A. Shultz.

which gradually lead to accepted modifications. Other times, gear is developed with specific goals in mind – for example, to avoid "ghost fishing" if gear is lost, or to take advantage of new technologies. Regardless of how the process occurs, improving fishing gear is often a collaborative effort with managers, researchers, and the fishing community all playing important roles, bringing scientific and practical knowledge to bear on real-world problems and opportunities.

Fishermen, scientists, and decision-makers have worked effectively together toward improving efficiency and reducing bycatch in many fisheries. The Oregon pink shrimp (*Pandalus jordani*) fishery provides an excellent example of such collaborative work. Although metal grate bycatch reduction devices (BRD) have been used for decades and are effective for significantly reducing bycatch, these devices cannot prevent the accidental catch of small fish while still allowing for the capture of pink shrimp. However, research has shown that fish respond to visual stimuli in a way that shrimp do not. This information was used to design a study to investigate the feasibility and effectiveness of adding pressure-activated LED lights to shrimp trawl nets. Working directly with fishermen and their gear during the fishing season, researchers investigated the impacts of adding LED lights at various locations along trawl nets. Results showed that LED lights *added to the footropes* dramatically decreased bycatch of small and juvenile fish without negatively impacting shrimp catch (Hannah et al., 2015; Oregon Department of Fish and Wildlife, 2019). Recognizing the benefits of this efficient and inexpensive change, most Oregon pink shrimp vessels voluntarily adopted the use of LED lights in 2014; they were made mandatory in Oregon and Washington in 2018 (Oregon Department of Fish and Wildlife, 2019) (see Figure 14.3).

New Markets for Seafood

Seafood markets link local seafood producers to local, regional, and global communities. Throughout the United States, there are communities that have an identity largely connected to a particular industry: mining towns, manufacturing towns, or lumber towns. The same is true for many coastal communities, where the fishing industry is central to their identity. While many of the practical partnerships that form in coastal communities revolve around communities of interest, they can support communities of place as well – and often, communities of place and communities of interest along the coast are inextricably linked.

Despite similarities that may exist among fishing communities throughout the United States, the fishing industry itself is remarkably heterogeneous. Fishing operations range widely from large trawlers with large crews (like the ones operating offshore Alaska) to small vessels with crews of just one or two people. These smaller operations comprise the vast majority of commercial fishing operations in the United States and globally. Fisheries use a plethora of gear types to target a wide range of species that may be sold live, fresh, or processed into a variety of seafood products. Aquaculture in the United States is also diverse, with operations ranging from small producers of boutique products such as dulse or caviar, to large-scale facilities such as nearshore oyster farms in Northern California and the

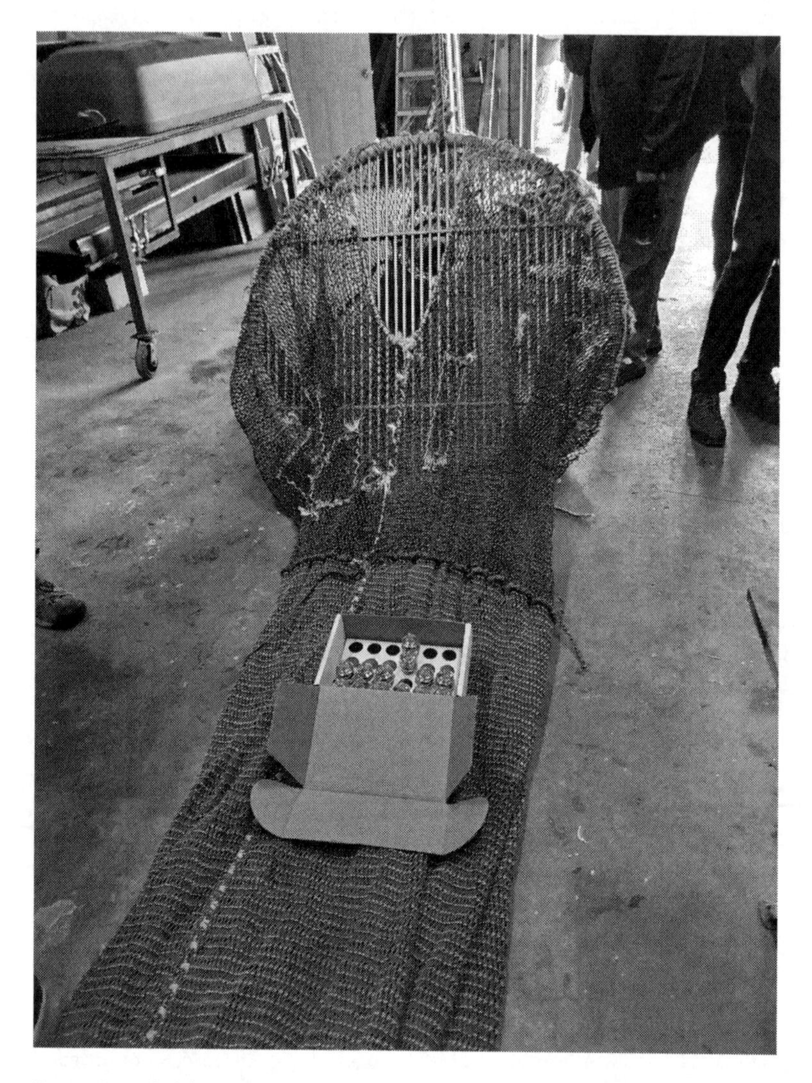

Figure 14.3 A fisherman displays a pink shrimp net with a bycatch reduction device and escape hole clearly visible. On top of the net are pressure-activated LED lights, which the fishing community uses to further reduce bycatch of small fish. Photo credit: A. Doerr.

Pacific Northwest. This diversity, together with regulatory, operational, environmental, social, and economic challenges, has led to growing interest in a varied mix of seafood marketing opportunities (Pomeroy et al., 2020). While the United States exports about 30% of its edible catch, the majority of the remaining 70% is consumed in restaurants or processed and sold in local retail stores (National Marine Fisheries Service, 2018). Realizing a potential opportunity to connect directly with local consumers, many in the fishing industry have been exploring new markets for seafood, such as community-supported fisheries (referred to as CSFs), dockside markets, a greater presence at farmers' markets, and pathways to bring local seafood to grade schools, hospitals, and college campuses.

In California, fishermen and growers were increasingly interested in how these marketing options might help mitigate the many challenges they faced. Recognizing this need, California Sea Grant Extension specialists sought to expand understanding of the ins and outs of various seafood marketing arrangements. Working in collaboration across the Sea Grant Extension network, the Extension specialists engaged with fishermen, seafood marketers, and others in several coastal communities to conduct field research with the goals of characterizing the different marketing options, the conditions required for establishing and maintaining them, and their implications for fishermen's operations and the well-being of both fishermen and fishing communities. They used their findings to develop a website called *Market Your Catch* (Culver et al., 2015), a clearinghouse for information resources and tools developed by several Sea Grant programs, and others. Consistent with Sea Grant's role as a neutral broker of information, the website provides information about different types of markets and customers, key considerations for evaluating the feasibility and utility of a market type for each individual fisherman's situation, and information on how to get started in or expand such seafood markets. Sea Grant Extension staff continue to share this information through workshops and presentations to fishing communities and others. Of particular interest is understanding the permits and licenses required, and the regulations that govern seafood direct marketing. As these vary by state and locality, Extension specialists have developed state-specific permit pages for the website that provide an overview of potential requirements and contact information for relevant agencies. With the growing interest in and expansion of US aquaculture, the website is also being updated to provide analogous information to people involved in aquaculture.

Change and Working Waterfronts

Building on the California working waterfront case study described above, with renewed and growing interest in ocean renewable energy and mariculture, the conversation has shifted to *supporting existing uses while accommodating new and expanded uses*. As questions emerge about how these new uses might fit within – and affect – working waterfronts, Sea Grant Extension is engaging with existing and new users, port managers, state agencies, and others to build shared understandings of diverse working waterfront needs along with potential conflicts, compatibilities, and synergies between existing and new uses. For example, Extension staff and colleagues are building on decades of community-based experience and knowledge in the Santa Barbara Channel region to develop StoryMaps and related resources to capture the common and divergent waterfront and marine space needs of fisheries and aquaculture. Capturing and conveying this information in ways that are accessible to the diverse groups involved requires understanding differences not only in how space is valued and used, but also in how people communicate those values and needs. The collection and translation of this information generated by commercial fishermen and aquaculturists will enable them, along with port managers and decision-makers, to better understand one another and support better-informed decisions.

Conclusion

There are multiple pathways available to individuals who would like to support coastal communities directly or indirectly. These pathways, however, share important commonalities. When working in coastal communities, it is important to acknowledge that all of these communities – whether of place or of interest – are coupled natural–human systems, and that their vulnerability, resilience, adaptive capacity, and wellbeing are dependent on recognizing these connections. Place-based and interest-based engagement are tried and true methods used by Extension agents. Why? Because, ultimately, continuous and cooperative learning and understanding are keys to addressing the tensions that can arise due to change; the marine environment has, and always will be, changing.

Coastal community development is not a one-time activity or process. Coastal communities are constantly changing: responding to environmental and demographic shifts, as well as changes in the economic and political landscapes within which they operate. Individuals seeking to work within these communities should be knowledgeable in recognizing and responding to these shifts, skilled at engaging with people both positively and negatively impacted, and able to use their skills and knowledge to confront new challenges and embrace new opportunities.

Professional Pathways

There are many careers that directly or indirectly support building resiliency and adaptive capacity in coastal communities. Such careers may be found within local government, such as city planning, community management, or county finance. Additionally, many communities have Development or Services offices. Outside of government, a plethora of nonprofit organizations work within communities to enhance both individual and collective capabilities. And there are careers through colleges and universities (like the Sea Grant Extension programs as mentioned in detail above) that connect community members, managers, scientists, and educators, facilitating communication and collaboration to identify and address the particular challenges and opportunities coastal communities face.

When one decides that they want to do professional work that supports coastal communities, they are committing themselves to building the knowledge, skills, and abilities to create or strengthen bridges between groups of people, between people and the marine environment, and between use and conservation of these places. The effort needed to fulfill this commitment might extend beyond their university training, as it is often not available through many traditional academic programs. Table 14.1 highlights a subset of knowledge types, along with associated skills and abilities, which are necessary or valuable to a career in community development, outreach, and engagement.

Questions for Reflection

1. Why is it important to consider both natural and social systems when working in coastal communities?
2. People who serve as educational liaisons and relationship and/or knowledge bridge-builders, such as Extension agents, who work with a variety of research, industry, and policy-making partners, are expected to be neutral. What does it mean to be neutral, and why is it important?
3. This chapter discusses several practical partnerships, where industry, scientists, and managers come together to develop solutions to problems facing coastal communities. Can you think of other examples of practical partnerships? Who was involved, and what was the outcome?
4. Describe the difference between communities of place and communities of interest, and why considering both is important in marine resource decision-making.
5. Why is it important to bring together scientific and practical knowledge to address challenges and opportunities for coastal community development? Can you provide a real-world example where this has been done – or a hypothetical situation where it could be done?

Note

1 Extension typically involves some form of field-based educational training or exchange program developed with, and for, rural people that is aimed at supporting them in the context of changing needs (Oakley & Garforth, 1985).

References

Adams, A. J., & Cooke, S. J. (2015). Advancing the science and management of flats fisheries for bonefish, tarpon, and permit. *Environmental Biology of Fishes*, 98, 2123–2131. https://doi.org/10.1007/s10641-015-0446-9.

Berkes, F., Colding, J., & Folke, C. (Eds.). (2003). *Navigating social-ecological systems: Building resilience for complexity and change*. Cambridge University Press.

Carpenter, S., Walker, B., Anderies, M., & Abel, N. (2001). From metaphor to measurement: Resilience of what to what? *Ecosystems*, 4(8), 765–781. https://doi.org/10.1007/s10021-001-0045-9

Chavez, F. P., Costello, C., Aseltine-Neilson, D., Doremus, H., Field, J. C., Gaines, D. D., Hall-Arber, M., Mantua, N. J., McCovey, J., Pomeroy, C., Sievanen, L., Sydeman, W., & Wheeler, S. A. (2017). *Readying California fisheries for climate change*. California Ocean Science Trust. Retrieved June 30, 2022 from https://www.oceansciencetrust.org/wp-content/uploads/2016/06/Climate-and-Fisheries_GuidanceDoc.pdf

Clay, R., & Jones, S. (2009). A brief history of community economic development. *Journal of Affordable Housing & Community Development Law*, 18(3), 257–267.

Culver, C., Richards, J., & Pomeroy, C. (2007). *Commercial fisheries of the Santa Barbara Channel and associated infrastructure needs*. Report to the Ventura Port District and the Santa Barbara Harbor District. California Sea Grant College.

Culver, C., Stroud, A., Pomeroy, C., Doyle, J., Von Harten, A., & Georgilas, N. (2015). Market Your Catch website. https://marketyourcatch.msi.ucsb.edu/

Fedler, T. (2019). *The 2018 economic impact of flats fishing in The Bahamas*. The Bonefish and Tarpon Trust. Retrieved June 30, 2022 from https://www.bonefishtarpontrust.org/downloads/research-reports/stories/bahamas-flats-economic-impact-report.pdf

Folke, C. (2006). Resilience: The emergence of a perspective for social – ecological systems analyses. *Global Environmental Change*, *16*(3), 253–67. https://doi.org/10.1016/j.gloenvcha.2006.04.002

Hannah, R. W., Lomeli, M. J.M., & Jones, S. A. (2015). Tests of artificial light for bycatch reduction in an ocean shrimp (Pandalus jordani) trawl: Strong but opposite effects at the footrope and near the bycatch reduction device. *Fisheries Research*, *170*, 60–67. https://doi.org/10.1016/j.fishres.2015.05.010

Holling, C. S. (1973). Resilience and stability of ecological systems. *Annual Review of Ecology and Systematics*, *4*, 1–23. https://doi.org/10.1146/annurev.es.04.110173.000245

Industrial Economics, Inc. (2012). *Identification of outer Continental Shelf renewable energy space-use conflicts and analysis of potential mitigation measures*. US Department of the Interior, Bureau of Ocean Energy Management: 414p.

Nelson, D., Adger, N., & Brown, K. (2007). Adaptation to environmental change: Contributions of a resilience framework. *Annual Review of Environment and Resources*, *32*(1), 395–419. https://doi.org/10.1146/annurev.energy.32.051807.090348

National Marine Fisheries Service. (2018). Fisheries of the United States, 2017 Report. Retrieved November 2, 2020 from https://www.fisheries.noaa.gov/resource/document/fisheries-united-states-2017-report

Oakley, P., & Garforth, C. (1985). *Guide to extension training (No. 11)*. Food & Agriculture Organization of the United Nations. Retrieved June 30, 2022 from https://www.fao.org/3/t0060e/t0060e.pdf

Oregon Department of Fish and Wildlife. (2019). *30th annual (diamond edition) Oregon Department of Fish and Wildlife Marine Resources Program pink shrimp review* [Newsletter]. Retrieved November 2, 2020 from https://www.dfw.state.or.us/MRP/shellfish/commercial/shrimp/docs/30th_APSR_2019.pdf

Peterson, G. (2000). Political ecology and ecological resilience: An integration of human and ecological dynamics. *Ecological Economics*, *35*, 323–336. https://doi.org/10.1016/S0921-8009(00)00217-2

Pomeroy, C., Rice, S., Culver, C., & Baker, V. (2020). Seafood direct marketing: Supporting critical decision making in Alaska and California. In J. Zelasney, A. Ford, L. Westlund, A. Ward, & O. Riego Peñarubia (Eds.), *Securing sustainable small-scale fisheries: Showcasing applied practices in value chains, post-harvest operations and trade* (pp. 85–103). FAO Fisheries and Aquaculture Technical Paper No. 652. FAO. https://www.fao.org/documents/card/en/c/cb0472en/

Pomeroy, C., Thomson, C, & Stevens, M. M. (2010). *California's North coast fishing communities: Historical perspective and recent trends*. California Sea Grant and NOAA Fisheries Southwest Fisheries Science Center.

Shackeroff, J., Hazen, E., & Crowder, L. (2009). The oceans as peopled landscapes. In K. L. McLeod & H. M. Leslie (Eds.), *Ecosystem-Based management for the oceans* (pp. 33–51). Island Press.

Sherman, K. D., Shultz, A. D., Dahlgren, C. P., Thomas, C., Brooks, E., Brookes, A., Brumbaugh, D. R., Gittens, L., & Murchie, K. J. (2018). Contemporary and emerging fisheries in The Bahamas – Conservation and management challenges, achievements and future directions. *Fisheries Management and Ecology*, *25*(5), 319–331. https://doi.org/10.1111/fme.12299

Walker, B., & Salt, D. (2006) *Resilience thinking: Sustaining ecosystems and people in a changing world*. Island Press.

Whitney, C. K., Bennett, N. J., Ban, N. C., Allison, E. H., Armitage, D., Blythe, J. L., Burt, J. M., Cheung, W., Finkbeiner, E. M., Kaplan-Hallam, M., Perry, I., Turner, N. J., & Yumagulova, L. (2017). Adaptive capacity: From assessment to action in coastal social-ecological systems. *Ecology and Society*, *22*(2), 22. https://doi.org/10.5751/ES-09325-220222

Wiseman, J., & Brasher, K. (2008). Community wellbeing in an unwell world: Trends, challenges, and possibilities. *Journal of Public Health Policy*, *29*(3), 353–366. https://doi.org/10.1057/jphp.2008.16

15 Marine Entrepreneurship

A. N. Doerr, L. Anderson, and J. Scorse

Introduction

According to the World Bank, the blue economy "*is the sustainable use of ocean resources for economic growth, improved livelihood and jobs, and ocean ecosystem health*" (World Bank, 2017). In the context of climate change, increasing demand for marine products, and rapidly degrading marine ecosystems the world over, it is important that humanity move away from the unsustainable use of ocean and coastal resources. The blue economy replaces fossil fuels with renewable energy, such as offshore wind power and wave and tidal energy systems. It promotes sustainable wild seafood harvest, carbon-neutral aquaculture production, and new seafood technologies (both plant- and cell-based) that supplement demand for food from the sea and have the potential to reduce overfishing that exceeds scientifically determined maximum sustainable yields (MSY) and degrades marine ecosystems. Additionally, the blue economy encourages coastal development that makes use of living shorelines to provide climate resilience, while also restoring coastal ecosystems. Moreover, it provides opportunities for public recreation and ecotourism that educate people about ocean and coastal ecosystems. The blue economy both encourages and enables people to think creatively about jobs and career possibilities centered on the sustainable and renewable use of ocean resources.

Approximately 40% of the world's population and many of the world's largest cities can be found within 100 km of the coast (Maul & Duedall, 2019). This concentration of human population along the coast will only accelerate in the coming decades (Environmental Protection Agency, 2017; Maul & Duedall, 2019). Steering ocean and coastal economic activity in the blue direction is the work of policymakers, regulators, educators, activists, consumers, community members, and entrepreneurs. Additionally, it is often the business sector that can promote rapid change and help communities adopt new technologies at scale. These changes will have major ramifications not only for income, employment, and equity, but also for the mental and physical health of much of the world's population, since healthy waterways promote and support healthy humans (see, e.g., Korfmacher et al., 2015; Wheeler et al., 2012).

This transition to blue economic development and blue growth is going to require a new wave of entrepreneurs to discover innovative solutions to ocean and coastal resource needs and forge new industries. There is tremendous opportunity

DOI: 10.4324/9781003058151-18

in sectors as diverse as ocean energy systems, bioplastics, and new modes of seafood production. We strongly encourage students the world over to ride this new wave and help develop an equitable and sustainable blue economy for the 21st century that meets human needs while mitigating greenhouse gas emissions and restoring our most precious coastal ecosystems.

Opportunities for Marine-Based Entrepreneurship

Food from the Sea

Commercial Fishing

The oceans contain some of the most valuable natural capital on the planet: wild seafood. Commercial harvest of marine species dates back thousands of years, having increased rapidly in the early to mid-20th century, and remained relatively stable since the mid-1980s (Food and Agriculture Organization of the United Nations [FAO], 2020). Global catch has been level for the last 40 years while harvest and management methods continue to evolve. In this context, the future of commercial fishing does not necessarily involve catching more fish, faster; rather it is about increasing value, reducing waste, and managing responsibly – all of which require innovation and entrepreneurship.

In many countries, including parts of Africa, Asia, Latin America, and the Western Pacific, small-scale fisherfolk and aquaculture farmers dominate the seafood industry (FAO, 2020). Artisanal fisheries are commonplace, with small boats launching directly from shore, bringing in daily catch to local markets for immediate community consumption. In contrast, the vessels associated with industrialized fishing are typically much larger, technologically capable, and efficient at harvesting living marine resources.

Within both artisanal and industrial fishing, and the many sizes and scales of fishing in between, challenges with illegal, unreported, and unregulated (IUU) fishing still exist globally (Sumaila et al., 2020). While many fisheries throughout the world are sustainably harvested, proper management and new technologies will be needed to ensure wild-caught fisheries continue to prosper. Entrepreneurial individuals may be at the forefront of supporting such fisheries, selectively harvesting popular and emerging species, and making use of the latest science, technology, or socioeconomic systems to catch and distribute food from the sea.

BOX 15.1 Defining Sustainable Seafood

Monterey Bay Aquarium's Seafood Watch program, the Marine Stewardship Council (MSC), and FishWise all have developed rating systems based on environmental (and sometimes social) criteria that consumers, as well as retailers, may use to determine which fish to purchase. These rating systems have helped usher in major shifts in the fishing industry across the world; this demand-side intervention has sent powerful signals up the supply chain, resulting in the development of new markets and measurable improvements in many fisheries. But more work needs to be done since most seafood consumed in the world does not meet sustainability criteria (FAO, 2020), and even these certification systems do not address all dimensions of the sustainability challenges. In addition, due to persistent and widespread fish mislabeling, "impostor fish" represent a sizable segment of the seafood supply chain, which fundamentally undermines sustainability efforts (Kroetz et al., 2020; Luque & Donlan, 2019).

Aquaculture

Although there is potential to increase the volume of wild-caught fish by targeting new species and improving harvest technologies (Costello et al., 2020), much of the increase in future seafood consumption will come from aquaculture (FAO, 2020). Aquaculture currently comprises just more than half of the fish consumed by humans, much of it grown inland rather than in the ocean. The potential exists for expansion of both nearshore and offshore aquaculture for invertebrates (e.g., oysters, sea cucumbers), finfish (e.g., Atlantic salmon, snapper), and seaweeds (e.g., giant kelp, dulse); as well as sea ranching or "penning", where juvenile fish are caught and then raised in ocean pens for harvest as adults. However, valid concerns associated with marine aquaculture, particularly for finfish, include the spread of disease to wild species and the environmental impacts of raising large numbers of fish in small spaces. With continued growth in the importance of aquaculture, there is significant potential for the development of new technologies and practices to overcome these challenges and advance sustainable aquaculture.

BOX 15.2 Case Study of Aquaculture: Monterey Bay Seaweeds

Dr. Mike Graham is a busy man, and not only because he has eight children. He is a professor of phycology at Moss Landing Marine Lab on California's Central Coast, where he primarily teaches courses on seaweeds, including their history and ecology, and is also the Vice President of the California Aquaculture Association (CAA). Never one to miss an opportunity to put theory into practice, in 2015 he embarked on a business venture to grow and market native Monterey Bay Seaweed to restaurants. The business is called Monterey Bay Seaweeds. He was not interested in mass-producing seaweed in large ocean-based farms, but instead, cultivating a niche product that would attract the attention of top chefs and connoisseurs. First, though, he had to obtain a permit to remove sea water from the Monterey Bay National Marine Sanctuary (MBNMS).

These seaweeds are grown on land in large tubs where the water is continuously circulated, and conditions can be precisely monitored and managed. His operation has the capacity to grow 1,000 lb of seaweed per year in a single eight-by-eight-ft area, surpassing production capabilities of other land-grown produce when comparing the necessary water and space requirements. After obtaining regulatory approval from the MBNMS, he began experimenting with different sea vegetable varieties, as well as a mixed system with the highly prized California abalone. Graham currently produces west coast dulse, sea lettuce, and ogo and ships them fresh around the country to some of the nation's top chefs, who use these vegetables in everything from appetizers and entrees to desserts and cocktails. Some of the plants can grow up to 15% in one day. The water that is discharged back into the bay is cleaner when it leaves his facility than when it entered. Except for a small amount of electricity that he uses for water circulation, his production method is energy-free, requiring no inputs.

Graham has a grand vision for US aquaculture, and California more specifically. He views sustainability as the driving force for the industry, with numerous opportunities for local production, including all sorts of delicious niche products like his own. For example, in 2015, Monterey Bay Seaweeds' popular dulse was profiled by *Bon Appétit* for its rich umami flavor, closely resembling bacon when fried. Next time you are in one of your favorite restaurants you may end up sampling some of his products as his reputation grows.

Seafood Markets

There are significant marine entrepreneurship opportunities in buying and selling seafood. In the United States, government figures estimate that in 2018 there were 3,314 seafood processing and wholesale plants that employed 63,005 people (National Marine Fisheries Service, 2020). China has one of the world's largest seafood processing sectors, with close to 10,000 processing plants as of 2014 (FAO, 2017). However, buyers vary in size from large, vertically integrated, multinational seafood companies, all the way to small fishing cooperatives[1] and direct-to-consumer dock sales.[2] Food Hubs[3] and Food Aggregators also support small independent fishing businesses without the legal ties of a cooperative.

Purchasing, processing, packaging, and distribution of seafood are separate activities that can be carried out by multiple businesses, although these activities are often vertically integrated within a single firm. Internationally, we have seen a consolidation in seafood processing, due in part to the high capital cost of infrastructure required for this business venture, including physical plants, docks, hoists, ice production, refrigeration, processing equipment, and trucks. For many wild-caught and farmed species, seafood processing has become a global enterprise, with products caught in one country and processed in another before being exported to yet another (Gephart et al., 2019). However, market disruptions, such as the global economic crisis of the late 2000s and the COVID-19 pandemic in the 2020s, have demonstrated the challenges with such a dispersed production chain, and alternative systems have emerged, including direct marketing of seafood and the development of local markets. There is strong evidence that such alternative processing and sale options will continue to be important and may provide opportunity for enterprising individuals (Stoll et al., 2021).

BOX 15.3 Case Study of Seafood Market Disruptors

As global population and wealth increase, demand for wild fish and farm-raised varieties will remain strong. However, it would be wise to keep an eye on the industries described below. They are emerging now, poised for growth, and likely to challenge the traditional notion of seafood:

1. Cell-based seafood production uses cells taken from live fish to grow exact genetic replicas of their flesh in brewery-like conditions far away from the oceans. While this type of seafood production may sound like science fiction, it is fast becoming reality. In 2020, Singapore became the first country to grant regulatory approval for the commercial sale of cell-based meat (in this case for chicken), and many other countries are looking to move forward on this front as well. (For more on this, see Lucas, 2020.)
2. Sea vegetable aquaculture is also gaining momentum in the United States. Sea vegetables such as kelp, dulse, and sea lettuce can grow quickly, can purify the water as they grow, need no fertilizers, and are highly nutritious. Sea vegetables are used as food additives for humans and livestock – potentially reducing methane emissions in cows – as well as in cosmetics, fertilizers, and wastewater treatment (Dundas et al., 2020). There are also industrial applications for sea vegetables, with innovations occurring in chemicals, fuels, and other manufacturing inputs. (See Box 15.2: Case Study of Aquaculture: Monterey Bay Seaweeds.)
3. While not 100% seafood, there are many companies that are producing plant-based versions of popular seafood products that resemble the real thing in both taste and texture. Plant-based tuna, crab cakes, and fish burgers have garnered so much international attention that the World Sustainability Organization will soon begin certifying plant-based seafood products under a Gold Level version of their Friend of the Sea certification system (Lamy, 2020). One company, in particular, to keep an eye on is Finless Foods, which is producing both plant- and cell-based seafood products.

For fishers and small-fish buyers in areas typically dominated by wholesalers, selling directly to consumers has been eased through the advancement of online shopping sites, packaging improvements, and third-party shipping services. These marketing models have capitalized on cultural movements aimed at being knowledgeable about one's food source, such as the "Locavore" and "Know your Farmer/Fisher" movements. An example of this might be a community-supported fishery (CSF), a program modeled after community-supported agriculture (CSA), where consumers pay fishers for products in advance. With a CSF, the consumer is effectively buying a subscription, or a share, and one or more fishers provide each consumer with his or her share of the catch on a prearranged schedule over a specified period. CSFs allow fishermen to distribute less-familiar products to customers, which they may be hesitant to purchase at a restaurant or retail market. CSFs also provide smaller-scale fishers with the security of stable demand, while also allowing them to fish more sustainably by focusing on seasonally available or less-commonly caught species.

In other parts of the world, such as the Caribbean and Latin America, some species are caught by industrial fleets for processing, while other species are targeted by artisanal fishers and sold via dock sales, at local markets, or through the integration of fishing, processing, and cooking at local fish stands or restaurants. In many small-scale and artisanal fisheries, women make up the bulk of the post-harvest workforce, playing crucial roles in processing and marketing (Gopal et al., 2020; Harper et al., 2020).

Boat, Gear, and Equipment Engineering and Design

Fishing-vessel and gear engineering, design, and building are also important elements of the industry. Boat owners are continually reinvesting in their operations to improve efficiency and profitability. For example, fishing gear, like trawl nets, is continually redesigned for species selectivity and to minimize habitat impacts. Boat owners are sometimes required to invest in new technologies, such as Automatic Identification System (AIS) and Vessel Monitoring System (VMS) satellite tracking systems, that keep them compliant with changing regulations, improve safety, protect the marine environment, and improve vessel monitoring requirements. The global emphasis on improving efficiency and sustainability in fisheries and aquaculture means there are countless opportunities for developing new technologies, both terrestrial and marine, for these industries.

Ocean and Coastal Tourism and Recreation

Sustainable tourism has been defined by the United Nations World Tourism Organization as tourism that considers its impacts on environmental resources, socio-cultural systems, and economic conditions, while providing benefits to local stakeholders and host nations (United Nations World Tourism Organization, 2016). Ecotourism, which builds upon the sustainable tourism mindset by further emphasizing greener travel and often a cultural or environmental education component, has continued to grow in market share, and many consumers intentionally opt for sustainable or "green" travel and tourism opportunities. The growth in the sustainable tourism market provides many opportunities for socially and environmentally minded entrepreneurs.

A variety of businesses directly rely on or benefit from the ocean but are non-consumptive in nature. Common non-consumptive businesses include diving and snorkeling companies, surf shops, whale watching tours, glass-bottom boat operators, and ocean recreation companies that rent kayaks, jet skis, and other marine vehicles. These ocean and coastal tourism industries are globally important, making up approximately 26% of the ocean's US$1.5 trillion contribution to the world economy (Organization for Economic Cooperation and Development, 2016). The National Oceanic and Atmospheric Administration (NOAA) estimates marine tourism and recreation industries in the United States employed more than 2 million people and added approximately $124 billion to the gross domestic product (GDP) in 2017 (National Oceanic and Atmospheric Administration [NOAA], 2020).

Many of these businesses are small, independent operations, and estimates for most of the industry project strong economic growth (Organization for Economic Cooperation and Development, 2016). Small differences can set apart similar businesses – allowing, for example, multiple diving shops or kayak tour operations in the same region. Although these businesses are using the ocean in more passive ways, they are still likely to experience some impacts from climate and market changes. For instance, shifting ocean conditions impact the behavior of marine mammals, affect the health of coral reefs, and can change typical current or wave patterns. Marine pollution can adversely impact these activities as well, potentially challenging the ability of a business to provide standard services, while regional or global changes to behavior, such as COVID-19, or market downturns, can significantly reduce demand for services.

It is important to note that not all of these non-consumptive industries have wholly benign impacts on the marine environment. Diving operations can inundate reefs with tourists and damage sensitive corals, and whale watching boats can harass and even sometimes directly injure whales. Stringent codes of conduct and enforcement measures are necessary to ensure that such industries do not contribute to ecosystem degradation, even if they are not engaged in extraction of marine resources.

There are also consumptive industries that cater primarily to coastal tourists. Charter fishing companies, which provide individuals an opportunity to fish in less-accessible areas or without purchasing their own gear, often operate both nearshore and offshore. Restaurants that provide local seafood, cooked using regional recipes or ingredients, are a staple in many coastal areas. Extractive businesses may be especially sensitive to fluctuations in resource availability, although diversifying services and products can help these businesses weather both expected and unexpected variability.

BOX 15.4 Case Study of Coastal Tourism: Local Ocean Seafoods Restaurant and Market

Local Ocean Seafoods (LOS) is a restaurant and fish market in Newport, Oregon, US, with an emphasis on local, seasonal catch; they purchase directly from about 60 boats. LOS buys and freezes large volumes of a given species during its peak season to offer local products year-round. Risks and benefits are inherent in this business's buying model. There are increased logistical costs associated with offloading, storing, and transporting products, and an upfront depletion of cash flow required to build inventory. Additionally, handling volumes of whole fish requires filleting space, specialty skills, and enhanced waste-management measures. However, this model also offers benefits, such as full traceability, the ability to tell the harvest story of the fish, and secured year-round inventory.

In March 2020, the COVID-19 pandemic had shuttered virtually every restaurant dining room in the nation. Markets for seafood froze up overnight. The entire $100 billion seafood industry network, including harvesters, processors, distributors, retailers, and restaurants, suffered a significant, sector-wide decline. This was largely the result of the closure and/or curtailed operations of eating establishments, where two-thirds of Americans consume their seafood (NOAA, 2021).

In response to this disruption, LOS created a new market channel, DockBox – a weekly make-at-home seafood meal kit. DockBox allowed the company to continue to move frozen and fresh inventory and keep staff employed during dining room closures. The market response was favorable; within five months the initial local market was expanded to 60 miles inland, then 200 miles to the major urban center, and finally 350 miles to a popular recreation destination.

When the company could only legally provide takeout dining, DockBox accounted for over half the company sales, with upward of 300 units sold per week. It allowed the company to maintain its core niche: the valuable relationships with local fishers, as well as the connection to a loyal customer base.

Ports as Blue Economy Incubators

Most of the goods consumed around the world travel to and from national ports. Traditionally, the maritime industry has not been especially innovative regarding sustainability, but this is beginning to change. In California, the ports of Los Angeles, Long Beach, and San Diego all have ambitious pollution reduction programs that are delivering very significant air quality benefits to the surrounding communities and are also innovators on the change to zero-emissions vehicles for transporting cargo (California Air Resources Board, 2018). The innovation does not stop there, however; the Port of San Diego has an Aquaculture and Blue Technology incubator program that helps jumpstart new blue economy businesses in the region, including marine debris removal technology, bio-enhancing shoreline armoring technology, and a new approach to soil remediation in marine environments.

North of San Diego, the Port of Los Angeles's AltaSea campus is nearing completion. This state-of-the-art facility will include workspace for many blue economy startups, a science center bringing together the region's top marine talent, along with an education center open to the public, all situated on a historic pier with easy access to the deep ocean. These are examples of ports not only taking the needs of their adjacent frontline communities seriously (by taking steps to improve air quality) but helping to catalyze a whole new set of industries that can provide local employment opportunities – all while seeking solutions to the unique challenges plaguing marine environments.

BOX 15.5 Case Study of Ports as Incubators: Ocean Cluster Houses

Ocean cluster houses drive innovation by providing small and new ocean-related businesses with private or shared office space, meeting rooms, access to technology, and other business services that cater to new venture creation. Co-location and intellectual synergy result in a high turnover of ideas and knowledge contribution.

The concept originated in Reykjavik, Iceland, and as of 2018, the Icelandic Ocean Cluster House had more than 100 companies working at their offices (Den Hollander & Thorsteinsson, 2020). For example, one business is involved in fish processing for food consumption, while a related business is using fish organs for nutraceuticals, fish oils, and food supplements, and yet another is leathering the fish skins for apparel. The concept has sparked several spin-off organizations across North America and Europe.

The purpose of cluster houses is to create value by connecting entrepreneurs, businesses, and knowledge of the marine industries. Depending on its location, an ocean cluster can include leaders and newcomers in fisheries, aquaculture, defense, offshore resources, marine renewables, bioresources, shipping, and ocean technology.

Living Shorelines for Coastal Resilience

Even if humanity manages to significantly reduce greenhouse gas emissions and avert the more extreme future climate scenarios, the coming decades will still bring stronger, more frequent storms and sea level rise due to the CO_2 that we have already emitted. Therefore, protecting our coastal infrastructure will be paramount (and expensive). In some areas, retreat from the coasts will be necessary, as adaptation costs will be too high. However, there will be many possibilities where smart planning can provide sufficient coastal resilience to protect property, improve public safety, and even enhance marine biodiversity.

This affirmative coastal future will require significant advances in the development of living shorelines – natural (and hybrid) infrastructure that makes use of dunes, sea grasses, wetlands, mangroves, reefs, and other ecological systems to dampen climate impacts in coastal regions. These systems are already being built with great success around the world, but the industry is still in its infancy. Research at the Center for the Blue Economy in California suggests that living shorelines are already cost-competitive with traditional gray infrastructure, such as seawalls, and that the main impediment to their adoption is informational and political, rather than economic (Colgan & Newkirk, 2016). The technologies are new, and developers and municipalities do not have the types of detailed engineered specifications that exist for other forms of coastal armoring, making it difficult to compare the relative performance of different systems and to build confidence in their durability. Development of clear and transparent engineering standards for living shorelines is necessary, along with the accumulation of time-series data to document their functionality. Like most new technologies, living shorelines also benefit from peer effects – when neighbors see others nearby successfully implementing nature-based solutions to protect their property, they are much more likely to invest in their own.

Not only will the development of a robust living-shorelines industry help coastal communities effectively adapt to climate change, but ecosystems will benefit greatly in the process, as these systems provide critical habitats for many marine species. Moreover, this coastal restoration industry will employ many types of workers, including laborers, natural scientists, and engineers.

There are likely some coastal regions that will require hardened gray infrastructure, and living shorelines alone will not suffice, but hybrid systems that combine the best of both approaches will likely be well-suited to most environments. This exciting new field can help us reorient our relationship with the shorelines in ways that benefit both humans and non-humans in a symbiotic manner, which is not only more aesthetically pleasing, but can also provide tangible health and environmental benefits.

Entrepreneurship, Business Development, and Risk

Beyond Business as Usual

Entrepreneurship, whether in the ocean and coastal sectors or elsewhere, shares many characteristics with standard business planning. Market identification and development is crucial for a successful business, as is a sound financial plan that accounts for upfront and long-term costs, capital investment, and human resources. Entrepreneurs, however, often take business creation a step further, developing unique products, innovating product delivery, or expanding the current market in unexpected ways.

Our case studies highlight the entrepreneurial mindset. Monterey Bay Seaweeds, for example, identified a new market and developed a product to fit that niche; Local Oceans Seafoods encountered challenges with their standard business model due to COVID-19, and identified an innovative way to deliver their product to both existing and new customers, expanding their customer base in the process of adapting to a change in external drivers. Even within existing markets, significant room exists for innovation and expansion: developing new gear types in fisheries or providing educational kayak tours instead of simply renting the gear to tourists.

As innovation and risk are often closely associated with entrepreneurship, it is especially important for entrepreneurs to have well-developed business plans. Such plans need to include clear market identification and potential for market development – in fact, such development may be a key component of the entrepreneurial nature of the business – as well as highly detailed financial plans that include supply networks, infrastructure, depreciation (which may be especially relevant in marine and coastal businesses), and investors/investment potential. Relationships need to be built with potential customers. It can be helpful to identify potential staff or associates who strongly believe in the enterprise, because businesses may require some time before becoming financially successful.

Although finances are important for any new business, marine entrepreneurship does not need to be commercial in nature. New educational programs or organizations (such as aquariums or ocean exploration camps), environmental or conservation nonprofits, and research institutions have all been developed by marine entrepreneurs.

Nature and Risks of Marine Industries

The marine environment is naturally in a state of flux throughout the year, with natural events, such as seasonal temperature change, upwelling, storm events, and species migrations affecting marine-based businesses. Anthropogenic climate change has increased the rate and severity of these changes in some locations and has led to greater uncertainty about future ocean conditions globally. In this last section of this chapter, we highlight some of the common attributes of marine environments that entrepreneurs and business owners should consider.

Species Variability

If a business is dependent on marine species, it must consider both annual and long-term shifts in resource availability. Fishing and whale-watching industries, for example, must contend with seasonal variability in their market species, as well as day-to-day uncertainties in the location and availability of the animals. This uncertainty is compounded by climate change, which can result in long-term shifts in the location and behavior of marine species.

Environmental Variability

Shifts in sea surface temperature (SST), pH, dissolved oxygen (DO) levels, salinity, and the occurrence of marine heat waves, harmful algal blooms (HABs), or coral diseases can lead to prolonged or permanent changes in ocean ecosystems. A business dependent on marine species must consider both annual and long-term shifts in environmental parameters that may affect resource health and availability. Of course, daily and seasonal weather patterns will affect any business that requires access to the ocean. Thus, storms, swells, wind, and wave height will also require careful consideration for the safety of workers and customers.

Seasonality

Many coastal towns are dependent on seasonal tourism. Seafood restaurants and charter fishing operations have high and low seasons and need to anticipate visitation levels to ensure their good financial health, even when their customer base is low.

Property Rights

Ownership of marine resources may be limited or nonexistent in an ocean setting, with access to fish, fishing grounds, dive sites, whales, and other marine mammals largely dependent on being in the right place at the right time – and possibly limited to the first vessel at that location on a given day.

Corrosion

Salt water is known for its deleterious qualities, and anything on or near the ocean will require continuous monitoring and upkeep. Physical capital and depreciation are concerns in any business setting, but businesses dependent on seawater access must factor in extra funds for maintenance and repair.

Conclusion

The development of a blue economy for the 21st century is just getting underway, but already we are witnessing the emergence of new technologies and industries. To understand how important and exciting these advancements are, imagine coastal regions where:

- All the power is generated cleanly by the wind, the waves, and the tides.
- No fishers or coastal hotels ever have to worry about the scourge of potential oil spills.

- Residential and commercial properties are protected by restored dunes or reefs that provide critical habitats for threatened and endangered species.
- The nearby port is powered exclusively by zero-emissions vehicles (ZEVs), such that local air quality dramatically improves, and ships slow down to protect whales.
- Sustainable local food systems provide high-quality protein for coastal communities and export without damage to the marine environment.

Humans have long had a complicated and often unhealthy relationship with the oceans, using them as pollution sinks or over-exploiting their natural resources. A new blue economy holds the promise to renew our relationship with the three-quarters of the planet covered in salt water in ways that enable us all to thrive. There is plenty of work to do, so get ready to ride the wave!

Professional Pathways

Many of the technologies that will define this new blue economy have yet to be invented, perfected, and/or brought to scale. To prepare for these careers, students need to have both vision and interdisciplinary training. For instance, you might train as engineers, planners, biologists, business leaders, and policymakers who can set the right incentives and prioritize public investments to accelerate blue growth. However, entrepreneurial skills are not discipline-specific. Instead, they require creativity and a commitment to the ocean and to human wellbeing. While it is true that entrepreneurship can be risky, specific knowledge that might help prepare you for this career includes business planning and marketing, coupled with a solid understanding of the nature and character of the ocean and its resources.

Questions for Reflection

1. How do trends in dietary preferences influence the development and markets of plant- and cell-based seafood alternatives? How would you market such products to increase their reach and market share?
2. Offshore renewable energy has many potential benefits but may disproportionately impact some marine businesses. What are some of the businesses that may be impacted, and what are some of these trade-offs? How should offshore energy developers work with these businesses to mitigate and/or minimize impacts from these projects?
3. What policies should be put in place by cities to incentivize living shorelines? How should demonstration projects be used to build public support?
4. As a coastal tourist, what types of sustainable and/or eco-friendly options do you seek out when you travel? How can these services be better marketed to the general public to expand their adoption? What are potential negative impacts from increased ecotourism, and how can these be addressed?

Notes

1 A fishing cooperative allows smaller fishers and fish buyers to pool resources in activities like catching, distributing, and marketing. Fishing co-ops have been successful in multiple countries, such as Rwanda, Mexico, Chile, Sri Lanka, and Japan, although they are less common in the United States, other than in Alaska.
2 In most US states, individual fishers may purchase a license to sell their own catch directly from their vessel to the public or restaurants. Chefs buying fish directly from harvesters must balance the risks, including less consistent availability, reduced species variety, larger purchase requirements, and fileting of whole fish, which can cause waste management issues. These risks are balanced by the benefits of freshness, accountability, and supporting a local food system. This market channel is most desirable when both the catch and fish-buyer prices are low.
3 A food hub is a "business or organization that actively manages the aggregation, distribution, and marketing of source-identified food products primarily from local and regional producers to strengthen their ability to satisfy wholesale, retail, and institutional demand" (Barham et al., 2012).

References

Barham, J., Tropp, D., Enterline, K., Farbman, J., Fisk, J., & Kiraly, S. (2012, April). *Regional Food Hub resource guide.* U.S. Dept. of Agriculture. http://dx.doi.org/10.9752/MS046.04-2012

California Air Resources Board. (2018, September 26). *CARB announces more than $200 million in new funding for clean freight transportation.* Retrieved January 16, 2021 from https://ww2.arb.ca.gov/news/carb-announces-more-200-million-new-funding-clean-freight-transportation

Colgan, C. S., & Newkirk, S. G. (2016). *Tradable permits for shoreline protection: Reshaping regulation under the Coastal Act for the era of sea level rise.* Center for the Blue Economy. https://www.middlebury.edu/institute/sites/www.middlebury.edu.institute/files/2018-10/10.12.18.Shoreline%20Tradable%20Permits%20Working%20Paper--originalpublishdate--Sept.2016.pdf

Costello, C., Cao, L., Gelcich, S., Cisneros-Mata, M. Á., Free, C. M., Froehlich, H. E., Golden, C. D., Ishimura, G., Maier, J., Macadam-Somer, I., Mangin, T., Melnychuk, M. C., Miyahara, M., de Moor, C. L., Naylor, R., Nøstbakken, L., Ojea, E., O'Reilly, E., Parma, A. M. … Lubchenco, J. (2020). The future of food from the sea. *Nature*, 588, 95–100. https://doi.org/10.1038/s41586-020-2616-y

Den Hollander, N., & Thorsteinsson, T. V. (2020). *A systematic approach to analyze industrial clusters: A case study of the Iceland ocean cluster* (Publication No. TRITA ITM-EX 2020:198) [Master's Thesis, KTH Royal Institute of Technology]. DiVA Portal. http://urn.kb.se/resolve?urn=urn%3Anbn%3Ase%3Akth%3Adiva-279730

Dundas, S. J., Levine, A. S., Lewison, R. L., Doerr, A. N., White, C., Galloway, A. W. E., Garza, C., Hazen, E. L., Padilla-Gamiño, J., Samhouri, J. F., Spalding, A., Stier, A., Hill, T., & White, J. W. (2020). Integrating oceans into climate policy: Any green new deal needs a splash of blue. *Conservation Letters, 13*(5), e12716. https://doi.org/10.1111/conl.12716

Environmental Protection Agency. (2017). *Climate impacts on coastal areas.* Retrieved January 12, 2021 from https://19january2017snapshot.epa.gov/climate-impacts/climate-impacts-coastal-areas_.html#Overview ()

Food and Agriculture Organization of the United Nations. (2017). *Fishery and aquaculture country profiles: The People's Republic of China.* Retrieved August 24, 2021 from http://www.fao.org/fishery/facp/CHN/en#CountrySector-PostHarvest

Food and Agriculture Organization of the United Nations. (2020). *The state of world fisheries and aquaculture 2020. Sustainability in action.* https://doi.org/10.4060/ca9229en

Gephart, J. A., Froehlich, H. E., & Branch, T. A. (2019). To create sustainable seafood industries, the United States needs a better accounting of imports and exports. *Proceedings of the National Academy of Sciences, 116*(19), 9142–9146. https://doi.org/10.1073/pnas.1905650116

Gopal, N., Hapke, H. M., Kusakabe, K., Rajaratnam, S., & Williams, M. J. (2020). Expanding the horizons for women in fisheries and aquaculture. *Gender, Technology and Development, 24*(1), 1–9. https://doi.org/10.1080/09718524.2020.1736353

Harper, S., Adshade, M., Lam, V. W. Y., Pauly, D., & Sumaila, U. R. (2020). Valuing invisible catches: Estimating the global contribution by women to small-scale marine capture fisheries production. *PLoS ONE, 15*(3), e0228912. https://doi.org/10.1371/journal.pone.0228912

Korfmacher, K. S., Aviles, K., Cummings, B. J., Daniell, W., Erdmann, J., & Garrison, V. (2015). Health impact assessment of urban waterway decisions. *International Journal of Environmental Research and Public Health, 12*(1), 300–321. https://doi.org/10.3390/ijerph120100300

Kroetz, K., Luque, G. M., Gephart, J. A., Jardine, S. L., Lee, P., Chicojay Moore, K., Cole, C., Steinkruger, A., & Donlan, C. J. (2020). Consequences of seafood mislabeling for marine populations and fisheries management. *Proceedings of the National Academy of Sciences, 117*(48), 30318–30323. https://doi.org/10.1073/pnas.2003741117

Lamy, J. (2020, October 2). *WSO's Friend of the Sea certification matters for the plant-based seafood industry. Here's why.* Good Food Institute. Retrieved February 1, 2021 from https://gfi.org/blog/friend-of-the-sea-certification/

Lucas, A. (2020, December 1). *Singapore issues first regulatory approval for lab-grown meat to Eat Just.* CNBC. https://www.cnbc.com/2020/12/01/singapore-issues-first-regulatory-approval-for-lab-grown-meat-to-eat-just.html

Luque, G. M., & Donlan, C. J. (2019). The characterization of seafood mislabeling: A global meta-analysis. *Biological Conservation, 236*, 556–570. https://doi.org/10.1016/j.biocon.2019.04.006.

Maul, G. A., & Duedall, I. W. (2019). Demography of coastal populations. In C. W. Finkl & C. Makowski (Eds.), *Encyclopedia of coastal science.* Encyclopedia of Earth Sciences Series. Springer. https://doi.org/10.1007/978-3-319-93806-6_115

National Marine Fisheries Service. (2020). *Fisheries of the United States, 2018* (Current fishery statistics No. 2018). U.S. Department of Commerce. Retrieved February 1, 2021 from https://www.fisheries.noaa.gov/national/commercial-fishing/fisheries-united-states-2018

National Oceanic and Atmospheric Administration. (2020). *Fast facts: Tourism and recreation.* Retrieved January 12, 2021 from https://coast.noaa.gov/states/fast-facts/tourism-and-recreation.html#:~:text=Workers%20in%20the%20ocean%2Dbased,the%20national%20economy%20each%20year

National Oceanic and Atmospheric Administration. (2021). *NOAA fisheries updated impact assessment of the COVID-19 crisis on the U.S. commercial seafood and recreational for-hire/charter industries.* U.S. Department of Commerce. Retrieved February 1, 2021 from https://media.fisheries.noaa.gov/2021-01/Updated-COVID-19-Impact-Assessment.pdf

Organization for Economic Cooperation and Development. (2016). *The ocean economy in 2030.* OECD Publishing. https://doi.org/10.1787/9789264251724-en

Stoll J. S., Harrison, H. L., De Sousa, E., Callaway, D., Collier, M., Harrell, K., Jones, B., Kastlunger, J., Kramer, E., Kurian, S., Lovewell, M. A., Strobel, S., Sylvester, T., Tolley, B., Tomlinson, A., White, E. R., Young, T., & Loring, P. A. (2021). Alternative seafood networks during COVID-19: Implications for resilience and sustainability. *Frontiers in Sustainable Food Systems*, 5, 614368. https://doi.org/10.3389/fsufs.2021.614368

Sumaila, U. R., Zeller, D., Hood, L., Palomares, M. L. D., Li, Y., & Pauly, D. (2020). Illicit trade in marine fish catch and its effects on ecosystems and people worldwide. *Science Advances*, 6(9), eaaz3801. https://doi.org/10.1126/sciadv.aaz3801

United Nations World Tourism Organization. (2016). *Measuring sustainable tourism: Developing a statistical framework for sustainable tourism*. Retrieved January 26, 2021 from https://webunwto.s3-eu-west-1.amazonaws.com/2019-08/mstoverviewrev1.pdf

Wheeler, B. W., White, M., Stahl-Timmins, W., & Depledge, M. H. (2012). Does living by the coast improve health and wellbeing? *Health & Place*, *18*(5), 1198–1201. https://doi.org/10.1016/j.healthplace.2012.06.015.

World Bank. (2017). What is the blue economy? Retrieved August 12, 2021, from https://www.worldbank.org/en/news/infographic/2017/06/06/blue-economy

16 Measuring Progress toward a Blue Economy

A Critical Overview of Development Indicators

A. M. Cisneros-Montemayor

Introduction: Oceans, Indicators, and a Blue Economy

The oceans have been a crucial part of human culture, subsistence, and economies for millennia, and they are increasingly viewed as a new frontier in the context of modern economic growth and globalized markets (Organization for Economic Cooperation and Development [OECD], 2016; World Bank & United Nations Department of Economic and Social Affairs [UNDESA], 2017). Oceans are estimated to contribute US$1.5 trillion per year to the global economy, supporting hundreds of millions of jobs (OECD, 2016). In the future, investment in ocean sectors, like offshore wind energy, fisheries and aquaculture, mangrove restoration, and decarbonization of shipping, could provide an additional US$15.5 trillion from 2020 to 2050, with economic returns estimated to be five times higher than investment costs (Konar & Ding, 2020).

Beyond financial projections, oceans are vitally important for human communities and livelihoods, including 200–300 million artisanal fishers and processors (90% of total employment related to oceans) (Teh et al., 2014), the provision of essential micronutrients to 2.1 billion people (Golden et al., 2016), and the maintenance of coastal Indigenous cultures (Cisneros-Montemayor et al., 2016). Many ocean policies do not take these social benefits into account (Voyer et al., 2018), and this is often due to development assumptions that are rarely stated but are reflected in the choice of progress objectives and indicators for ocean and coastal development. So, our choice of indicators – how and what we measure to keep track of policy outcomes – is really crucial, both as a transparent reflection of our development goals and assumptions, and as a way to make sure we are prioritizing what we intended to accomplish.

This chapter provides a critical review of development indicators as they relate to a Blue Economy approach, currently the most commonly used term in ocean development policy and one which ideally prioritizes social benefits and equity. The chapter begins by defining Blue Economy and how it differs from other ocean development approaches and describing the role of indicators in sustainable development. The second section examines the particular challenges for tracking policy progress in ocean systems and uses the UN Sustainable Development Goal 14 (Life Below Water) as an example. The third section discusses the process of considering and selecting indicators for a Blue Economy, including those related to social equity, environmental sustainability, and economic viability.

DOI: 10.4324/9781003058151-19

What Is a Blue Economy?

The term "Blue Economy" was first proposed by Small Island Developing States (SIDS) – a group of island nations with enormous marine territory relative to land area – to promote sustainable economic growth based on ocean resources, but specifically emphasizing that this growth be equitable and appropriate given local cultures, contexts, and development objectives (UNDESA, 2014). Based on that, here we define Blue Economy as the establishment of ocean sectors that are socially equitable, environmentally sustainable, and economically viable. Even though all three of these components are important, the prioritization of social equity is what sets a Blue Economy apart as a new and transformative approach (Cisneros-Montemayor et al., 2021; see Figure 16.1).

Since it was first proposed, the Blue Economy term has sometimes been used in quite different ways by different interest groups (Voyer et al., 2018), though the focus on social equity should ideally distinguish the Blue Economy from other related approaches (Cisneros-Montemayor et al., 2019). For example, "Blue Growth" refers to the economic expansion of the ocean economy (that is, any and all industries linked to marine and coastal areas; Colgan, 2016), with environmental sustainability implied but not always emphasized, and social equity even less so (Cisneros-Montemayor et al., 2019; Voyer et al., 2018). Another term, "Sustainable Ocean Economy", explicitly emphasizes environmental sustainability, and increasingly addresses issues of social equity (see, for example, the "Blue Papers" of the High-Level Panel for a Sustainable Ocean Economy: https://www.oceanpanel.org/blue-papers), although it still aims for expansion of ocean industries to increase total financial benefits. As mentioned above, how we think about ocean development and its ultimate goals (e.g., environmental conservation, social well-being, economic efficiency) will inform what indicators we choose.

What Are Development Indicators?

Simply put, an indicator is an observable characteristic that reflects the current state of a system or its components. Indicators can be direct – such as temperature as a reflection of thermal energy, or height as a reflection of size – or indirect – such as retail sales as a reflection of economic activity, or goals scored as a reflection of how good a soccer player is. Notice that each of these examples of indicators could have different units (e.g., °C or °F, centimeters or inches), and we must be aware that these may actually measure slightly different things (e.g., total sales or total value of sales). Most matters related to sustainable development – including the Blue Economy – are very complex and almost always need indirect indicators to gauge progress on the various overarching policy goals.

The challenges posed by this complexity have led to much research into indicator design and selection, and there are hundreds of development indicators in use throughout the world at national and local levels (see Hass et al., 2002, for an excellent review). Many of these indicators are highly useful for local contexts, either directly or in modified ways (Elfes et al., 2014), although there are a number of pertinent critiques regarding their unstated assumptions and uncritical application (see, for example, Bell & Morse, 2004; Merry, 2011). As an example, one of the most widely recognized and used development indicators is the United Nations Development Programme's Human Development Index (HDI). (An index is an indicator that is re-scaled, say from 0 to 1, and usually represents an average or composite of several indicators.) The HDI is an indicator of human development comprising health, wealth, and education. Respectively, its three sub-indicators are national average life expectancy, income per capita, and years of schooling (United Nations Development Programme, 2020). Notice how, even if we agree that health, wealth, and education are key aspects of human well-being, we could question if other aspects should also be included, whether the specific indicators we chose are appropriate (for example, average income does not capture inequality), or if national-level averages reflect local disparities within nations. This is not meant as a criticism of HDI, specifically, but it is important to think about these types of issues when using any indicators.

Context and Goals for a Blue Economy

In the context of a Blue Economy, it is important to recognize that oceans are a unique social–ecological system with very particular challenges for development that go beyond the general underlying tensions and tradeoffs in advancing social, ecological, and economic goals (Halpern et al., 2013). A unique challenge in ocean systems is that many living marine resources are very difficult to observe and measure directly. This means that basic information on species and ecosystems, such as preferred habitats, food web relationships, and local abundances, must be inferred from a patchwork of field samples and estimations (Hilborn & Mangel, 1997). Ocean currents and species migrations and movement mean that pollution or other types of human impacts can quickly extend beyond points of origin, and again are difficult to track (Fabricius, 2005).

In terms of governance, ownership and tenure of marine areas and resources are much less well-defined than in terrestrial systems (Guggisberg, 2019), which can lead to increased conflicts over resources (Spijkers et al., 2018) and worsen social inequities and power dynamics (Vierros et al., 2020). In this sense, the oceans and ocean industries are outliers in modern market systems – which usually focus on private property and individual profits – because there is still a much stronger emphasis on the benefits of common resources, traditional uses, and freedom of movement (Seto & Campbell, 2019). Indeed, it has been specifically argued that Blue Economy approaches should not be used to impose legal and market systems that could disenfranchise and marginalize traditional ocean users like artisanal and subsistence fishers (Bennett et al., 2015; Seto & Campbell, 2019).

Given the complexity of ocean systems that a Blue Economy aims to address, a fundamental first step is determining which goals or objectives must be achieved. An entire literature exists that is devoted to best practices in eliciting, formulating, and agreeing upon these goals. In general terms, we must first recognize who will be affected by ocean development – like that envisioned by a Blue Economy – and fully involve them in a transparent policy process. To be clear, choosing which indicators to use is only one part of a larger process that tries to include a diversity of perspectives to agree on goals, objectives, and the strategies needed to achieve them (Valentin & Spangenberg, 2000).

Goals, Targets, and Indicators in Ocean Policy

To ground the discussion above, we can use as an example the most prominent, current set of global ocean goals included within the UN Sustainable Development Goals (SDGs). The SDGs comprise 17 goals and 169 individual targets which have been agreed upon by all UN member nations under the 2030 Agenda for Sustainable Development (United Nations [UN], 2015); the oceans are specifically addressed in SDG 14 (Life Below Water) and its targets. It is important to remember that SDG 14 does not equal a Blue Economy, and there are other examples of ocean policies, but it is discussed here as a concrete example of existing goals and indicators related to oceans. Furthermore, SDG 14 has been widely adopted by marine agencies in their own national ocean plans, so future Blue Economy plans will probably have to work within its commitments (Table 16.1).

Table 16.1 UN Sustainable Development Goal (SDG) 14 and targets (UN, 2015). The associated indicators adopted by the UN (UN, 2017) are in italics

SDG 14: Life Below Water
14.1 By 2025, prevent and significantly reduce marine pollution of all kinds, in particular from land-based activities, including marine debris and nutrient pollution.
14.1.1 Index of coastal eutrophication and floating plastic debris density.
14.2 By 2020, sustainably manage and protect marine and coastal ecosystems to avoid significant adverse impacts, including by strengthening their resilience, and take action for their restoration in order to achieve healthy and productive oceans.
14.2.1 Proportion of national exclusive economic zones managed using ecosystem-based approaches.
14.3 Minimize and address the impacts of ocean acidification, including through enhanced scientific cooperation at all levels.
14.3.1 Average marine acidity (pH) measured at agreed suite of representative sampling stations.
14.4 By 2020, effectively regulate harvesting and end overfishing, illegal, unreported, and unregulated fishing and destructive fishing practices and implement science-based management plans, in order to restore fish stocks in the shortest time feasible, at least to levels that can produce maximum sustainable yield as determined by their biological characteristics.
14.4.1 Proportion of fish stocks within biologically sustainable levels.
14.5 By 2020, conserve at least 10% of coastal and marine areas, consistent with national and international law and based on the best available scientific information.
14.5.1 Coverage of protected areas in relation to marine areas.
14.6 By 2020, prohibit certain forms of fisheries subsidies which contribute to overcapacity and overfishing, eliminate subsidies that contribute to illegal, unreported, and unregulated fishing and refrain from introducing new such subsidies, recognizing that appropriate and effective special and differential treatment for developing and least developed countries should be an integral part of the World Trade Organization fisheries subsidies negotiation.
14.6.1 Progress by countries in the degree of implementation of international instruments aiming to combat illegal, unreported, and unregulated fishing.

(Continued)

Table 16.1 Continued

SDG 14: Life Below Water
14.7 By 2030, increase the economic benefits to small island developing States and least developed countries from the sustainable use of marine resources, including through sustainable management of fisheries, aquaculture, and tourism.
14.7.1 Sustainable fisheries as a percentage of GDP in small island developing States, least developed countries, and all countries.
14.a Increase scientific knowledge, develop research capacity, and transfer marine technology, taking into account the Intergovernmental Oceanographic Commission Criteria and Guidelines on the Transfer of Marine Technology, in order to improve ocean health and to enhance the contribution of marine biodiversity to the development of developing countries, in particular small island developing States and least developed countries.
14.a.1 Proportion of total research budget allocated to research in the field of marine technology.
14.b Provide access for small-scale artisanal fishers to marine resources and markets.
14.b.1 Progress by countries in the degree of application of a legal/regulatory/policy/institutional framework which recognizes and protects access rights for small-scale fisheries.
14.c Enhance the conservation and sustainable use of oceans and their resources by implementing international law as reflected in UNCLOS, which provides the legal framework for the conservation and sustainable use of oceans and their resources, as recalled in paragraph 158 of The Future We Want.
14.c.1 Number of countries making progress in ratifying, accepting, and implementing through legal, policy and institutional frameworks, ocean-related instruments that implement international law, as reflected in the United Nation Convention on the Law of the Sea, for the conservation and sustainable use of the oceans and their resources.

A significant challenge with this set of targets is that, while they are relatively comprehensive, they have few specific benchmarks for success (Cormier & Elliott, 2017); this is a shortcoming which must be addressed when designing indicators (as discussed below). The one exception is 14.5, aiming to "conserve at least 10% of coastal and marine areas", but even here there is no specific indicator on the quality of protection (Table 16.1). This can lead to "paper parks" with poor implementation (Gill et al., 2017), or that impact marginalized communities (including Indigenous peoples; Cisneros-Montemayor et al., 2018b), or the declaration of protected areas where there is little human activity to begin with (Devillers et al., 2020). There are ways to evaluate progress toward similar unspecific goals (Andriantiatsaholiniaina et al., 2004; Cisneros-Montemayor et al., 2018b), but to achieve a Blue Economy, objectives and goals must match local needs – and, ideally, include quantifiable indicators and benchmarks to track progress and adapt policy strategies to local contexts.

Another issue is that most SDG 14 targets focus on environmental protection (Table 16.1), even if it is linked to subsequent economic and social benefits. This may follow from a common development model which assumes that the state of the environment and its natural capital are the base for development and must be prioritized accordingly, with social systems and governance institutions as a secondary layer that will lead to a final layer of economic production and benefits (a "layer cake" model; Folke et al., 2016). A growing body of literature argues for a different approach in which social inequities and power dynamics must be addressed first, given that they are at the root of both inequitable distributions of economic benefits and unsustainable use of natural resources (Bennett et al., 2019).

As seen through the example of SDG 14, indicators are essential for tracking true progress toward policy goals and can also highlight possible gaps in goals themselves. In the case of the Blue Economy, a specific aim to prioritize equitable outcomes requires actions – and associated progress indicators – beyond the state of the environment. For example, a fish stock can be recovered to a target abundance given decreases in fishing mortality, but we can simultaneously address existing or historical inequities in the distribution of catch, such that we reduce impacts and increase benefits for marginalized fishers (a "just transition" to sustainability; Bennett et al., 2019). If they do not explicitly account for equity in implementation, even well-intentioned conservation or sustainability goals – including reducing pollution or decreasing illegal fishing – can have negative effects on human wellbeing, usually impacting marginalized populations the most (Bennett et al., 2019; Cisneros-Montemayor et al., 2018a).

Measuring Progress toward a Blue Economy

General Principles for Blue Economy Indicators

Indicators for a Blue Economy – prioritizing social equity while promoting environmental sustainability and economic viability – must be able to directly measure progress on many different aspects of development (Cisneros-Montemayor et al., 2021). For example, if decreasing poverty is the ultimate objective of development for a coastal community, one would wish to track indicators of poverty and purchasing power itself, not only the state of local fish stocks or even per capita income (since local costs of living can change as a result of interventions). One must be aware of local contexts (e.g., are there local power dynamics or group inequities?), ultimate drivers (e.g., is poverty directly related to fishing, or to other factors?), and underlying assumptions of development (e.g., will recovering fish stocks lead to higher local incomes?) when designing strategies and their indicators, making sure to track progress on the desired objectives as directly as possible.

An influential set of guidelines for designing indicators are the Bellagio Principles (Pintér et al., 2012), which continue to be used in many national planning strategies (Hass et al., 2002). Notably, the first principle outlines the assumed goal, "delivering well-being within the capacity of the biosphere to sustain it for future generations"; agreement on what is understood as wellbeing will of course be an important first step for applying this to a specific place (Pintér et al., 2012). Goals must be clearly defined and informed by diverse perspectives, recognizing social and ecological relationships and how they can change. Progress can then be assessed using indicators which are adequately scoped (e.g., with useful time frames and geographic scales); transparently collected, interpreted, and communicated; part of a framework of interrelated indicators and goals; and involve broad public participation and a capacity to update our strategies as we create new information (Pintér et al., 2012; Valentin & Spangenberg, 2000).

Similar to the Bellagio Principles but closer to the implementation stage, a common framework for choosing useful indicators (or goals) is the SMART criteria. That is, an indicator should be Specific, Measurable, Achievable, Relevant, and Time-bound (Cormier & Elliott, 2017). The most important aspects of the

SMART framework include explicit and transparent discussion of how indicators specifically reflect intended goals; whether they can be reliably and continuously collected; and whether indicators change and can be measured fast enough to allow for adaptive policy responses (i.e., changing as needed based on outcomes). Of course, measuring more indicators could provide more details about our progress, especially if we are working in very complex systems and have multiple interrelated goals (both of which are almost always true for oceans). However, we also have to consider tradeoffs, including added costs of monitoring and the challenges of interpreting many different types of data. In any case, composite indicators that aggregate multiple indicators into one or only a few – such as the HDI discussed above, or the Ocean Health Index (Halpern et al., 2012) – are easier to communicate to politicians and the general public. In practice, these "headline" indicators (Hass et al., 2002) can be used together with more detailed technical indicators as long as they are all collected in a formal and transparent way that makes it easier to see how they reflect overarching policy goals.

Following these general principles, in this section we will discuss potential indicators that can be used to gauge our policy progress and help direct our management strategies to achieve and maintain a Blue Economy. For clarity, we separately address indicators related to social equity, environmental sustainability, and economic viability, but remember that these aspects of development and their corresponding indicators will almost always be very closely interconnected (Figure 16.1).

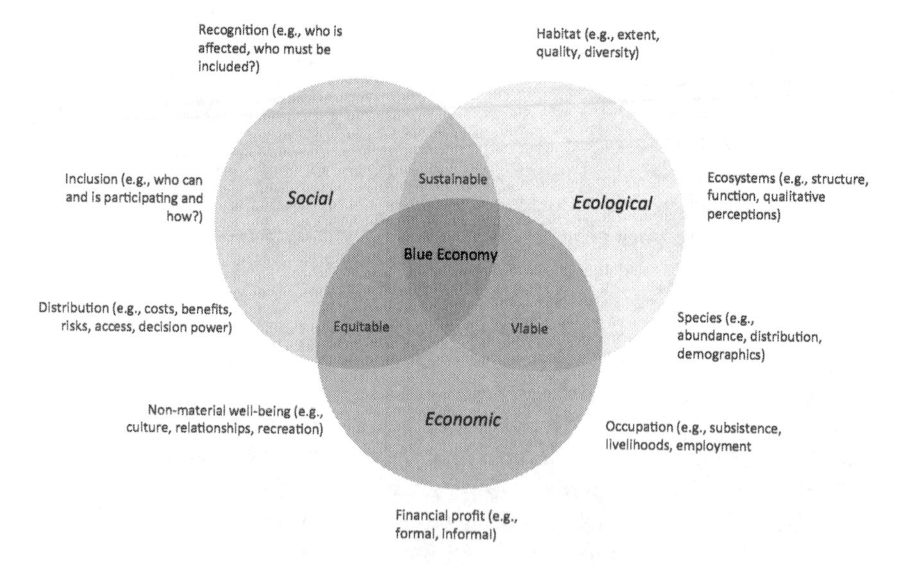

Figure 16.1 Blue Economy conceptual diagram (adapted from Cisneros-Montemayor et al., 2019) including key components and examples of corresponding indicators. This list is not meant to be exhaustive and there are many alternative or additional indicators; their selection must be carefully considered given the specific place, context, and objectives.

Social Equity

Social equity does not have a unique definition but can be synonymous with fairness in human relationships. From a simplified but useful perspective, "[to] discuss the equity of public policies is to discuss who gets what and by what rules" (Blanchard, 1986). Three aspects are generally recognized as central in the context of a Blue Economy. *Recognitional equity* refers to identifying and considering all of the different human groups affected, directly or indirectly, by ocean development (the "who" in the quote above). *Procedural equity* refers to the way the strategies of development and rules concerning the distribution of benefits and costs are formulated, ideally fully incorporating the perspectives of all of the prior recognized groups. *Distributional equity* refers to outcomes of development and the subsequent share of these groups in benefits, costs, or risks.

Distribution is probably the most intuitive aspect of equity for the majority of policy practitioners and natural science researchers and is relatively straightforward to measure, yet is not always specifically considered. Common indicators of distribution include those of income equality, such as the Gini index and other related measures (World Bank, 2016). Note, however, that these indicators measure (in)equality in distribution, not equity. Sometimes it may be desirable for some groups to receive a larger share of benefits – for example, if an ocean industry aims to prioritize increased income for specific marginalized populations. Furthermore, distribution of income is only one small part of equity. Other indicators can include ease of access to resources and public services (e.g., distance or time from places of work, education, and recreation), relative purchasing power (i.e., accounting for changes in local prices of goods and services), or employment of different groups within institutional or company hierarchies (e.g., managerial versus entry-level positions).

To better help address equity outcomes, any indicator related to a development must be collected in a way that allows for comparison across different groups of people. This can include material benefits (e.g., income, employment, food), negative impacts (e.g., health, insecurity), risks (real or perceived), and qualitative perceptions of wellbeing and satisfaction (e.g., through individual surveys). Depending on the context, monitoring specific outcomes for different groups during the measurement of these indicators requires accounting for gender, sex, age, nationality, ethnicity, religion, political affiliation, Indigeneity, profession, location, etc. The fundamental assumption is that ocean development impacts people in different ways, so ensuring equitable outcomes requires an understanding of these differences so that appropriate policies can address pertinent issues and actively intervene to ensure that vulnerable human groups are protected and supported (Bennett et al., 2019).

It is very important to remember that adequately accounting for distributional equity requires that recognitional and procedural equity concerns be addressed first. This is what will determine the appropriate human groups that must be included, the types of benefits or impacts to consider, and the social and power dynamics that could influence distribution. In practice, distributional concerns are an inherent part of any political process, but recognitional and procedural ones often are not, because marginalized populations, by definition, require specific actions to bring into policy discussions. Similar to distribution, these indicators may include, for example, the age, gender, ethnicity, and residence of participants in meetings; the types of existing processes for determining ownership (formal or informal) and access to resources; and the types and appropriateness of our own efforts at contacting and including individuals given local customs. Active collaboration with social scientists is essential for designing indicators of social equity, regardless of whether they are linked to ecological or economic outcomes (Hicks et al., 2016). A first step toward this must be for teams tasked with monitoring policy progress to include sociologists, anthropologists, historians, etc., with the tools and expertise to define and monitor equity outcomes and help us to leave no one behind in the global push for human wellbeing (UN, 2015).

Ecological Sustainability

Sustainability, in an ecological context, implies using natural resources in such a way that ecosystem functions necessary to produce them, and other ecosystem services, are maintained through time. A common misconception is that ecological sustainability is inherently at odds with maximizing financial benefits. For natural resources, including those underpinning ocean sectors, unsustainable exploitation actively decreases financial benefits, both currently and for any future industries (Daly, 1990).

In oceans, the most common approach to monitoring ecological sustainability is stock assessment models used for most marine fisheries. Most agencies focus on determining the necessary exploitation rate required to achieve the maximum yearly production that can be sustained over time (maximum sustainable yield, or MSY; Food and Agriculture Organization of the United Nations, 2018), and the yearly state of the fishery in relation to this reference point. While there are many possible reference points that can be drawn from such models, including the maximum economic yield (which maximizes profit instead of production), the vast majority of models focus on stocks (a distinct population) of single species and generally do not explicitly account for ecological dynamics or variability. Nevertheless, single-species indicators can be quite useful to monitor ecological state, whether of exploited species or of others including seabirds, coastal vegetation, marine mammals, etc. These indicators can be related to total production, as in most fisheries, but can also include measures of observed or estimated abundance, distribution, extractive effort, production rate (e.g., fish catch per boat), or population structure (e.g., size, age, and sex ratios). A common criticism of these approaches is that they require large amounts of data that may not be available; while this is true for more sophisticated models, there are many alternative indicators designed to provide credible advice in data-limited situations (see, for example, Carruthers & Hordyk, 2018; Cope & Punt, 2009; Kleisner et al., 2013).

The widely recognized concept of ecosystem services, or nature's contributions to people (Díaz et al., 2018), focuses on functioning ecosystems – not species in isolation – as the providers of human benefits. These include market benefits, such as food, livelihoods, income, and shelter, and non-market benefits, such as culture, inspiration, kinship, and recreation, among many others (Allison et al., 2020). Recognizing that marine ecosystems underpin human benefits means that we must account for ecological functions beyond the single-species indicators mentioned above, which can be done qualitatively or quantitatively (Arreguín-Sánchez & Ruiz-Barreiro, 2014). For example, quantifying the effects of ocean sectors on ecosystems can be done through ecosystem models such as Ecopath with Ecosim, Atlantis, or others (Fulton, 2010; Plagányi, 2007). These allow users to represent a full marine ecosystem and its trophic webs, and track potential impacts on marine species of, for example, fishing pressure, habitat modification, pollution, shipping noise, or conservation and restoration actions. Ecosystem indicators include ratios of species abundances (reflecting ecosystem structure), rates of energy transfer (pointing to potential large-scale changes), and the relative abundances of key species (including primary producers, keystone species, or apex predators) (Coll et al., 2016; Perrings et al., 2011).

In some cases, quantitative marine ecosystem models can be built using local information provided by fishers or other ocean users (Bentley et al., 2019; Cisneros-Montemayor et al., 2020). This has the added benefit of requiring close collaboration between different stakeholders, potentially encouraging trust, and further cooperative management actions. Even conceptual ecosystem models based on scientific and/or local knowledge can be highly useful to anticipate potential ecosystem effects from specific ocean sectors, identify ecological or cultural keystone species, and highlight specific ecological relationships that could lead to significant ecosystem changes (such as cascading effects or alternate stable states) and therefore require further research (Cisneros-Montemayor et al., 2020).

In addition to species and ecosystems, indicators for a Blue Economy must monitor marine habitats. This includes physical habitat, such as sea ice, kelp forests, mangroves, and coral reefs, and water quality more broadly. Importantly, it is not only the extent of coverage that must be monitored, but the quality of habitat. The simplest example is that of living versus dead coral reefs, where the latter can still provide habitat for marine species but with a vastly lower overall productivity. Similarly, mangrove forests can be highly resilient to human disturbance and pollution, but the species associated with them may not be.

Economic Viability

Although economics deals with a wide range of issues related to resource allocation and utility, we focus in this subsection on what is termed the "ocean economy", which refers to the financial economic activity supported by or linked to ocean resources. This includes sectors based on renewable and/or living marine resources, such as fisheries, ecotourism, blue carbon, mariculture, bioprospecting, and offshore wind and tidal energy (Cisneros-Montemayor et al., 2019). The ocean economy is not limited to these sectors, however, or explicitly tied to sustainability or equity, but to an accounting of economic activity related to the ocean. It therefore also includes sectors such as offshore oil and gas, shipping and port services, marine technology and infrastructure development, mass coastal tourism, coastal development, etc. (World Bank & UNDESA, 2017). These would not be considered within a Blue Economy strictly defined as requiring sustainability, except perhaps as part of temporary transition plans toward renewable resource sectors, yet they must be accounted for within indicators of the ocean economy.

A wider consideration of economic benefits (both market and non-market) from multiple ocean sectors means that we must take special care to avoid double-counting values. This is important to help policy decisions comparing ocean industries to each other, but also to be able to compare the ocean economy as a whole to other sectors of the economy (Colgan, 2016). In general terms, an accurate accounting of the ocean economy involves: (1) categorizing types of ocean sectors depending on their degree of dependence on oceans and/or their living resources; (2) considering the geographic location and corresponding jurisdiction of ocean industries; (3) ensuring appropriate confidentiality and transparency in data collection; (4) including both natural capital (value of resources "in the water") and flows (value of harvested resources); and calculating value added (the economic activity related to post-harvest processing, marketing, and industry services) for marine sectors as completely as possible. Some measures of economic viability account for the fact that some industries receive subsidies to operate, and these subsidies should be subtracted to estimate societal returns (Schuhbauer et al., 2019a). For example, a fishery that receives subsidies to lower operating costs may bring a lot of societal benefits but should not include public subsidies in estimates of these benefits; rather, subsidies are costs to society that may or may not be used for other types of investments, including for ocean sectors with more direct benefits to local coastal communities (Schuhbauer et al., 2019b).

Accurate financial accounting can be challenging in the context of oceans and their associated sectors, but there are already data and frameworks specifically designed to do so (Colgan, 2016; Fenichel et al., 2020; Wang, 2016). However, because a Blue Economy is intended to provide local benefits, national-level indicators such as GDP will most often not be appropriate for reflecting progress on ocean development goals (Fenichel et al., 2020). Furthermore, oceans support very large informal economies and livelihoods, and indicators should account for this regardless of whether they are officially recognized as part of the formal economy. An important first step in doing so must be to identify and include all stakeholder groups as previously discussed in the Social Equity section. This can help to more fully appreciate the breadth of economic indicators necessary to reflect local contexts, objectives, and dynamics.

The most widely used measure of economic viability, especially in a development context, is Net Present Value (NPV), which estimates the net benefits of a public or private investment over the length of time it is expected to operate (or over a long time span, say 30 or 50 years, representing multiple generations). This calculation is often used to gauge return on investment (ROI) and assumes that humans place more value on benefits (or costs) today than in the future – i.e., the future is discounted. Higher discount rates mean that we value the future less;

therefore, we would be less willing to invest in a project expected to provide benefits in the future but known to have high costs today. (Reluctance to take actions on climate change is one good example of this).

In order to account for benefits and costs in NPV, discount rates are often calculated by economic planning departments or investment banks. But one should consider their use very carefully because these rates (usually between 3 and 10% per year) have been shown to be potentially vastly different from those measured directly through field surveys of ocean users. One field survey of artisanal fishers, by far the largest group of marine employment throughout the world, found discount rates of well over 200% (Teh et al., 2014). At a deeper level, it has been argued that by assuming high discount rates we are implicitly and wrongfully discounting the needs of future generations. A more equitable approach involves "resetting" the NPV clock so that benefits to future generations are better captured, known as intergenerational discounting (Sumaila & Walters, 2005). Discount rates can also be lowered by reducing the urgency to use up natural resources today, which can be supported by promoting alternative livelihoods and education on responsible consumption, decreasing environmental impacts of marine industries, and promoting more equitable sharing of current benefits from marine resources. Like any indicator, the use of NPV is not inherently flawed, but it requires critical and careful consideration of its assumptions, parameters, and limitations for advising policy.

The various economic indicators discussed above are intended to be direct measurements of economic activity, but not all economic values from ocean ecosystem services can be measured directly. These ecosystem services (or nature's contributions to people) include provisioning of material goods, regulation of climate and environmental systems, and a wide array of cultural, inspirational, and recreational contributions (Díaz et al., 2018). Some, such as shadow values (e.g., the approximate market value of fish that are caught and eaten, not sold) or regulating services (e.g., storm protection by mangrove forests or reefs), can be indirectly calculated based on the market value of substitutes (in the prior examples, fish sold at a market or the construction and maintenance of a seawall). Other methods for indirectly calculating economic indicators related to ocean development and ecosystem services include hedonic pricing, benefit-transfer, and contingent valuation (e.g., surveys of willingness-to-pay or -accept) (Austen et al., 2019). Within a Blue Economy approach, analyzing economic viability must recognize that economic values go far beyond prices and income, so indicators must try as much as possible to reflect deeper changes in human wellbeing (Allison et al., 2020).

Conclusion

There are hundreds of potential indicators that have been proposed, adapted, and used (Hass et al., 2002), but local contexts and objectives should always dictate which ones are relevant for a specific application. There are many seemingly useful indicators that are very difficult to actually collect, and which result in theoretically sound frameworks with little hard data to evaluate progress. Conversely, we must ensure that the indicators that are easier to collect over time or across regions truly reflect policy outcomes and progress on objectives (Fernandez-Macho, 2016). Finally, when evaluating any indicators of sustainable development – particularly in complex systems such as oceans – it is important to not let the measures become the goal.

To use a final example to bring together the discussion above, SDG 14.7 aims to increase benefits to SIDS from ocean sectors including fisheries, tourism, and mariculture (Table 16.1). However, the corresponding indicator (14.7.1) aims for increasing the proportion of fish stocks assessed as sustainable (e.g., at MSY) and their percent contribution to national GDP. Ideally, this indicator is intended to reflect the overall state of local marine ecosystems, helping direct efforts to protect and recover fish stocks and thus contribute to GDP growth. However, one could also make progress on this indicator by modifying reference points or assessment methods to make it easier for stocks to be deemed sustainable. If local fishers are displaced to reduce overfishing, loss of access to food and cultural practices would not be accounted for in the measurement of benefits by the indicator. Conversely, if a nation sees significant economic growth from investments in ecotourism or the technology sector, the relative contribution of fisheries to GDP would decrease, showing a negative trend in the indicator even if stocks are sustainable. Any of these possible drivers of indicator trends would clearly not reflect progress on the intended goal.

The example above is intended as one more illustration to emphasize the power of indicators to facilitate or impede achievement of a Blue Economy or other development goals, and we should now be able to think of many others. Any and all social, ecological, and economic aspects considered in an objective can easily be lost depending on the choice of indicator, which is why great care must be taken in designing them (Valentin & Spangenberg, 2000). Social systems, ecosystems, and economies are complex and interconnected, so indicators must not be evaluated in isolation (Cisneros-Montemayor et al., 2021). As this chapter highlights, designing, choosing, and collecting indicators for a Blue Economy cannot be a purely technical process, but requires transparent and inclusive collaboration within a dedicated policy process to make sure that we are advancing our intended goals.

Professional Pathways

Students interested in working in the context of the Blue Economy can follow a variety of paths. Typical fields of study would likely include economics, political science, sociology, or other interdisciplinary human dimension fields such as political economy, marine studies, or development studies. Specific entities to explore for career opportunities might include regional or national natural-resource planning institutions; fisheries and ocean management departments at the international (e.g., FAO), national (e.g., national governments), or local (e.g., state or counties) level; intergovernmental organizations related to sustainable development such as the UN agencies or the World Bank; marine industry associations (e.g., shipping, tourism, logistics); or intergovernmental, non-governmental, or national marine and environmental conservation institutes.

Questions for Reflection

1. What is a Blue Economy?
2. How is the Blue Economy framework linked to equitable and sustainable development?
3. What are development indicators, and how do we choose them for a Blue Economy?
4. How can we consider social equity in choosing indicators – even ones related to financial or environmental outcomes?

Acknowledgments

AMCM gratefully acknowledges comments and suggestions of the editors and three anonymous reviewers. AMCM is supported by the Nippon Foundation Ocean Nexus Center at EarthLab, University of Washington, in collaboration with the University of British Columbia.

References

Allison, E. H., Kurien, J., Ota, Y., Adhuri, D. S., Bavinck, J. M., Cisneros-Montemayor, A. M., Fabinyi, M., Jentoft, S., Lau, S., Mallory, T. G., Olukoju, A., van Putten, I., Stacey, N., Voyer, M., & Weeratunge, N. (2020). *The human relationship with our ocean planet* (p. 80). World Resources Institute. https://oceanpanel.org/publication/the-human-relationship-with-our-ocean-planet/

Andriantiatsaholiniaina, L. A., Kouikoglou, V. S., & Phillis, Y. A. (2004). Evaluating strategies for sustainable development: Fuzzy logic reasoning and sensitivity analysis. *Ecological Economics*, 48(2), 149–172. https://doi.org/10.1016/j.ecolecon.2003.08.009

Arreguín-Sánchez, F., & Ruiz-Barreiro, T. M. (2014). Approaching a functional measure of vulnerability in marine ecosystems. *Ecological Indicators*, *45*, 130–138. https://doi.org/10.1016/j.ecolind.2014.04.009

Austen, M., Andersen, P., Armstrong, C., Döring, R., Hynes, S., Levrel, H., Oinonen, S., Ressurreição, A., & Coopman, J. (2019). *Valuing marine ecosystems – Taking into account the value of ecosystem benefits in the Blue Economy* [Preprint]. MarXiv. https://doi.org/10.31230/osf.io/vy3kp

Bell, S., & Morse, S. (2004). Experiences with sustainability indicators and stakeholder participation: A case study relating to a "blue plan" project in Malta. *Sustainable Development*, *12*(1), 1–14. https://doi.org/10.1002/sd.225

Bennett, N. J., Blythe, J., Cisneros-Montemayor, A. M., Singh, G. G., & Sumaila, U. R. (2019). Just transformations to sustainability. *Sustainability*, *11*(14), 3881. https://doi.org/10.3390/su11143881

Bennett, N. J., Govan, H., & Satterfield, T. (2015). Ocean grabbing. *Marine Policy*, *57*, 61–68. https://doi.org/10.1016/j.marpol.2015.03.026

Bentley, J. W., Hines, D., Borrett, S., Serpetti, N., Fox, C., Reid, D. G., & Heymans, J. J. (2019). Diet uncertainty analysis strengthens model-derived indicators of food web structure and function. *Ecological Indicators*, 98, 239–250. https://doi.org/10.1016/j.ecolind.2018.11.008

Blanchard, W. (1986). Evaluating social equity: What does fairness mean and can we measure it? *Policy Studies Journal*, *15*(1), 29–54. https://doi.org/10.1111/j.1541-0072.1986.tb00442.x

Carruthers, T. R., & Hordyk, A. R. (2018). The data-limited methods toolkit (DLM tool): An R package for informing management of data-limited populations. *Methods in Ecology and Evolution*, 9(12), 2388–2395. https://doi.org/10.1111/2041-210X.13081

Cisneros-Montemayor, A. M., Cashion, T., Miller, D. D., Tai, T. C., Talloni-Álvarez, N., Weiskel, H. W., & Sumaila, U. R. (2018a). Achieving sustainable and equitable fisheries requires nuanced policies not silver bullets. *Nature Ecology & Evolution*, *2*(9), 1334. https://doi.org/10.1038/s41559-018-0633-0

Cisneros-Montemayor, A. M., Moreno-Báez, M., Reygondeau, G., Cheung, W. W. L., Crosman, K. M., González-Espinosa, P. C., Lam, V. W. Y., Oyinlola, M. A., Singh, G. G., Swartz, W., Zheng, C., & Ota, Y. (2021). Enabling conditions for an equitable and sustainable blue economy. *Nature*, *591*(7850), 396–401. https://doi.org/10.1038/s41586-021-03327-3

Cisneros-Montemayor, A. M., Moreno-Báez, M., Voyer, M., Allison, E. H., Cheung, W. W. L., Hessing-Lewis, M., Oyinlola, M. A., Singh, G. G., Swartz, W., & Ota, Y. (2019). Social equity and benefits as the nexus of a transformative blue economy: A sectoral review of implications. *Marine Policy*, *109*, 103702. https://doi.org/10.1016/j.marpol.2019.103702

Cisneros-Montemayor, A. M., Pauly, D., Weatherdon, L. V., & Ota, Y. (2016). A global estimate of seafood consumption by coastal Indigenous peoples. *PLoS ONE*, *11*(12), e0166681. https://doi.org/10.1371/journal.pone.0166681

Cisneros-Montemayor, A. M., Singh, G. G., & Cheung, W. W. L. (2018b). A fuzzy logic expert system for evaluating policy progress towards sustainability goals. *Ambio*, *47*(5), 595–607. https://doi.org/10.1007/s13280-017-0998-3

Cisneros-Montemayor, A. M., Zetina-Rejón, M. J., Espinosa-Romero, M. J., Cisneros-Mata, M. A., Singh, G. G., & Fernández-Rivera Melo, F. J. (2020). Evaluating ecosystem impacts of data-limited artisanal fisheries through ecosystem modelling and traditional fisher knowledge. *Ocean & Coastal Management*, *195*, 105291. https://doi.org/10.1016/j.ocecoaman.2020.105291

Colgan, C. S. (2016). Measurement of the ocean economy from national income accounts to the sustainable blue economy. *Journal of Ocean and Coastal Economics*, *2*(2), 12. https://doi.org/10.15351/2373-8456.1061

Coll, M., Shannon, L. J., Kleisner, K. M., Juan-Jordá, M. J., Bundy, A., Akoglu, A. G., Banaru, D., Boldt, J. L., Borges, M. F., Cook, A., Diallo, I., Fu, C., Fox, C., Gascuel, D., Gurney, L. J., Hattab, T., Heymans, J. J., Jouffre, D., Knight, B. R. … Shin, Y.-J. (2016). Ecological indicators to capture the effects of fishing on biodiversity and conservation status of marine ecosystems. *Ecological Indicators*, *60*, 947–962. https://doi.org/10.1016/j.ecolind.2015.08.048

Cope, J. M., & Punt, A. E. (2009). Length-based reference points for data-limited situations: Applications and restrictions. *Marine and Coastal Fisheries: Dynamics, Management, and Ecosystem Science*, *1*(1), 169–186. https://doi.org/10.1577/C08-025.1

Cormier, R., & Elliott, M. (2017). SMART marine goals, targets and management – Is SDG 14 operational or aspirational, is "Life Below Water" sinking or swimming? *Marine Pollution Bulletin*, *123*(1–2), 28–33. http://dx.doi.org/10.1016/j.marpolbul.2017.07.060

Daly, H. E. (1990). Toward some operational principles of sustainable development. *Ecological Economics*, *2*(1), 1–6. https://doi.org/10.1016/0921-8009(90)90010-R

Devillers, R., Pressey, R. L., Ward, T. J., Grech, A., Kittinger, J. N., Edgar, G. J., & Watson, R. A. (2020). Residual marine protected areas five years on: Are we still favouring ease of establishment over need for protection? *Aquatic Conservation: Marine and Freshwater Ecosystems*, *30*(9), 1758–1764. https://doi.org/10.1002/aqc.3374

Díaz, S., Pascual, U., Stenseke, M., Martín-López, B., Watson, R. T., Molnár, Z., Hill, R., Chan, K. M. A., Baste, I. A., Brauman, K. A., Polasky, S., Church, A., Lonsdale, M.,

Larigauderie, A., Leadley, P. W., van Oudenhoven, A. P. E., van der Plaat, F., Schröter, M., Lavorel, S. … Shirayama, Y. (2018). Assessing nature's contributions to people. *Science*, *359*(6373), 270–272. https://doi.org/10.1126/science.aap8826

Elfes, C. T., Longo, C., Halpern, B. S., Hardy, D., Scarborough, C., Best, B. D., Pinheiro, T., & Dutra, G. F. (2014). A regional-scale ocean health index for Brazil. *PLoS ONE*, 9(4), e92589. https://doi.org/10.1371/journal.pone.0092589

Fabricius, K. E. (2005). Effects of terrestrial runoff on the ecology of corals and coral reefs: Review and synthesis. *Marine Pollution Bulletin*, *50*(2), 125–146. https://doi.org/10.1016/j.marpolbul.2004.11.028

Food and Agriculture Organization of the United Nations. (2018). *The state of world fisheries and aquaculture 2018. Meeting the Sustainable Development Goals* (p. 227). FAO.

Fenichel, E. P., Milligan, B., Porras, I., Addicott, E. T., Árnasson, R., Bordt, M., Djavidnia, S., Dvarskas, A., Goldman, E., Grimsrud, K., Lange, G.-M., Matuszak, J., Muawanah, U., Quaas, M., Soulard, F., & Zhang, J. (2020). *National accounting for the ocean and ocean economy* (High Level Panel for a Sustainable Ocean Economy, p. 48). World Resources Institute.

Fernandez-Macho, J. (2016). A statistical assessment of maritime socioeconomic indicators for the European Atlantic area. *Journal of Ocean and Coastal Economics*, *2*(2), 4. https://doi.org/10.15351/2373-8456.1047

Folke, C., Biggs, R., Norström, A. V., Reyers, B., & Rockström, J. (2016). Social-ecological resilience and biosphere-based sustainability science. *Ecology and Society*, *21*(3), 41. https://www.jstor.org/stable/26269981

Fulton, E. A. (2010). Approaches to end-to-end ecosystem models. *Journal of Marine Systems*, *81*(1–2), 171–183. https://doi.org/10.1016/j.jmarsys.2009.12.012

Gill, D. A., Mascia, M. B., Ahmadia, G. N., Glew, L., Lester, S. E., Barnes, M., Craigie, I., Darling, E. S., Free, C. M., Geldmann, J., Holst, S., Jensen, O. P., White, A. T., Basurto, X., Coad, L., Gates, R. D., Guannel, G., Mumby, P. J., Thomas, H., … Fox, H. E. (2017). Capacity shortfalls hinder the performance of marine protected areas globally. *Nature*, *543*(7647), 665–669. https://doi.org/10.1038/nature21708

Golden, C., Allison, E. H., Cheung, W. W., Dey, M. M., Halpern, B. S., McCauley, D. J., Smith, M., Vaitla, B., Zeller, D., & Myers, S. S. (2016). Fall in fish catch threatens human health. *Nature*, *534*(7607), 317–320. https://doi.org/10.1038/534317a

Guggisberg, S. (2019). The roles of nongovernmental actors in improving compliance with fisheries regulations. *Review of European, Comparative & International Environmental Law*, *28*(3), 314–327. https://doi.org/10.1111/reel.12304

Halpern, B. S., Klein, C. J., Brown, C. J., Beger, M., Grantham, H. S., Mangubhai, S., Ruckelshaus, M., Tulloch, V. J., Watts, M., White, C., & Possingham, H. P. (2013). Achieving the triple bottom line in the face of inherent trade-offs among social equity, economic return, and conservation. *Proceedings of the National Academy of Sciences*, *110*(15), 6229–6234. https://doi.org/10.1073/pnas.1217689110

Halpern, B. S., Longo, C., Hardy, D., McLeod, K. L., Samhouri, J. F., Katona, S. K., Kleisner, K., Lester, S. E., O'Leary, J., Ranelletti, M., Rosenberg, A. A., Scarborough, C., Selig, E. R., Best, B. D., Brumbaugh, D. R., Chapin, F. S., Crowder, L. B., Daly, K. L., Doney, S. C. … Zeller, D. (2012). An index to assess the health and benefits of the global ocean. *Nature*, 488(7413), 615–620. https://doi.org/10.1038/nature11397

Hass, J., Brunvoll, F., & Hoie, H. (2002). Overview of sustainable development indicators used by national and international agencies (OECD Statistics Working Papers, No. 2002/02). OECD Publishing. https://doi.org/10.1787/838562874641

Hicks, C. C., Levine, A., Agrawal, A., Basurto, X., Breslow, S. J., Carothers, C., Charnley, S., Coulthard, S., Dolsak, N., Donatuto, J., Garcia-Quijano, C., Mascia, M. B., Norman,

K., Poe, M. R., Satterfield, T., St. Martin, K., & Levin. P. S. (2016). Engage key social concepts for sustainability. *Science, 352*(6281), 38–40.

Hilborn, R., & Mangel, M. (1997). *The ecological detective: Confronting models with data.* Princeton University Press.

Kleisner, K., Zeller, D., Froese, R., & Pauly, D. (2013). Using global catch data for inferences on the world's marine fisheries: Catch data for fisheries inferences. *Fish and Fisheries, 14*(3), 293–311. https://doi.org/10.1111/j.1467-2979.2012.00469.x

Konar, M., & Ding, H. (2020). *A sustainable ocean economy for 2050: Approximating its benefits and costs* (p. 62). High Level Panel for a Sustainable Ocean Economy. Retrieved July 11, 2022 from https://www.readkong.com/page/a-sustainable-ocean-economy-for-2050-approximating-its-7989226

Merry, S. E. (2011). Measuring the world: Indicators, human rights, and global governance. *Current Anthropology, 52*(S3), S83–S95. https://doi.org/10.1086/657241

OECD [Organisation for Economic Co-operation and Development]. (2016). *The ocean economy in 2030.* OECD Publishing. https://doi.org/10.1787/9789264251724-en

Perrings, C., Naeem, S., Ahrestani, F. S., Bunker, D. E., Burkill, P., Canziani, G., Elmqvist, T., Fuhrman, J. A., Jaksic, F. M., Kawabata, Z., Kinzig, A., Mace, G. M., Mooney, H., Prieur-Richard, A.-H., Tschirhart, J., & Weisser, W. (2011). Ecosystem services, targets, and indicators for the conservation and sustainable use of biodiversity. *Frontiers in Ecology and the Environment,* 9(9), 512–520. https://doi.org/10.1890/100212

Pintér, L., Hardi, P., Martinuzzi, A., & Hall, J. (2012). Bellagio STAMP: Principles for sustainability assessment and measurement. *Ecological Indicators, 17,* 20–28. https://doi.org/10.1016/j.ecolind.2011.07.001

Plagányi, É. E. (2007). *Models for an ecosystem approach to fisheries* (FAO Fisheries Technical Paper No. 477; p. 108). FAO. http://www.fao.org/docrep/010/a1149e/a1149e00.htm

Schuhbauer, A., Cisneros-Montemayor, A. M., Chuenpagdee, R., & Sumaila, U. R. (2019b). Assessing the economic viability of small-scale fisheries: An example from Mexico. *Marine Ecology Progress Series, 617–618,* 365–376. https://doi.org/10.3354/meps12942

Schuhbauer, A., Cisneros-Montemayor, A. M., & Sumaila, U. R. (2019a). Economic viability of small-scale fisheries: A transdisciplinary evaluation approach. In R. Chuenpagdee & S. Jentoft (Eds.), *Transdisciplinarity for small-scale fisheries governance* (Vol. 21, pp. 93–117). Springer International Publishing. https://doi.org/10.1007/978-3-319-94938-3_6

Seto, K., & Campbell, B. (2019). The last commons: (Re)constructing an ocean future. In A. M. Cisneros-Montemayor, W. W. L. Cheung, & Y. Ota (Eds.), *Predicting future oceans: Sustainability of ocean and human systems amidst global environmental change* (pp. 365–376). Elsevier.

Spijkers, J., Morrison, T. H., Blasiak, R., Cumming, G. S., Osborne, M., Watson, J., & Österblom, H. (2018). Marine fisheries and future ocean conflict. *Fish and Fisheries, 19*(5), 798–806. https://doi.org/10.1111/faf.12291

Sumaila, U. R., & Walters, C. (2005). Intergenerational discounting: A new intuitive approach. *Ecological Economics, 52*(2), 135–142. https://doi.org/10.1016/j.ecolecon.2003.11.012

Teh, L. S. L., Teh, L. C. L., & Sumaila, U. R. (2014). Time preference of small-scale fishers in open access and traditionally managed reef fisheries. *Marine Policy, 44,* 222–231. https://doi.org/10.1016/j.marpol.2013.08.028

United Nations. (2015). *Transforming our world: The 2030 agenda for sustainable development* (A/Res/70/1, p. 35). United Nations General Assembly. https://www.ceeol.com/search/article-detail?id=304581

United Nations. (2017). *Work of the statistical commission pertaining to the 2030 agenda for sustainable development* (A/RES/71/313). United Nations General Assembly.

United Nations Department of Economic and Social Affairs. (2014). *Blue economy concept paper*. United Nations Department of Economic and Social Affairs. https://sustainabledevelopment.un.org/content/documents/2978BEconcept.pdf

United Nations Development Programme. (2020, December 15). *Human development report 2020. The next frontier: Human development and the Anthropocene*. United Nations Development Programme. http://hdr.undp.org/en/content/human-development-report-2020

Valentin, A., & Spangenberg, J. H. (2000). A guide to community sustainability indicators. *Environmental Impact Assessment Review*, 20(3), 381–392. https://doi.org/10.1016/S0195-9255(00)00049-4

Vierros, M. K., Harrison, A. L., Sloat, M. R., Crespo, G. O., Moore, J. W., Dunn, D. C., Ota, Y., Cisneros-Montemayor, A. M., Shillinger, G. L., Watson, T. K., & Govan, H. (2020). Considering indigenous peoples and local communities in governance of the global ocean commons. *Marine Policy*, *119*, 104039. https://doi.org/10.1016/j.marpol.2020.104039

Voyer, M., Quirk, G., McIlgorm, A., & Azmi, K. (2018). Shades of blue: What do competing interpretations of the blue economy mean for oceans governance? *Journal of Environmental Policy & Planning*, 20(5), 595–616. https://doi.org/10.1080/1523908X.2018.1473153

Wang, X. (2016). The ocean economic statistical system of China and understanding of the blue economy. *Journal of Ocean and Coastal Economics*, 2(2). https://doi.org/10.15351/2373-8456.1055

World Bank. (2016). *Poverty and shared prosperity 2016: Taking on inequality*. The World Bank. https://doi.org/10.1596/978-1-4648-0958-3

World Bank, & United Nations Department of Economic and Social Affairs. (2017). *The potential of the Blue Economy: Increasing long-term benefits of the sustainable use of marine resources for Small Island Developing States and Coastal Least Developed Countries* (p. 50). World Bank. https://openknowledge.worldbank.org/bitstream/handle/10986/26843/115545.pdf?sequence=1&isAllowed=y

17 Conclusion

Marine Studies as an Interdisciplinary Approach to the Study of the Coupled Natural–Human Ocean System

D. O. Suman and A. K. Spalding

The ocean covers more than 70% of the Earth's surface. About 97% of the water on Earth is found in the Ocean. Between 50 and 80% of oxygen produced on Earth comes from the ocean, mostly from phytoplankton photosynthesis. Marine fish provide about 15% of protein consumed by humans. The marine realm provides many services to humans, including provision of food, energy, and minerals; climate regulation; recreational opportunities, spiritual and cultural meanings, creative inspiration, maritime transportation routes, opportunities for scientific study, carbon sequestration, oxygen production, among others. We use the Ocean in so many ways!

However, despite its great importance to humans, the Ocean is in trouble. One-third of fish stocks are overfished, and only 7% have any room for growth. Society has used the Ocean as a waste bin. By 2050, some researchers estimate that the weight of plastics in the ocean could exceed the weight of fish. Plastic waste has been found on the deep Mariana Trench, and Antarctic waters and deep-sea sediments contain microplastics. Nutrient pollution plagues the world's coastal waters. Recent dramatic oil spills from maritime accidents and offshore oil operations add to the substantial amounts of hydrocarbons that are released to marine waters from numerous smaller operational sources. Invasive species of algae and mollusks carried around the world in ballast waters create havoc in marine and freshwater ecosystems. Due to global climate change, the average surface ocean temperature may increase between 0.7°C and 3°C by 2100, impacting the distribution of living marine resources and causing rising sea levels that have already begun to impact low-lying coastal areas. Predictions for global sea level rise by 2100 range from 43 to 84 cm relative to 1986–2005 levels.

Human activities are ultimately responsible for these adverse impacts to the Ocean and its resources. At the same time, the degradation of the marine environment also creates feedback loops that affect human societies in many ways. This volume has stressed the importance of considering the Ocean as a Coupled Natural–Human System, where people are an integral part of the marine environment. Therefore, our efforts to study and protect the Ocean must fully integrate human dimensions approaches, including social sciences, arts, and humanities. The chapters in this volume do just that. Part I laid the foundation for what we

DOI: 10.4324/9781003058151-20

have called "Marine Studies", outlining the various human dimension approaches and the natural and physical processes that define the marine environment. Part II examined some of the grand challenges that the global Ocean faces, namely, overfishing, climate change, marine pollution, international ocean governance; and the various ways in which humans have contributed to and been affected by these challenges. Limitations of space have made it impossible to discuss additional challenges, such as inadequate coastal planning, consumption patterns, deep-sea mining, globalization of trade and its impacts on maritime transportation and port development, among others. Yet, as you find your niche in Marine Studies, we invite you to further explore other current and emerging ocean issues. Finally, Part III described a suite of approaches that could be used to address and explore the Ocean's human dimensions. Taken together, the essays describe some of the interlocking pieces that comprise the Coupled Natural–Human Ocean System. Table 17.1 illustrates some of the linkages between the challenges (fisheries and aquaculture, renewable energy, climate change, High Seas biological diversity) and several of the human dimension topics (social justice, international ocean governance, and marine conservation/marine protected areas).

Integrating the Human Dimension

Solutions to marine environmental problems must integrate people if they are to be effective. In particular, the chapters in Part III of this volume offer a variety of perspectives on how to approach this integration. For instance, the chapter on **ocean governance** introduces the international ocean regime by outlining how regulating and managing human activities in ocean and coastal spaces requires governance of this global commons. This is not a simple issue because governance depends entirely on the voluntary goodwill and cooperation of Nations, enforcement mechanisms are weak, and the levels of development of Nations vary widely. Specifically, it describes how ocean governance has evolved from a focus on zones and boundaries (spatial governance), exemplified by the UN Convention on the Law of the Sea (UNCLOS), to a more nuanced approach that is based on issues (functional governance), such as biological diversity, climate change, marine pollution, among others.

Marine protected areas (MPAs) are an important tool for ocean governance and conservation. The chapter on MPAs and marine reserves shows how this tool is gaining prominence as a way to address overfishing and biodiversity loss. It also introduces how MPAs call for human dimension considerations, such as the need for social inclusion and participation of all stakeholders in the planning, development, and management of MPAs; consideration of the potential adverse socio-economic impacts of conservation measures and the need for compensation; implementation of management plans and sustainable funding mechanisms; consideration of the cultural and economic needs of fishers who may be restricted from traditional fishing grounds; and the need to coordinate management of MPAs and other conservation measures into broader coastal and watershed management, as well as other marine sectors through marine spatial planning.

Table 17.1 Relevance of human dimension considerations to challenges faced by the ocean

		Human Dimension Approaches		
		Conservation/MPAs	*Social Justice*	*International Ocean Governance*
Challenges	**Fisheries**	Siting of MPAs in best locations to protect and enhance fish stocks; socio-economic impacts of MPAs; enforcement issues	Allocation of coastal space between recreational fisheries and tourism facilities; allocation of quotas between industrial and artisanal fisheries; beneficiaries and losers of ITQ systems	Governance and management of marine living resources on the High Seas; management of transboundary stocks; promotion of sustainable fisheries; High Seas MPAs
	Ocean Aquaculture	Uses of restorative aquaculture to improve water quality	Conflicts between capture fisheries and aquaculture operations; labor conditions of aquaculture workers	Permitting, siting, and management of ocean aquaculture operations in national waters and High Seas; transboundary impacts
	Climate Change	MPAs as an adaptive strategy to climate change; new and mobile MPAs that account for displacement of marine living resources from climate change	Sea level rise and climate gentrification; social vulnerabilities from climate change impacts; allocations of displaced fish stocks	Management of the Ocean to promote climate change mitigation; linkages between climate change and biodiversity conventions
	Marine Pollution	Impacts of marine pollution on MPAs and conservation efforts; conservation efforts to assist recovery from marine pollution events	Social vulnerabilities from pollution impacts	Implementation of measures to reduce marine pollution from vessels, land-based sources, and invasive species; emerging pollutants
	Renewable Ocean Energy	Synergies between renewable energy facilities and MPAs	Siting of facilities and space conflicts with other marine uses; energy pricing; ownership of operations	Siting of energy facilities and subsea cables; contingency planning
	High Seas Biodiversity	High Seas MPAs (designation, implementation, management, enforcement)	Sharing of benefits from uses of High Seas biodiversity; transfer of new technologies	Governance frameworks for High Seas MPAs, marine genetic resources, environmental assessments, and transfer of technologies

The chapter on **social justice** expands on the tradeoffs illustrated in using MPAs as a governance tool by showcasing how human activities in the Ocean benefit some socio-economic groups and disadvantage others or how some maritime activities deprive certain users of the marine space or resources that they have traditionally used and enjoyed. Examples include industrial fisheries that usurp living marine resources that are important for small-scale fishers, shrimp aquaculture ponds constructed in mangroves that block access for traditional users of mangrove resources, exclusive coastal residential or tourist developments that displace vulnerable coastal communities. Specifically, this chapter explores these issues through the lens of a coastal fishing community in southern Oregon in the United States that examines the intersection of gender, race, and socio-economic class. Other case studies discussed in the essays mention the increased vulnerability of disadvantaged communities to sea level rise and other climate change impacts, and the squeeze experienced by small-scale fishers from expanding coastal tourism.

Those who work in marine spaces must further recognize and respect different ways of knowledge about the natural environment. Marine resource users have great knowledge about the marine environment, even though it may be different from what we have been taught is "scientific". The chapter on **traditional ecological knowledge (TEK)** advances this concept further, epitomizing the marine space as a Coupled Natural–Human System. Indigenous peoples consider themselves to be part of the Earth home and have a spiritual and ceremonial relationship with the environment: important lessons for all of us in these challenging times of a climate change crisis. TEK is rooted in the land and waters, and the chapter asks important questions about whether this traditional knowledge can exist when Native peoples have been colonized, massacred, and displaced from their lands. The essay encourages us to give Indigenous peoples that space and respect to develop and fortify their traditional knowledge base and gain the tribal sovereignty that they desire. Importantly, we should never assume or expect that TEK can solve global climate change, marine pollution, or overfishing. Indeed, we invite you to explore how a TEK perspective can help us develop our own environmental ethic and more integrated way of living.

Coastal communities maintain special relationships with the adjacent Ocean that evolve due to natural and human-caused changes in the social, economic, and natural environments. The chapter on **Community development** defines coastal communities and outlines how a community-centered approach aims to strengthen community wellbeing, resilience, and their capacity to adapt to changing conditions. Coastal communities are undoubtedly on the frontline of the environmental crisis of our time – from climate change, fish stock fluctuations and displacements, gentrification and increasing coastal tourism, as well as the rise and fall of industries. How well communities adapt to change while maintaining their identities and structures depends on knowledge and training of its residents, social capital (trust and cohesion within the community), financial and infrastructure resources, and natural capital or ecosystem resources. The chapter illustrates how those of us who are helping coastal communities strengthen their resilience must understand coastal and marine ecosystem processes, while maintaining a keen sensitivity to the human dimensions of the community and its residents.

In addition to the people and environment-centered approaches of the previous chapters, the chapter on **marine entrepreneurship** showcases how the business sector offers numerous opportunities for innovation that broadly promote sustainability and as well address some of the challenges that the Ocean faces. Clearly, successful marine business ventures require well thought-out business and financial plans, as well as market identification and evaluation. The chapter also points out how the marine environment presents unique risks of environmental variability, difficult environmental conditions, seasonality, insecure property rights, among others for which an entrepreneur must prepare. Examples include sustainable fishing operations, carbon-neutral aquaculture, renewable energy technologies, new seafood technologies and products, and innovative marketing strategies that create tighter links between the fisher and the consumer. Non-consumptive activities, such as community-based coastal tourism, diving and surfing businesses, and charter fishing operations, also show room for growth.

The final chapter of this volume explores the concept of the **Blue Economy** as an approach that has the potential to support the integration of natural and social sciences through its particular emphasis on equity and justice. Although different definitions exist for Blue Economy, this chapter defines it as the application of Sustainable Development to the Ocean with particular emphasis on social justice and proposes indicators for the three branches of sustainability – economic, social, and environmental.

These seven topics showcase that to understand the Ocean and address the grand challenges outlined in Part II of this book, we must include a broad suite of lens and disciplines – drawing from the natural and social sciences, as well as the humanities. However, the treatment must be holistic and transdisciplinary and not focused only on fisheries biology, physical and chemical oceanography, economics/business, anthropology, political science, or law. The Coupled Natural–Human Ocean System must draw on all these areas of knowledge, ways of learning, and the lens through which we view the Earth and its residents.

Role of Marine Studies – Ocean Literacy, Capacity-Building, Education

Our responsibility as citizens, and particularly as marine studies students and professionals, is to act in ways that minimize our adverse impacts on the Ocean, protect and restore its environmental quality, promote social justice, and find and implement solutions to the challenges that we have created for the Ocean's health and, ultimately, for society. There are many ways we can further these goals: critically examining the ways people use the Ocean; being sensitive in our professions to the human dimensions of the marine environment; recognizing how gender, race, and social class are reflected in relations between stakeholders; becoming involved with movements to protect the Ocean; voting, encouraging our representatives to support sustainable policies, and participating in governmental initiatives; and making sustainable choices as consumers, among others.

Our efforts to study and understand the Ocean must fully integrate human dimensions; viewing the Ocean as a Coupled Natural–Human System is essential for creating meaningful, innovative, adaptive, equitable, and effective responses and policies to these challenges. We hope that this book has provided readers with an appreciation of the importance of this integration.

Index